Parasites and Wildlife

Parasites and Wildlife

Editors

Rafael Calero-Bernal
Ignacio García-Bocanegra

MDPI • Basel • Beijing • Wuhan • Barcelona • Belgrade • Manchester • Tokyo • Cluj • Tianjin

Editors
Rafael Calero-Bernal
Complutense University of Madrid
Spain

Ignacio García-Bocanegra
Universidad de Córdoba
Spain

Editorial Office
MDPI
St. Alban-Anlage 66
4052 Basel, Switzerland

This is a reprint of articles from the Special Issue published online in the open access journal *Animals* (ISSN 2076-2615) (available at: https://www.mdpi.com/journal/animals/special_issues/parasite_wildlife).

For citation purposes, cite each article independently as indicated on the article page online and as indicated below:

LastName, A.A.; LastName, B.B.; LastName, C.C. Article Title. *Journal Name* **Year**, *Volume Number*, Page Range.

ISBN 978-3-0365-6852-2 (Hbk)
ISBN 978-3-0365-6853-9 (PDF)

Cover image courtesy of Ignacio García-Bocanegra

Contents

About the Editors . . . vii

Preface to "Parasites and Wildlife" . . . ix

Rafael Calero-Bernal and Ignacio García-Bocanegra
Parasites and Wildlife
Reprinted from: *Animals* **2023**, *13*, 628, doi:10.3390/ani13040628 . . . 1

Pamela C. Köster, Justinn Renelies-Hamilton, Laia Dotras, Manuel Llana, Celia Vinagre-Izquierdo, Petras Prakas, et al.
Molecular Detection and Characterization of Intestinal and Blood Parasites in Wild Chimpanzees (*Pan troglodytes verus*) in Senegal
Reprinted from: *Animals* **2021**, *11*, 3291, doi:10.3390/ani11113291 . . . 5

Dalytė Mardosaitė-Busaitienė, Jana Radzijevskaja, Linas Balčiauskas and Algimantas Paulauskas
Babesia microti in Rodents from Different Habitats of Lithuania
Reprinted from: *Animals* **2021**, *11*, 1707, doi:10.3390/ani11061707 . . . 29

Federica Giorda, Umberto Romani-Cremaschi, Antoinette E. Marsh, Carla Grattarola, Barbara Iulini, Alessandra Pautasso, et al.
Evidence for Unknown *Sarcocystis*-Like Infection in Stranded Striped Dolphins (*Stenella coeruleoalba*) from the Ligurian Sea, Italy
Reprinted from: *Animals* **2021**, *11*, 1201, doi:10.3390/ani11051201 . . . 39

Petras Prakas, Linas Balčiauskas, Evelina Juozaitytė-Ngugu and Dalius Butkauskas
The Role of Mustelids in the Transmission of *Sarcocystis* spp. Using Cattle as Intermediate Hosts
Reprinted from: *Animals* **2021**, *11*, 822, doi:10.3390/ani11030822 . . . 57

Age Kärssin, Liidia Häkkinen, Annika Vilem, Pikka Jokelainen and Brian Lassen
Trichinella spp. in Wild Boars (*Sus scrofa*), Brown Bears (*Ursus arctos*), Eurasian Lynxes (*Lynx lynx*) and Badgers (*Meles meles*) in Estonia, 2007–2014
Reprinted from: *Animals* **2021**, *11*, 183, doi:10.3390/ani11010183 . . . 67

Peng-Wu Yin, Xian-Guo Guo, Dao-Chao Jin, Rong Fan, Cheng-Fu Zhao, Zhi-Wei Zhang, et al.
Distribution and Host Selection of Tropical Rat Mite, *Ornithonyssus bacoti*, in Yunnan Province of Southwest China
Reprinted from: *Animals* **2021**, *11*, 110, doi:10.3390/ani11010110 . . . 81

Katarzyna Skrzypek, Jacek Karamon, Małgorzata Samorek-Pieróg, Joanna Dabrowska, Maciej Kochanowski, Jacek Sroka, et al.
Comparison of Two DNA Extraction Methods and Two PCRs for Detection of *Echinococcus multilocularis* in the Stool Samples of Naturally Infected Red Foxes
Reprinted from: *Animals* **2020**, *10*, 2381, doi:10.3390/ani10122381 . . . 97

Patricia Barroso, Ignacio García-Bocanegra, Pelayo Acevedo, Pablo Palencia, Francisco Carro, Saúl Jiménez-Ruiz, et al.
Long-Term Determinants of the Seroprevalence of *Toxoplasma gondii* in a Wild Ungulate Community
Reprinted from: *Animals* **2020**, *10*, 2349, doi:10.3390/ani10122349 . . . 105

Maira Riquelme, Rodrigo Salgado, Javier A. Simonetti, Carlos Landaeta-Aqueveque, Fernando Fredes and André V. Rubio
Intestinal Helminths in Wild Rodents from Native Forest and Exotic Pine Plantations (*Pinus radiata*) in Central Chile
Reprinted from: *Animals* **2021**, *11*, 384, doi:10.3390/ani11020384 . . . . **127**

Igor C. L. Acosta, Solange M. Gennari, Horwald A. B. Llano, Sebastián Muñoz-Leal and Rodrigo M. Soares
Molecular Characterization of New Haplotype of Genus *Sarcocystis* in Seabirds from Magdalena Island, Southern Chile
Reprinted from: *Animals* **2021**, *11*, 245, doi:10.3390/ani11020245 . . . . **137**

Anabel Martínez-Padilla, Javier Caballero-Gómez, Ángela Magnet, Félix Gómez-Guillamón, Fernando Izquierdo, Leonor Camacho-Sillero, et al.
Zoonotic Microsporidia in Wild Lagomorphs in Southern Spain
Reprinted from: *Animals* **2020**, *10*, 2218, doi:10.3390/ani10122218 . . . . **149**

Sándor Hornok, Kristin Mühldorfer, Nóra Takács, Regina Hofmann-Lehmann, Marina L. Meli, Miklós Gyuranecz, et al.
Broad Range Screening of Vector-Borne Pathogens in Arctic Foxes (*Vulpes lagopus*) in Iceland
Reprinted from: *Animals* **2020**, *10*, 2031, doi:10.3390/ani10112031 . . . . **157**

About the Editors

Rafael Calero-Bernal

DVM (2006), MSc (2007) and PhD (2011) at the University of Extremadura (Spain). Dr. Calero-Bernal has spent research training periods in recognized research institutions in Portugal (INSA, Lisbon), Italy (ISS, Rome), Spain (ISCIII, Madrid), Brazil (FIOCRUZ, Manaus), Equatorial Guinea (CRCE, Bata) and USA (USDA, Beltsville), with special aim in deepen the knowledge on One Health approach of parasitic diseases caused by Apicomplexan agents. Currently, he is member of the SALUVET research group, and Assistant Professor at the Animal Health Department, Veterinary Faculty, Complutense University of Madrid, Spain.

Ignacio García-Bocanegra

DVM (2001), DFST (2003), PhD (2005), MSc (2009) and Dip ECZM (Wildlife Population Health) (2016). Dr. García-Bocanegra is full professor in the Department of Animal Health, Veterinary Faculty, University of Córdoba, Spain. He is currently head of the Animal Health and Zoonosis Research Group (GISAZ). His main topics of research are the epidemiology and control of infectious diseases shared between wildife and livestock under the One Health approach.

Preface to "Parasites and Wildlife"

In a changing world, most emerging infectious diseases have been propagated from the wilderness to anthropized areas. Today, research on parasites that may affect the biology and population equilibrium of their hosts is of major interest, especially when a One Health perspective is considered.

The present Special Issue aims to provide a collection of comprehensive investigations falling within several areas of interest, such as epidemiology, diagnosis, emerging zoonoses, food safety, conservation issues, parasite–host interactions, and pathology, in infections caused by parasites in wild host species.

Rafael Calero-Bernal and Ignacio García-Bocanegra
Editors

animals

MDPI

Editorial

Parasites and Wildlife

Rafael Calero-Bernal [1,*] and Ignacio García-Bocanegra [2,*]

1 Salud Veterinaria y Zoonosis (SALUVET), Department of Animal Health, Faculty of Veterinary, Complutense University of Madrid, 28040 Madrid, Spain
2 Grupo de Investigación en Sanidad Animal y Zoonosis (GISAZ), Departamento de Sanidad Animal, UIC Zoonosis y Enfermedades Emergentes ENZOEM, Universidad de Córdoba, 14014 Córdoba, Spain
* Correspondence: r.calero@ucm.es (R.C.-B.); nacho.garcia@uco.es (I.G.-B.)

Macro and micro-parasites are integrated into ecosystems worldwide and are considered important elements of biodiversity. Close relationships between parasites and vertebrate hosts cause evolutionary changes in each, which are sometimes driven by the heterogeneity of elements composing the ecosystems. Knowledge of parasites affecting human, domestic, and wild animal species requires, therefore, a deep understanding of the complex network created by the interactions between these pathogens, host species, and ecosystems. Nowadays, researchers face the challenge of studying the unpredictable change of these interactions as consequences of natural and anthropic disequilibrium.

In a changing world, most transboundary diseases have been propagated from the wilderness to anthropized areas. During the last decades, societies are increasingly concerned about the relevance of wildlife as a fundamental part of Global Health. In this respect, research on parasites that may affect the biology and population equilibrium of wildlife is of major interest, especially when a One Health perspective is considered.

The present Special Issue "Parasites and Wildlife" was aimed at providing a collection of comprehensive investigations falling within a number of areas of interest such as epidemiology, diagnosis, emerging zoonoses, food safety, conservation issues, parasite–host interactions, and pathology in infections caused by parasites in wild host species.

In sum, eight articles, two communications, and two brief reports from over 103 authors based in field areas of nine countries have been compiled. The pretended international dimension of the proposal was achieved given the fact that 44 institutions from 18 countries were involved (Figure 1a). Investigations were focused on 51 wild, including avian and mammal, host species and were focused on several topics represented by the keywords summarized in Figure 1b.

Citation: Calero-Bernal, R.; García-Bocanegra, I. Parasites and Wildlife. *Animals* **2023**, *13*, 628. https://doi.org/10.3390/ani13040628

Received: 1 February 2023
Revised: 8 February 2023
Accepted: 9 February 2023
Published: 10 February 2023

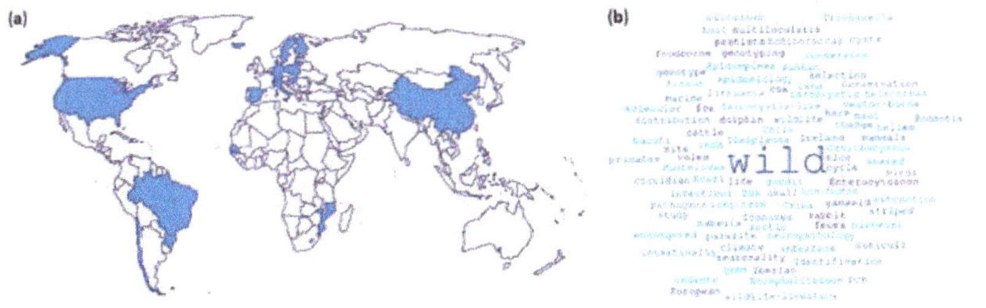

Figure 1. (**a**) Worldwide research institutions (n = 44) from 18 countries participating in the interdisciplinary Special Issue on Parasites and Wildlife. (**b**) Word cloud with the keywords utilized by each research paper. Brazil (n = 2): University of São Paulo, University Santo Amaro; Chile (n = 2): Universidad de Chile, Universidad de Concepción; China (n = 3): Guizhou University, Provincial Key Laboratory for Agricultural Pest Management in Mountainous Region, Dali University; Denmark

(*n* = 3): University of Copenhagen, Technical University of Denmark, Statens Serum Institut; Estonia (*n* = 2): Estonian Veterinary and Food Laboratory, Estonian University of Life Sciences; Finland (*n* = 1): University of Helsinki; Germany (*n* = 2): Leibniz Institute for Zoo and Wildlife Research, Freie Universität Berlin; Hungary (*n* = 2): University of Veterinary Medicine, Centre for Agricultural Research; Iceland (*n* = 1): Icelandic Institute of Natural History; Italy (*n* = 5): Istituto Zooprofilattico Sperimentale del Piemonte, Liguria e Valle d'Aosta, Local Veterinary Services in Imperia, University of Siena, University of Teramo, Local Veterinary Services in Asti; Lithuania (*n* = 2): Vytautas Magnus University, Nature Research Centre; Mozambique (*n* = 1): Universidade Licungo; Poland (*n* = 1): National Veterinary Research Institute; Senegal (*n* = 1): Jane Goodall Institute Senegal; Spain (*n* = 11): Institutes of Health Carlos III, Jane Goodall Institute Spain, Complutense University of Madrid, University of Las Palmas de Gran Canaria, Mundomar, Universidad de Córdoba, Universidad de Castilla-La Mancha, Consejo Superior de Investigaciones Científicas, Universidad de Málaga, Universidad San Pablo-CEU, Junta de Andalucía; Sweden (*n* = 1): Lund University; Switzerland (*n* = 1): University of Zurich; USA (*n* = 3): Ohio State University, Department of Health and Human Services, Food and Drug Administration, United States Department of Agriculture.

An overview of the research compiled according to the taxonomic organization and complexity of the target organisms (fungi, protozoa, helminths, and arthropods) is presented below.

In recent years, investigations on protozoan parasites have significantly increased and this is reflected in the present collection. Köster et al. [1] investigated, by means of molecular methods, the occurrence and genetic diversity of intestinal and blood protozoa as well as filariae in faecal samples from wild chimpanzees (*Pan troglodytes verus*) in the Dindefelo Community Nature Reserve, Senegal. Findings suggested that chimpanzees might play a more complex role in the epidemiology of pathogenic and commensal fungal, protozoan, and nematode species than initially anticipated, and two important facts contributed to the interest of such research: (i) wild chimpanzee populations in West Africa have dramatically decreased as a direct consequence of anthropogenic activities and infectious diseases, and (ii) findings suggested potential cross-species transmission between wild chimpanzees and humans in areas where both species overlap. Indeed, the presence of zoonotic Microsporidia was investigated by Martínez-Padilla et al. [2] in kidney samples of European wild rabbit (*Oryctolagus cuniculus*) and Iberian hare (*Lepus granatensis*) consumed by humans in Spain. Molecular methods allowed the detection of *Enterocytozoon bieneusi* and *Encephalitozoon intestinalis* but not *E. hellem* nor *E. cuniculi*. Despite the fact that the presence of the above agents may pose a public health concern, additional studies are required to define the frequency and characterize the real potential associated with zoonotic risk.

The group of Apicomplexan parasites have been extensively addressed in this Special Issue. Prakas et al. [3] searched for the role of mustelids as definitive hosts in the transmission of various *Sarcocystis* spp. in Lithuania. Intestinal scraping of five species of mustelids was examined by molecular methods and four species of *Sarcocystis* including the zoonotic *S. hominis* were detected. The study provided strong evidence that members of the Mustelidae family serve as definitive hosts for *Sarcocystis* spp. using cattle as intermediate hosts and raising important connections between carnivore wildlife and livestock. In a similar line, Giorda et al. [4] investigated the presence of tissue cysts of an acknowledged *Sarcocystis*-like organism in the brain and muscle of two stranded striped dolphins (*Stenella coeruleoalba*) in the Ligurian coast of Italy. Non-suppurative meningoencephalitis was associated with the presence of genetic variants that showed the highest homology to *Sarcocystis* spp. infecting the Bovidae family. The present study added valuable information on the complex epidemiology of the Sarcocystis genus. Evidence of sarcocystid infection was also investigated by Acosta et al. [5] in three species of seabirds from Magdalena Island, Chile. Findings of ITS1, 18S, and cox1 nucleotide sequences from muscle tissues of two skuas, revealed closely related homologous sequences of *Sarcocystis halieti*, that is a species found in seabirds of the northern hemisphere. Further studies are needed to understand the epidemiology of the infection and its impact on the health of marine fauna.

Barroso et al. [6] carried out a serosurvey to investigate the potential environmental and host-dependent variables that may affect the prevalence of antibodies of the zoonotic protozoan *Toxoplasma gondii* in wild boar (*Sus scrofa*), red deer (*Cervus elaphus*), and fallow deer (*Dama dama*) present in the Doñana National Park, Spain. The high seroprevalence values detected in the three wild ungulate species analyzed suggested that the complex interplay of hosts and eco-epidemiological niches, along with the optimal climatic conditions for the oocysts' survival, may favor the spread of the parasite in its host community in the study area. The concomitance of effects among the species indicated that relevant drivers of risk operated at the community level.

The zoonotic character and the complex epidemiology of some helminthic genuses motivates continuous research. In this context, Skrzypek et al. [7] investigated, by means of molecular methods, the presence of the important zoonotic *Echinococcus multilocularis* cestode in fecal samples of red foxes (*Vulpes vulpes*) from Poland. Two DNA extraction methods were compared aiming at improving the sensitivity of the diagnostic assay. High prevalence figures detected indicated the importance of the host in the life cycle of the parasite. In addition, these figures implied that both extraction methods showed similar efficiency in DNA isolation and dealing with inhibitors. The number of worms detected in the intestines had no influence on the PCR results. Additional aspects related to the degradation of genetic material potentially harming the reliability of the diagnostic methods used on field-collected samples were highlighted.

Land-use changes are one of the most important drivers of zoonotic disease risk in humans, including parasites of wildlife origin. The effect of this anthropogenic land-use change on the parasitism (presence and prevalence) of intestinal helminths in wild rodents was investigated by Riquelme et al. [8] in central Chile. Despite the fact that the overall helminth prevalence was 16.95%, and some zoonotic species (*Hymenolepis* spp.) were present, the effect of habitat type, native forests, and adult and young pine plantations, the prevalence was not observed, while other factors, such as rodent species and season of the year, were relevant to explain changes in helminth prevalence. Therefore, additional investigations focused on habitat alterations and potential changes in parasitofauna are encouraged. Kärssin et al. [9] carried out a survey on the presence of *Trichinella* spp. in four wild free-ranging host species, including wild boars, brown bears, Eurasian lynxes, and badgers, in Estonia. All four European *Trichinella* species, including non-encapsulated *T. pseudospiralis*, were detected, and results indicated high infection pressure in the sylvatic cycles across the years—illustrating the continuous risk of spillover to domestic cycles and of transmission to humans.

Currently, the evolving arthropod populations and their impact as disease vectors are undeniable. Yin et al. [10] investigated the potential host selection of the tropical rat gamasid mite (*Ornithonyssus bacoti*) on different animal hosts and the distribution in different environmental gradients in Yunnan Province of Southwest China. Different parameters including mite abundance were evaluated in 15 host species including rodents and other small mammals. In sum, the main reservoirs of *O. bacoti* were the synanthropic rat species, *Rattus tanezumi*, and *R. norvegicus*, and along with the observed major distribution in the flatland landscape and indoor habitats raises important questions on the impact on the public and on the health of animals, as well as the interconnections of the sylvatic and domestic life cycles of the mite. Two studies approached the occurrence of vector-borne pathogens in mammal wild hosts; Hornok et al. [11] molecularly screened for vector-borne protozoan parasites and bacteria potentially present in liver tissues of the only native terrestrial mammal in Iceland, the arctic fox (*Vulpes lagopus*). In this pioneering study in Europe, and considering the importance of the vector tick species *Ixodes ricinus*, only DNA of *Anaplasma phagocytophilum* was detected. Results provide valuable information as baseline data for comparison in the tfuture monitoring of the emergence of ticks and tick-borne diseases, especially when considering the incoming warming climate and the predictable increase of the presence of *I. ricinus* in the area. In addition, Mardosaitė-Busaitienė et al. [12] investigated the role of the reservoir host of eight species of rodents for emerging *Babesia microti* in Lithuania. The remarkable prevalence of *B. microti* was detected in the species investigated and parasites identified presented molecular homology with zoonotic strains; in addition, a comprehensive ecological study allowed the identification of a gradient of prevalences in different habitats compatible with the sustainability of the life cycle.

Overall, the papers in this Special Issue reveal different perspectives of current research on parasitic diseases and their relationship with the wildlife compartment; it is clear that it is necessary to continue these field-based studies to unravel the intricate interactions between pathogens and wildlife, livestock, and humans, confirming the three compartments of the One Health approach.

Funding: TED2021-132599B-C22 project, funded by MCIN/AEI/10.13039/501100011033 and by the European Union "NextGenerationEU"/PRTR. Recovery, Transformation, and Resilience Plan—funded by the European Union—NextGenerationEU. CIBER—Consorcio Centro de Investigación Biomédica en Red- (CB 2021/13/00083), Instituto de Salud Carlos III, Ministerio de Ciencia e Innovación and Unión Europea—NextGenerationEU. Project LagoHealth; reference PID2019-111080RB-C21, Spanish Ministry of Science and Innovation.

Acknowledgments: We would like to thank all of the authors who contributed their papers to this Special Issue, and the reviewers for their helpful recommendations. Finally, we are also grateful to Moisés Gonzálvez for providing the cover image.

Conflicts of Interest: The authors declare no conflict of interest.

References

1. Köster, P.C.; Renelies-Hamilton, J.; Dotras, L.; Llana, M.; Vinagre-Izquierdo, C.; Prakas, P.; Sneideris, D.; Dashti, A.; Bailo, B.; Lanza, M.; et al. Molecular Detection and Characterization of Intestinal and Blood Parasites in Wild Chimpanzees (*Pan troglodytes verus*) in Senegal. *Animals* **2021**, *11*, 3291. [CrossRef] [PubMed]
2. Martínez-Padilla, A.; Caballero-Gómez, J.; Magnet, Á.; Gómez-Guillamón, F.; Izquierdo, F.; Camacho-Sillero, L.; Jiménez-Ruiz, S.; del Águila, C.; García-Bocanegra, I. Zoonotic Microsporidia in Wild Lagomorphs in Southern Spain. *Animals* **2020**, *10*, 2218. [CrossRef] [PubMed]
3. Prakas, P.; Balčiauskas, L.; Juozaitytė-Ngugu, E.; Butkauskas, D. The Role of Mustelids in the Transmission of *Sarcocystis* spp. Using Cattle as Intermediate Hosts. *Animals* **2021**, *11*, 822. [CrossRef] [PubMed]
4. Giorda, F.; Romani-Cremaschi, U.; Marsh, A.E.; Grattarola, C.; Iulini, B.; Pautasso, A.; Varello, K.; Berio, E.; Gazzuola, P.; Marsili, L.; et al. Evidence for Unknown *Sarcocystis*-Like Infection in Stranded Striped Dolphins (*Stenella coeruleoalba*) from the Ligurian Sea, Italy. *Animals* **2021**, *11*, 1201. [CrossRef] [PubMed]
5. Acosta, I.C.L.; Gennari, S.M.; Llano, H.A.B.; Muñoz-Leal, S.; Soares, R.M. Molecular Characterization of New Haplotype of Genus *Sarcocystis* in Seabirds from Magdalena Island, Southern Chile. *Animals* **2021**, *11*, 245. [CrossRef] [PubMed]
6. Barroso, P.; García-Bocanegra, I.; Acevedo, P.; Palencia, P.; Carro, F.; Jiménez-Ruiz, S.; Almería, S.; Dubey, J.P.; Cano-Terriza, D.; Vicente, J. Long-Term Determinants of the Seroprevalence of *Toxoplasma gondii* in a Wild Ungulate Community. *Animals* **2020**, *10*, 2349. [CrossRef] [PubMed]
7. Skrzypek, K.; Karamon, J.; Samorek-Pieróg, M.; Dąbrowska, J.; Kochanowski, M.; Sroka, J.; Bilska-Zając, E.; Cencek, T. Comparison of Two DNA Extraction Methods and Two PCRs for Detection of *Echinococcus multilocularis* in the Stool Samples of Naturally Infected Red Foxes. *Animals* **2020**, *10*, 2381. [CrossRef]
8. Riquelme, M.; Salgado, R.; Simonetti, J.A.; Landaeta-Aqueveque, C.; Fredes, F.; Rubio, A.V. Intestinal Helminths in Wild Rodents from Native Forest and Exotic Pine Plantations (*Pinus radiata*) in Central Chile. *Animals* **2021**, *11*, 384. [CrossRef]
9. Kärssin, A.; Häkkinen, L.; Vilem, A.; Jokelainen, P.; Lassen, B. *Trichinella* spp. in Wild Boars (*Sus scrofa*), Brown Bears (*Ursus arctos*), Eurasian Lynxes (*Lynx lynx*) and Badgers (*Meles meles*) in Estonia, 2007–2014. *Animals* **2021**, *11*, 183. [CrossRef] [PubMed]
10. Yin, P.-W.; Guo, X.-G.; Jin, D.-C.; Fan, R.; Zhao, C.-F.; Zhang, Z.-W.; Huang, X.-B.; Mao, K.-Y. Distribution and Host Selection of Tropical Rat Mite, *Ornithonyssus bacoti*, in Yunnan Province of Southwest China. *Animals* **2021**, *11*, 110. [CrossRef] [PubMed]
11. Hornok, S.; Mühldorfer, K.; Takács, N.; Hofmann-Lehmann, R.; Meli, M.L.; Gyuranecz, M.; Unnsteinsdóttir, E.R.; Greenwood, A.D.; Czirják, G.Á. Broad Range Screening of Vector-Borne Pathogens in Arctic Foxes (*Vulpes lagopus*) in Iceland. *Animals* **2020**, *10*, 2031. [CrossRef] [PubMed]
12. Mardosaitė-Busaitienė, D.; Radzijevskaja, J.; Balčiauskas, L.; Paulauskas, A. *Babesia microti* in Rodents from Different Habitats of Lithuania. *Animals* **2021**, *11*, 1707. [CrossRef] [PubMed]

animals

MDPI

Article

Molecular Detection and Characterization of Intestinal and Blood Parasites in Wild Chimpanzees (*Pan troglodytes verus*) in Senegal

Pamela C. Köster [1,†], Justinn Renelies-Hamilton [2,†], Laia Dotras [3], Manuel Llana [3], Celia Vinagre-Izquierdo [4], Petras Prakas [5], Donatas Sneideris [5], Alejandro Dashti [1], Begoña Bailo [1], Marta Lanza [1], Alejandra Jiménez-Mejías [1], Carlota Muñoz-García [1], Aly S. Muadica [1,6], David González-Barrio [1], José M. Rubio [1], Isabel Fuentes [1], Francisco Ponce-Gordo [7], Rafael Calero-Bernal [8,*] and David Carmena [1,*]

1 Parasitology Reference and Research Laboratory, Spanish National Centre for Microbiology, 28220 Majadahonda, Spain; pamelakster@yahoo.com (P.C.K.); dashti.alejandro@gmail.com (A.D.); BEGOBB@isciii.es (B.B.); mlanza@isciii.es (M.L.); alejandrajimenezmejias@gmail.com (A.J.-M.); carlota.mg114@gmail.com (C.M.-G.); muadica@gmail.com (A.S.M.); dgonzalezbarrio@gmail.com (D.G.-B.); jmrubio@isciii.es (J.M.R.); ifuentes@isciii.es (I.F.)
2 Section for Ecology and Evolution, Department of Biology, University of Copenhagen, DK-1165 Copenhagen, Denmark; claxon71@gmail.com
3 Jane Goodall Institute Spain and Senegal, Dindefelo Biological Station, Dindefelo, Kedougou, Senegal; laia.dotras@janegoodall.es (L.D.); manuel.llana@janegoodall.es (M.L.)
4 Evolutionary Ecology, Biology Department, Lund University, 223 62 Lund, Sweden; celia.vinagre.izquierdo@gmail.com
5 Nature Research Centre, LT-08412 Vilnius, Lithuania; prakaspetras@gmail.com (P.P.); donatas.sneideris@gmail.com (D.S.)
6 Departamento de Ciências e Tecnologia, Universidade Licungo, Quelimane 106, Mozambique
7 Department of Microbiology and Parasitology, Faculty of Pharmacy, Complutense University of Madrid, 28040 Madrid, Spain; pponce@ucm.es
8 Salud Veterinaria y Zoonosis (SALUVET), Department of Animal Health, Faculty of Veterinary, Complutense University of Madrid, 28040 Madrid, Spain
* Correspondence: r.calero@ucm.es (R.C.-B.); dacarmena@isciii.es (D.C.)
† These two authors contributed equally to this work.

Citation: Köster, P.C.; Renelies-Hamilton, J.; Dotras, L.; Llana, M.; Vinagre-Izquierdo, C.; Prakas, P.; Sneideris, D.; Dashti, A.; Bailo, B.; Lanza, M.; et al. Molecular Detection and Characterization of Intestinal and Blood Parasites in Wild Chimpanzees (*Pan troglodytes verus*) in Senegal. *Animals* **2021**, *11*, 3291. https://doi.org/10.3390/ani11113291

Academic Editor: Theo De Waal

Received: 16 June 2021
Accepted: 11 November 2021
Published: 17 November 2021

Simple Summary: Western chimpanzees are currently listed as a Critically Endangered subspecies. Human encroachment has taken a toll on this great ape due to fragmented habitat and the exchange of pathogens. This epidemiological study investigated the occurrence and genetic diversity of intestinal and blood parasites in faecal samples from wild chimpanzees living in the Dindefelo Community Nature Reserve, Senegal. We paid special attention to potential human-driven sources of infection and transmission pathways. Potential diarrhoea-causing protist parasites (e.g., *Cryptosporidium* spp., *Giardia duodenalis*, *Entamoeba histolytica*) were detected at low infection rates (and densities) or absent, whereas commensals (*Entamoeba dispar*) or protist of uncertain pathogenicity (*Blastocystis* sp.) were far more abundant. We detected *Sarcocystis* spp. in chimpanzee faeces. Blood protist parasites such as *Plasmodium* spp. and *Trypanosoma brucei* spp. (the etiological agents of malaria and sleeping sickness, respectively, in humans) were also found at low prevalences, but microfilariae of the nematode *Mansonella perstans* were frequently found. Molecular analyses primarily revealed host-adapted species/genotypes and an apparent absence of gastrointestinal clinical manifestations in infected chimpanzees. Zoonotic events of still unknown frequency and directionality may have taken part between wild chimpanzees and humans sharing natural habitats and resources.

Abstract: Wild chimpanzee populations in West Africa (*Pan troglodytes verus*) have dramatically decreased as a direct consequence of anthropogenic activities and infectious diseases. Little information is currently available on the epidemiology, pathogenic significance, and zoonotic potential of protist species in wild chimpanzees. This study investigates the occurrence and genetic diversity of intestinal and blood protists as well as filariae in faecal samples (n = 234) from wild chimpanzees in the Dindefelo Community Nature Reserve, Senegal. PCR-based results revealed the presence of intestinal potential pathogens (*Sarcocystis* spp.: 11.5%; *Giardia duodenalis*: 2.1%; *Cryptosporidium hominis*:

0.9%), protist of uncertain pathogenicity (*Blastocystis* sp.: 5.6%), and commensal species (*Entamoeba dispar*: 18.4%; *Troglodytella abrassarti*: 5.6%). *Entamoeba histolytica*, *Enterocytozoon bieneusi*, and *Balantioides coli* were undetected. Blood protists including *Plasmodium malariae* (0.4%), *Trypanosoma brucei* (1.3%), and *Mansonella perstans* (9.8%) were also identified. Sanger sequencing analyses revealed host-adapted genetic variants within *Blastocystis*, but other parasitic pathogens (*C. hominis*, *P. malariae*, *T. brucei*, *M. perstans*) have zoonotic potential, suggesting that cross-species transmission between wild chimpanzees and humans is possible in areas where both species overlap. Additionally, we explored potential interactions between intestinal/blood protist species and seasonality and climate variables. Chimpanzees seem to play a more complex role on the epidemiology of pathogenic and commensal protist and nematode species than initially anticipated.

Keywords: protists; non-human primates; endangered; conservation; PCR; genotyping; epidemiology; zoonoses; seasonality

1. Introduction

Western chimpanzee populations (*Pan troglodytes verus*) have decreased by 80% between 1990 and 2014 in West Africa [1]. A combination of anthropogenic (agriculture, poaching, extractive industries, infrastructure development, human-chimpanzee interactions, lack of law enforcement) and natural (infectious diseases) threats are putting chimpanzees at the brink of extinction [2]. The breaking down of natural barriers has increased the likelihood of cross-species transmission of several zoonotic viral (rabies virus, Herpes B virus, Ebola virus, Yellow fever virus), bacterial (*Campylobacter jejuni*, enterotoxigenic *Escherichia coli*, *Klebsiella pneumoniae*, *Streptococcus pneumoniae*, *Streptococcus pyogenes*, *Staphylococcus aureus*, *Mycobacterium tuberculosis*) and parasitic (*Trypanosoma* spp., *Toxoplasma gondii*, *Babesia* spp., Coccidia, nematodes, and cestodes) infections that pose a threat both to endangered species (such as chimpanzees) and humans living nearby [3].

Little information is currently available on the epidemiology and potential health impact of zoonotic intestinal and blood protist species in wild chimpanzee populations. These include diarrhoea-causing intestinal protist species such as *Cryptosporidium* spp., *Giardia duodenalis*, and *Entamoeba histolytica* [4–7], and the microsporidia *Enterocytozoon bieneusi* [8]. Other protist species have uncertain or limited pathogenic significance, such as the stramenopile *Blastocystis* sp. [9] or the ciliate *Balantioides coli* [10]. Non-pathogenic protists species include *Entamoeba dispar* or the ape-adapted ciliate *Troglodytella abrassarti* [11]. Regarding blood parasites, malaria infections by at least seven *Plasmodium* species exhibiting strict host specificity are known to circulate in African great apes [12]. Thus, *Plasmodium reichenowi*, *Plasmodium billcollinsi*, *Plasmodium billbrayi* and *Plasmodium gaboni* infect only chimpanzees, while *Plasmodium praefalciparum*, *Plasmodium blacklocki* and *Plasmodium adleri* are restricted to gorillas. However, recent molecular studies have demonstrated that transfer of *Plasmodium* from humans towards chimpanzees is possible under confinement conditions [13], raising the question of whether cross-species transmission is also feasible under natural environmental conditions. Similarly, wild chimpanzees have been demonstrated to be suitable hosts for *Trypanosoma brucei*, the etiological agent of human and animal African trypanosomiasis [14]. However, it should be noted that assessing the role of non-human primates (NHP) in the epidemiology of *Plasmodium* and *Trypanosoma* parasites is severely hampered by the intrinsic difficulty of obtaining blood and tissue samples from the investigated animal populations, particularly in natural habitats. This issue is extensive also to filarial nematode parasites, whose microfilarial stages are circulating in the blood stream of the infected host. In this regard, DNA from *Mansonella* spp. has been identified in chimpanzee faecal samples from Gabon and Cameroon [15].

The protist enteroparasites *Cryptosporidium* spp., *G. duodenalis*, *Blastocystis* sp., and the microsporidia *E. bieneusi* exhibit extensive intra-species genetic diversity, allowing the identification of several genotypes/subtypes with marked differences in host specificity and range. *Cryptosporidium* encompasses at least 45 valid species, with *C. hominis* and *C. parvum* causing most of the infections reported in humans and NHP globally [16]. *Giardia duodenalis* is currently regarded as a multi-species complex comprising eight (A to H) distinct assemblages, of which zoonotic assemblages A and B are commonly reported to infect humans and other mammal species [17]. At least 25 subtypes (ST) have been identified within *Blastocystis* sp., of which ST1–9 and ST12 have been reported in humans [18]. Finally, more than 500 *E. bieneusi* genotypes have been defined and grouped in 11 phylogenetic groups. Group 1 and Group 2 include most of the potentially zoonotic genotypes, whereas the rest of the clusters display genotypes with strong host specificity [8].

Genotyping and subtyping methods are central in epidemiological studies to trace the origin of infections, understand the circulation of the pathogen in particular populations and geographical areas, provide information on transmission pathways, and assess the occurrence and directionality of potential zoonotic events. Data from this molecular-based epidemiological study aims to investigate the occurrence and genetic diversity of intestinal and blood parasites in faecal samples from wild chimpanzees living in the Dindefelo Community Nature Reserve (hereafter, Dindefelo) in Senegal, with special attention to potential human-driven sources of infection and cross-species transmission pathways.

2. Materials and Methods

2.1. Study Area

We collected faecal samples in Dindefelo, located in the south-eastern part of the Kedougou region, Senegal. This area is Sudano-Guinean woodland savannah mosaic intermixed with agricultural fields [19,20]. This highly seasonal habitat has a long dry season that lasts from November to May. Over 7000 people live in and around the reserve in 14 villages and hamlets [21]. In this anthropogenic landscape, the chimpanzees compete with the human population for food resources [22], making this area particularly well suited for transmission dynamics studies between humans and wild chimpanzees. The water scarcity during the dry season causes humans, livestock and wildlife, including chimpanzees, to use some of the same water sources [19].

2.2. Sampling and Data Collection

We collected 234 fresh (<24 h old) faecal samples of wild chimpanzee living in sympatry with humans. We were helped by experienced local field assistants to opportunistically identify and sample between November 2018 and March 2020. We obtained 5–10 g from the centre of faeces with sterile cotton swabs and stored them in 70% ethanol for preservation and transport to the lab. For each sample, we estimated the time in hours since defecation and noted faecal consistency using the Bristol Stool Scale (BSS), ranging 1–7 [23]. Additionally, we obtained daily rainfall data (GPM-3IMERGDF) [24] and surface air temperature (AIRS3STD; AIRS Science Team) [25] estimates from satellite data and averaged them across the 14,000 hectare study area. For rainfall data, we created a cumulative rainfall variable by adding the precipitation data from the 10 days before sampling.

2.3. DNA Extraction and Purification

We isolated genomic DNA from about 200 mg of each faecal specimen of chimpanzee by using the QIAamp DNA Stool Mini Kit (Qiagen, Hilden, Germany) according to the manufacturer's instructions, except that samples mixed with InhibitEX buffer were incubated for 10 min at 95 °C. Extracted and purified DNA samples were eluted in 200 µL of PCR-grade water and kept at 4 °C until further molecular analysis. We included a water extraction control in each sample batch processed.

2.4. Molecular Detection of Cryptosporidium spp.

We assessed the presence of *Cryptosporidium* spp. using a nested-PCR protocol to amplify a 587 bp fragment of the *ssu* rRNA gene of the parasite [26]. Amplification reactions (50 µL) included 3 µL of DNA sample and 0.3 µM of the primer pairs CR-P1/CR-P2 in the primary reaction and CR-P3/CPB-DIAGR in the secondary reaction (Table S1). Both PCR reactions were carried out as follows: one step of 94 °C for 3 min, followed by 35 cycles of 94 °C for 40 s, 50 °C for 40 s and 72 °C for 1 min, concluding with a final extension of 72 °C for 10 min. Sub-typing of the isolates identified as *C. hominis* was attempted at the *gp60* gene using the AL-3531/AL-3535 and AL-3532/AL-3534 primer pairs [27].

2.5. Molecular Differential Detection of Entamoeba histolytica and Entamoeba dispar

We carried out detection and differential diagnosis between pathogenic *E. histolytica* and non-pathogenic *E. dispar* by a qPCR method targeting a 172 bp fragment of the gene codifying the *ssu* rRNA gene of *E. histolytica* and *E. dispar* [28,29]. Amplification reactions (25 μL) consisted of 3 μL template DNA, 12.5 pmol of the primer set Ehd-239F/Ehd-88R, 5 pmol of each specific TaqMan® probe (Table S1), and TaqMan® Gene Expression Master Mix (Applied Biosystems). Detection of parasitic DNA was performed on a Corbett Rotor GeneTM 6000 real-time PCR system (QIAGEN) using an amplification protocol consisting of an initial hold step of 2 min at 55 °C and 15 min at 95 °C followed by 45 cycles of 15 s at 95 °C and 1 min at 60 °C. We included molecular biology grade water (no-template, negative Nzytech, Lisbon, Portugal) and genomic DNA (positive) controls in each PCR run.

2.6. Molecular Detection and Characterization of Giardia Duodenalis

We conducted *G. duodenalis* DNA detection using a real-time PCR (qPCR) method targeting a 62 bp region of the gene codifying the small subunit ribosomal RNA (*ssu* rRNA) of the parasite [30]. Amplification reactions (25 μL) consisted of 3 μL of template DNA, 0.5 μM of each primer Gd-80F and Gd-127R, 0.4 μM of probe (Table S1), and 12.5 μL TaqMan® Gene Expression Master Mix (Applied Biosystems, Waltham, MA, USA). Cycling conditions and data analysis were as described above for the detection of *E. histolytica/E. dispar*.

We subsequently assessed *G. duodenalis* isolates that tested positive by qPCR by sequence-based multi-locus genotyping of the genes encoding for the glutamate dehydrogenase (*gdh*) [31], β-giardin (*bg*) [32], and triose phosphate (*tpi*) [33] proteins of the parasite. We conducted amplifications by semi-nested and nested PCR protocols using specific primer pairs (Table S1).

2.7. Molecular Detection of Sarcocystis spp.

We detected *Sarcocystis* spp. by nested PCR amplifying a 550–600 bp fragment of the *ssu* rRNA gene. Final volumes (25 μL) of reaction mixtures included 3 μL of DNA sample and 0.4 μM of the outer Sgrau183/Sgrau182 and the inner Spri1/Spri2 primer sets (Table S1). Cycling conditions were as follows: enzyme activation at 94 °C for 5 min, followed by 30 cycles of 94 °C for 45 s, 59 °C for 50 s and 72 °C for 1 min, concluding with a final extension step of 72 °C for 10 min. PCR conditions for primary and secondary reactions were the same.

2.8. Molecular Detection and Characterization of Blastocystis sp.

We identified *Blastocystis* sp. by a direct PCR protocol targeting the *ssu* rRNA gene of the parasite [34]. The assay uses the pan-*Blastocystis*, barcode primer pair RD5/BhRDr to amplify a PCR product of ~600 bp. Amplification reactions (25 μL) included 5 μL of template DNA and 0.5 μM of each primer (Table S1). Amplification conditions consisted of one step of 95 °C for 3 min, followed by 30 cycles of 1 min each at 94, 59 and 72 °C, with an additional 2 min final extension at 72 °C.

2.9. Molecular Detection and Characterization of Enterocytozoon bieneusi

We conducted *E. bieneusi* detection by a nested PCR protocol to amplify the internal transcribed spacer (ITS) region as well as portions of the flanking large and small subunit of the ribosomal RNA gene as previously described [35]. We used the outer EBITS3/EBTIS4 and inner EBITS1/EBITS2.4 primer sets (Table S1) to generate a final PCR product of 390 bp, respectively. PCR reactions (50 μL) consisted of 1 μL of template DNA and 0.2 μM of each primer. Cycling conditions for the primary PCR consisted of one step of 94 °C for 3 min, followed by 35 cycles of amplification (denaturation at 94 °C for 30 s, annealing at 57 °C for 30 s, and elongation at 72 °C for 40 s), with a final extension at 72 °C for 10 min. Conditions for the secondary PCR were identical to the primary PCR except only 30 cycles were carried out with an annealing temperature of 55 °C.

2.10. Molecular Detection of Balantioides coli

We attempted *B. coli* detection by a direct PCR assay to amplify the complete ITS1–5.8s-rRNA–ITS2 region and the last 117 bp (3′ end) of the *ssu*-rRNA sequence of this ciliate using the primer set B5D/B5RC [36]. PCR reactions (25 µL) consisted of 2 µL of template DNA and 0.4 µM of each primer (Table S1). PCR conditions were as follows: 94 °C for 10 min; 30 cycles of 94 °C for 1 min, 60 °C for 1 min, 72 °C for 1 min, and a final extension for 5 min at 72 °C.

2.11. Molecular Detection of Troglodytella spp.

We carried out this ciliate mutualist's detection by a direct PCR method targeting a 401 bp fragment of the ITS region of the rDNA (ITS1-5.8S rDNA-ITS2) [37]. PCR reactions (25 µL) contained 2 µL of template DNA and 0.8 µM of each primer SSU-end/LSU-start (Table S1). Conditions of PCR for ITS amplification were initial denaturation for 2 min at 94 °C, 35 cycles of 45 s at 94 °C, 45 s at 50 °C, and 90 s at 72 °C, and terminal elongation for 5 min at 72 °C.

2.12. Molecular Detection of Plasmodium spp.

We performed *Plasmodium* spp. detection using a qPCR protocol targeting the *ssu* rRNA. Expected amplicon sizes varied depending on the species of *Plasmodium*. In the case of those infective to humans, the range would be from 356 bp for *P. falciparum* to 417 bp for *P. malariae*. The PCR mixture consisted of 1× Quantimix Easy Probes (Biotools, B&M Labs SA, Madrid, Spain), 0.2 µM of each primer (universal primer forward JM-U-0011-L and *Plasmodium*-specific reverse primer PLR-1080, Table S1), 0.2 × EvaGreen® Dye (Biotium, Inc. Hayward, Fremont, CA, USA) and 5 µL of template DNA in a reaction volume of 20 µL. We used a Rotor-Gene Q (QIAGEN) to perform the amplification, beginning with 3 min at 95 °C, followed by 45 cycles of 15 s at 95 °C, and 45 s at 62 °C. A final step of the melt program consisting of stepwise temperature increases of 0.5 °C from 60 °C to 95 °C with fluorescence acquisition at each temperature transition was included. We determined positive samples by post-reaction analysis by melting temperature (Tm) curve (Tm = 77.5 °C ± 1.0 °C) and by 2% agarose or automatic gel electrophoresis and confirmed by sequencing.

2.13. Molecular Detection of Trypanosomatid Species

We performed trypanosomatidae detection using a qPCR targeting the *ssu* rRNA. Expected amplicon sizes varied depending on the species, being 521 bp for *T. brucei*. The PCR mixture consisted of 1× Quantimix Easy Probes (Biotools), 0.2 µM of each primer (universal primer forward JM-U-0011-L and trypanosomatidae-specific reverse primer JM-T-0012n-R, Table S1), 0.2 × EvaGreen® Dye (Biotium) and 5 µL of template DNA in a reaction volume of 20 µL. Cycling conditions, melt step, and electrophoretic procedures were as described in Section 2.12.

2.14. Molecular Detection of Filarial Species

We detected filarial species using a qPCR targeting the ITS1 region of the main filarial species infecting humans [38]. Expected amplicon sizes varied depending on the filarial species; within the genus *Mansonella*, the range would be from 250 bp for *M. streptocerca* to 312 bp for *M. perstans*. The PCR mixture consisted of 1 × Quantimix Easy Probes (Biotools), 0.2/0.375 µM of the primer set FIL2-F/FIL 2-Loa/FIL2-R (Table S1), 0.2 × EvaGreen® Dye (Biotium) and 5 µL of template DNA in a reaction volume of 20 µL. Conditions of PCR for

ITS1 amplification were initial denaturation for 3 min at 95 °C, followed by 45 cycles of 10 s at 95 °C, 15 s at 50 °C, and 15 s at 60 °C. A final step of melt program consisting of stepwise temperature increases of 0.5 °C from 60 °C to 95 °C with fluorescence acquisition at each temperature transition was included. We determined positive samples by post-reaction analysis by melting temperature (Tm) curve (Tm = 77.5 ± 1.0 °C) and by 2% agarose or automatic gel electrophoresis and confirmed by sequencing (see below).

2.15. Sequencing

We directly sequenced positive-PCR/qPCR products of the expected sizes in both directions using appropriate primer sets (Table S1). We used capillary DNA sequencing electrophoresis using the BigDye® Terminator chemistry (Applied Biosystems) on an ABI PRISM 3130 automated DNA sequencer.

We have deposited the sequences obtained in this study in GenBank under accession numbers MZ182323–MZ182324 (*C. hominis*), MZ182325–MZ182328 (*Blastocystis* sp.), MZ182329–MZ182352 (*Sarcocystis* spp.), MZ224016 (*T. abrassarti*), MZ272002 (*P. malariae*), MZ272003 (*Plasmodium* sp.), MZ272004–MZ272006 (*T. brucei*), and MZ285880–MZ285897 (*M. perstans*).

2.16. Phylogenetic Analysis

We inferred evolutionary relationships among the identified *Blastocystis* subtypes by a phylogenetic analysis using the neighbor-joining method in MEGA 6 [39]. The evolutionary distances were computed using the Kimura 2 parameter method and modelled with a gamma distribution. We estimated the reliability of the phylogenetic analyses at each branch node by the bootstrapping 1000 times. We retrieved *Blastocystis* sp. sequences identified in captive or free-living chimpanzees globally from the NCBI database and included in the phylogenetic analysis for reference and comparative purposes.

2.17. Statistical Analyses

We analysed the relationship between cumulative rainfall and temperature (independent variables) and each parasite's presence (for parasites with >5 positive samples; dependent variable) with generalized linear models using a binomial family (glm function in stats package v.3.6.1) [40].

Additionally, we divided the year into two main seasons: dry season (November through May) and rainy season (June through October). Again, we analysed the relationship between seasonality and parasite presence with generalized linear models in the same way as above.

We analysed the relationship between BSS (independent variable) and each parasite's presence (for parasites with >5 positive samples; dependent variable) with additional binomial generalized linear models (glm function in stats package v.3.6.1) [40]. Finally, we correlated rainfall and temperature with stool consistency by using a Spearman correlation [41].

2.18. Parasite Interactions

We analysed the pairwise relationship between parasites (by fitting Fisher´s tests [42] for all parasites with at least 10 positive samples. We corrected for multiple testing by using the Benjamini and Hochberg method [43]. Further, we plotted a Principal Component Analysis (PCA) using the function prcomp from the stats R package v.3.6.1. There, we included minimum surface air temperature averaged over 10 days before sampling, and cumulative rainfall totalled over 10 days before sampling and binomial variables for the six most prevalent parasites.

3. Results

We show the full dataset indicating main features of chimpanzee faecal samples at the time of collection and PCR and sequencing results for the intestinal and blood parasite and commensal species investigated in the present study in Table S2.

3.1. Prevalence of Intestinal and Blood Parasites

We report the PCR-based prevalence rates of intestinal and blood parasite species identified in the wild chimpanzee population under investigation in Table 1. Among intestinal protists, the most common pathogenic protist identified was *Sarcocystis* spp. (11.5%), followed by *G. duodenalis* (2.1%), and *C. hominis* (0.9%). *Entamoeba histolytica* was not detected. We found non-pathogenic *E. dispar* in 18.4% of samples, the stramenopile *Blastocystis* sp. at a prevalence rate of 5.6%, whereas the microsporidia *E. bieneusi* was undetected. Among ciliates, we found the commensal *T. abrassarti* in 5.6% of samples, but none of them tested positive for *B. coli*. We detected the DNA of blood protists belonging to the genera *Plasmodium* (*P. malariae*) and *Trypanosoma* (*T. brucei*) at low (<2%) infection rates. Finally, the filarial nematode *Mansonella perstans* was identified in 9.8% of samples.

Table 1. Prevalence of intestinal and blood parasite and commensal species identified in the wild chimpanzee population (n = 234) investigated in the present study.

Species	Positive (n)	Prevalence (%)	95% Confidence Interval
Intestinal protists			
Entamoeba dispar	43	18.4	13.6–23.9
Sarcocystis spp.	27	11.5	7.7–16.3
Blastocystis sp.	13	5.6	3.0–9.3
Troglodytella abrassarti	13	5.6	3.0–9.3
Giardia duodenalis	5	2.1	0.7–4.9
Cryptosporidium hominis	2	0.9	0.1–3.1
Balantioides coli	0	0.0	-
Entamoeba histolytica	0	0.0	-
Intestinal microsporidia			
Enterocytozoon bieneusi	0	0.0	-
Blood protists			
Trypanosoma brucei	3	1.3	0.27–3.7
Plasmodium malariae	1	0.4	0.01–2.4
Plasmodium spp.	1	0.4	0.01–2.4
Filarial nematodes			
Mansonella perstans	23	9.8	6.3–14.4

3.2. Molecular Characterization of Intestinal Protist Species

Sequence analyses revealed that the two *Cryptosporidium*-positive isolates identified in the present study corresponded to *C. hominis*. One of them had 100% homology with reference sequence AF108865, whereas the remaining isolate differed from it by a single nucleotide polymorphism (SNP) involving an ambiguous (double peak) site at position 805 (Table 2). Attempts to determine the genotype family of these isolates at the 60 kDa glycoprotein (*gp60*) locus repeatedly failed.

The five DNA isolates that yielded a positive result for *G. duodenalis* by qPCR generated cycle threshold (Ct) values ranging from 25.6–33.9 (median: 31.4). None of these isolates could be successfully amplified at the *gdh*, *bg*, or *tpi* loci, so the assemblage/sub-assemblage of the parasite involved in these infections remained unknown.

Table 2. Diversity, frequency, and molecular features of *Cryptosporidium hominis*, *Blastocystis* sp., and *Troglodytella abrassarti* isolates identified in the wild chimpanzee population investigated in the present study.

Species	Genotype	Sub-Genotype	No. Isolates	Locus	Reference Sequence	Stretch	Single Nucleotide Polymorphisms	GenBank ID
Cryptosporidium hominis	-	-	1	*ssu* rRNA	AF108865	574–997	None	MZ182324
	-	-	1	*ssu* rRNA	AF108865	579–983	C805Y	MZ182323
Blastocystis sp.	ST1	Allele 1	1	*ssu* rRNA	AB107968	79–568	A132G	MZ182325
	ST1	Allele 7	1	*ssu* rRNA	HQ286905	14–506	A476G	MZ182326
	ST1	Allele 8	10	*ssu* rRNA	HQ286907	1–553	T1A, 10_11DelAG, 71DelC, 513DelC, T551A, A553C	MZ182327
	ST1	Alleles 7 + 8	1	*ssu* rRNA	HQ286907	73–553	G141R, 513DelC, T551A, A553C	MZ182328
Troglodytella abrassarti	-	-	13	ITS	EU680311	1–418	T82C, G177A	MZ224016

Del: base deletion; ITS: Internal transcribed spacer; R: A/G; ssu rRNA: Small subunit ribosomal RNA; Y: C/T.

We detected DNA of *Sarcocystis* spp. in 27 samples. We obtained pure sequences in 24 cases, while we observed double peaks presumably indicating coexistence of two or more genetic variants of the parasite in three additional samples. One 526 bp sequence (MZ182329) showed 99.6–99.8% similarity with *S. alces* (EU282018, KF8312734) and up to 97% similarity with other *Sarcocystis* spp. Eight 552 bp sequences (MZ182330–MZ182337) shared 99.3–100% similarity with *S. gracilis* and <95% similarity with other *Sarcocystis* spp. Five 503 bp sequences (MZ182338–MZ182342) displayed 99.0–100% similarity with *S. morae* and <95% similarity with other *Sarcocystis* spp. Three 508 bp sequences (MZ182343–MZ182345) revealed highest similarity with *S. truncata* (98.1–100%) and *S. japonica* (97.7–99.6%). Five sequences (MZ182346–MZ182350) varying in length (503–509 bp) demonstrated highest similarity with *S. sinensis* (93.4–100%), *S. bovifelis* (96.1–99.8%) and *S. bovini* (96.7–99.2%). Finally, two 522 bp sequences (MZ182351–MZ182352) showed 97.1–99.6% and 98.1–99.6% similarity with *S. capracanis* and *S. tenella*, respectively. Thus, we identified at least six genetic variants of *Sarcocystis* genus in faecal samples of chimpanzees.

We confirmed a total of 13 isolates as *Blastocystis*-positive by Sanger sequencing. All of them were assigned to the subtype ST1 of the protist (Table 2). Allele 8 was the most common genetic variant found within ST1 (76.9%, 10/13), followed by allele 1 (7.7%, 1/13), allele 7 (7.7%, 1/13), and mixed alleles 7 + 8 (7.7%, 1/13). Phylogenetic analyses based on the neighbor-joining method confirmed that all ST1 sequences generated in the present study clustered together in well-supported clades with sequences of free-living chimpanzees from Senegal and Tanzania previously deposited in GenBank (Figure 1). Of note, an additional 42 isolates yielded amplicons of the expected size but in the form of faint bands on gel that produced poor quality, unreadable sequences. In the absence of Sanger sequencing confirmation, we conservatively considered these isolates *Blastocystis*-negative.

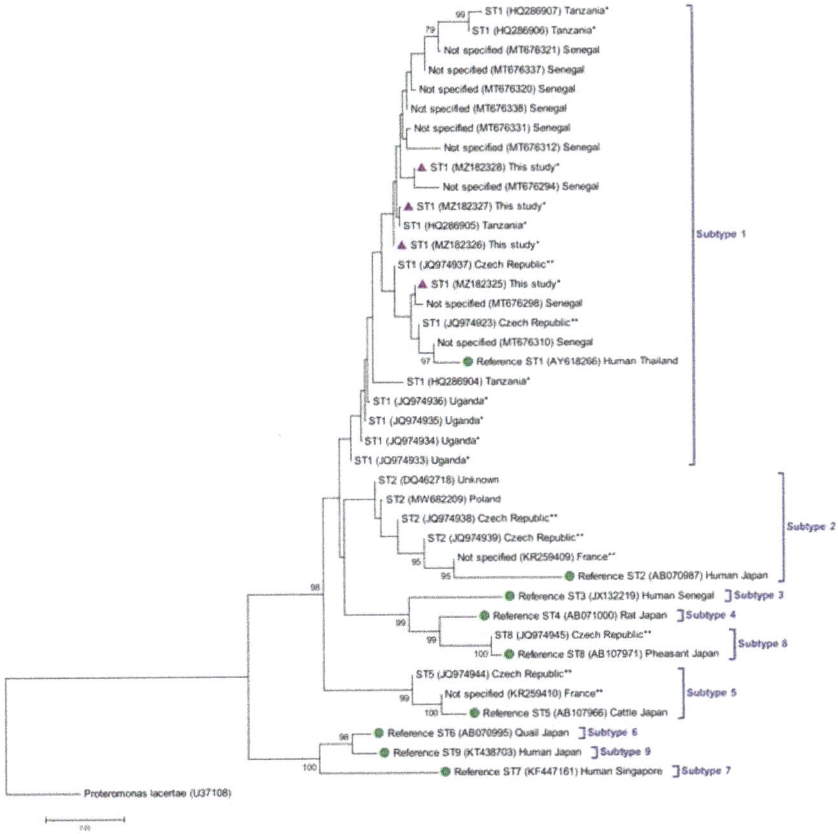

Figure 1. Phylogenetic relationships among *Blastocystis* sp. sequences identified in free-living and captive chimpanzees globally. GenBank accession numbers and country of origin are indicated. When known, sequences from free-living and captive chimpanzees were identified with superscript single and double asterisks, respectively. The analysis was conducted by a neighbor-joining method of the *ssu* rRNA gene. Genetic distances were calculated using the Kimura two-parameter model. Bootstrap values lower than 75% are not displayed. Purple-filled triangles represent sequences generated in the present study. Green filled dots represent reference sequences for subtypes ST1–ST9. *Proteromonas lacertae* was used as outgroup taxon to root the tree.

All 13 isolates yielding amplification products compatible with *T. abrassarti* were confirmed by Sanger sequencing. Representative sequences varied by two SNPs (T82C and G177A) with reference sequence EU680311 (Table 2).

3.3. Molecular Characterization of Blood Protist Parasites

Only two samples tested positive to *Plasmodium* spp. and an additional three to *T. brucei*. Plasmodia corresponded to *P. malariae* in one case and possibly also in the second (94% similarity with GenBank reference sequence XR_003751948.1). Phylogenetic analysis using the neighbor-joining method confirmed that both sequences belonged to the same cluster, although they were located in different tree branches (Figure S1). In the case of *T. brucei*, the three sequences generated were identical and fell into a well-defined cluster together with representative sequences of *T. brucei* (*T. brucei brucei*, *Tb rhodesiense*, *Tb gambiense*), *T. evansi*, and *T. equiperdum* retrieved from GenBank (Figure S2). Members of the above-mentioned species were integrated into the so-called brucei group [44]. The partial sequence of the *ssu* rRNA gene amplified by the PCR protocol used in the present study corresponded to a conserved region of the gene that does not allow for inter-species discrimination.

Sequence analyses of an additional 31 samples allowed the identification of other Trypanosomatida that are not considered parasites of primates. These include members of the genera *Phytomonas* ($n = 28$), *Herpetomonas* ($n = 2$) and *Lafontella* (previously *Herpetomonas*, $n = 1$), in addition to five cases of free-living Metakinetoplastin microorganisms of the orders Eubodonida ($n = 2$) and Neobodonida ($n = 3$). An ecological and epidemiological interpretation of these findings is provided in Table S1 (see text box on the upper right corner of the table).

3.4. Molecular Characterization of Filarial Parasites

We characterized 23 filarial nematodes by PCR and Sanger sequencing. All cases were homologous (≥99%) to *M. perstans*, although the sequences are not all the same as can be seen in the corresponding phylogenetic tree (Figure S3). In this case, all of them are integrated into the *M. perstans* cluster, close to the *M. streptocerca* cluster and further away from *M. ozzardi*.

3.5. Seasonality Effects on Parasites

Chimpanzee parasite prevalence differed by season (Figure 2; Table 3). *Mansonella perstans* had higher prevalence during the dry season than the rainy season ($p = 0.0241$), while the same is true for *E. dispar* ($p = 0.0202$).

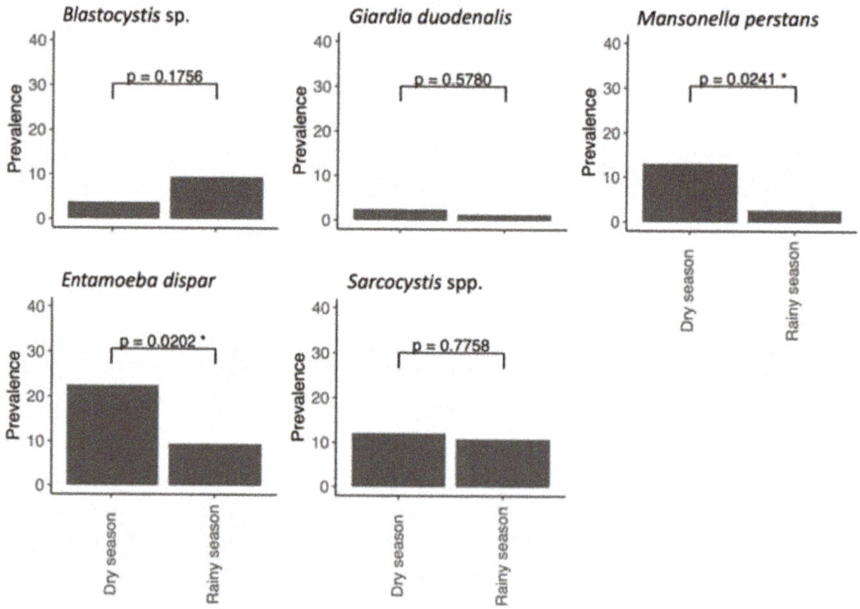

Figure 2. Chimpanzee parasite prevalence by season. Rainy season comprises months June through October, while dry season comprises November through May. See Table 3 for statistical differences between seasons. A superscript single asterisk denotes statistical significance.

Table 3. Results of the statistical analyses conducted to assess the relationship between the occurrence of the most frequent parasite species infecting chimpanzees and climatological (rainfall, temperature) or clinical (stool consistency) variables in the present study. For seasonality, estimates are given on the log odds ratio (not the response) scale. Statistically significant values are highlighted in bold.

Variable	Contrast	Estimate	SE	Z Ratio	*p* Value
Blastocystis sp.	Seasonality	0.4770	0.3520	1.355	0.1756
	Rainfall	0.0548	0.0235	2.330	**0.0198**
	Temperature min.	−0.0263	0.2118	−0.124	0.9010
	Stool consistency	0.1069	0.1632	0.655	0.5120
Giardia duodenalis	Seasonality	0.6270	1.1300	0.556	0.5780
	Rainfall	−0.4564	0.4465	−1.022	0.3067
	Temperature min.	1.7000	0.0754	2.255	**0.0241**
	Stool consistency	−0.0967	0.0460	−2.100	**0.0357**
Mansonella perstans	Seasonality	0.3569	0.1588	2.303	**0.0213**
	Rainfall	−0.0435	0.0237	−1.838	**0.0066**
	Temperature min.	−0.1081	0.1177	−0.918	0.3590
	Stool consistency	−0.1626	0.1151	−1.412	0.1579
Entamoeba dispar	Seasonality	−1.4600	0.5770	−2.525	0.0570
	Rainfall	0.2614	0.1591	0.164	0.1000
	Temperature min.	0.1270	0.4470	0.285	0.7758
	Stool consistency	0.0228	0.0200	1.141	0.2538
Sarcocystis spp.	Seasonality	1.7000	0.0754	2.255	**0.0241**
	Rainfall	−0.2865	0.1756	−1.631	0.1028
	Temperature min.	1.0200	0.4400	2.323	**0.0202**
	Stool consistency	−0.0435	0.0237	−1.838	**0.0066**
Bristol Stool Scale	Seasonality	−1.6200	0.2230	−7.285	**<0.0001**
	Rainfall *	$t = 5.961$	df = 210	-	**<0.0001**
	Temperature min. *	$t = 2.030$	df = 210	-	**0.04361**

* Rainfall data/temperature and stool consistency (as determined using the Bristol Stool Scale) are correlated with a Spearman correlation.

Climate variables also explained the variance in parasite prevalence in chimpanzee samples (Figure 3): *Blastocystis* sp. ($p = 0.0198$) was positively associated with the total rainfall during the 10 days before sampling, whereas *M. perstans* ($p = 0.0357$) and *E. dispar* ($p = 0.0066$) were negatively associated with it. *Mansonella perstans* ($p = 0.0213$) and *Sarcocystis* spp. ($p = 0.0338$) were positively and negatively associated with the mean minimum temperature averaged over the 10 days before sampling, respectively.

While the Bristol Stool Scale (BSS) shows that rainfall and temperature cause faecal samples to become more watery (higher levels on the BSS), the faecal consistency did not significantly explain variance in any parasite's prevalence (Table 3).

No negative or positive associations between parasites were detected (Table 4).

Table 4. Results of the statistical analyses conducted to assess potential relationships between prevalent parasites.

Parasite 1	Parasite 2	*p* Value	*p* Adjusted Value
Blastocystis sp.	*Giardia duodenalis*	1.0000	1.0000
Blastocystis sp.	*Mansonella perstans*	1.0000	1.0000
Blastocystis sp.	*Entamoeba dispar*	0.2654	1.0000
Blastocystis sp.	*Sarcocystis* spp.	1.0000	1.0000
Giardia duodenalis	*Mansonella perstans*	1.0000	1.0000
Giardia duodenalis	*Entamoeba dispar*	0.5875	1.0000
Giardia duodenalis	*Sarcocystis* spp.	0.4660	1.0000
Mansonella perstans	*Entamoeba dispar*	0.0874	0.8740
Mansonella perstans	*Sarcocystis* spp.	0.7384	1.0000
Entamoeba dispar	*Sarcocystis* spp.	0.7934	1.0000

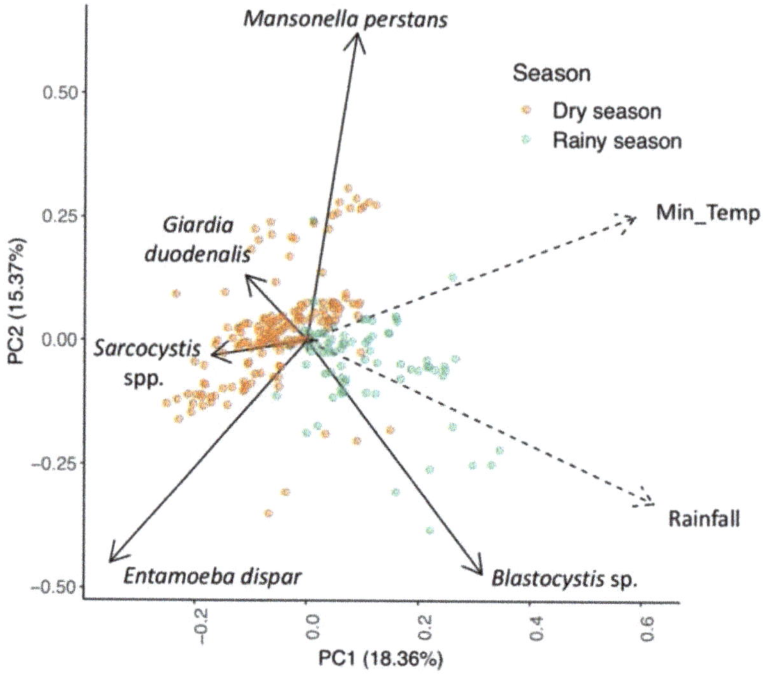

Figure 3. PCA plot of parasite presence in chimpanzee faecal samples. Each coloured dot indicates a single sample and is coloured by season. Solid arrows indicate parasite loading values, and dashed arrows indicate climate variable loading values. The closer two loading values are on the plot, the more likely they will correlate independently from other loading values. The distance from the coordinate centre (0, 0) can be used as a measure of strength of such associations. PC1 (positive values) captured high cumulative rainfall, while temperature was captured by a combination of PC1 and PC2 (positive values). Min_Temp represents minimum surface air temperature averaged over 10 days previous to sample collection. Rainfall represents cumulative rainfall totalled over the 10 days previous to sample collection.

4. Discussion

In this PCR-based epidemiological study, we investigated the occurrence and genetic diversity of six potentially pathogenic (*C. hominis*, *E. histolytica*, *G. duodenalis*, *Sarcocystis* spp., *E. bieneusi*, and *B. coli*) and three non-pathogenic (*Blastocystis* sp., *E. dispar*, and *T. abrassarti*) intestinal protist and microsporidia species, two blood protist parasites (*Plasmodium* spp., and *Trypanosoma* spp.), and a filarial nematode (*M. perstans*) in faecal samples from a wild western chimpanzee population living in Dindefelo, Senegal. Overall, this is one of the most complete molecular-based surveys on protist parasite species affecting endangered chimpanzees conducted in Africa to date. The main contributions of this study to the field include (i) the finding, for the first time, of *Sarcocystis* spp. DNA in faecal samples of wild chimpanzees globally, (ii) the confirmation that common diarrhoea-causing protist species including *C. hominis*, *E. histolytica*, *G. duodenalis* or the microsporidia *E. bieneusi* are present at low infection rates and parasitic intensities or absent in the surveyed chimpanzee population, (iii) the confirmation that wild chimpanzees may act as suitable hosts of *Plasmodium* and *Trypanosoma* species and filariae, playing a still not fully understood role in their epidemiology and transmission [14,45,46], and (iv) the provision of epidemiological evidence in support of the existence of overlapping domestic and sylvatic cycles and cross-species transmission suggesting that some of the infections detected in chimpanzees might be of anthropogenic nature.

Little information is currently available on the epidemiology of *Cryptosporidium* spp. in wild chimpanzees. Early epidemiological studies based on microscopy examination failed to detect the parasite in the Republic of Congo [47], whereas a prevalence of 9% was found in savannah chimpanzees in Tanzania [48]. Using PCR, *Cryptosporidium* spp. was undetected in wild chimpanzees from the same study area as the present study in Senegal [49], whereas an infection rate of 21% was reported in chimpanzees from the Greater Gombe Ecosystem in Tanzania [50]. In the present study, only 1% of chimpanzee samples harboured *Cryptosporidium* infections. Regarding molecular diversity, Tanzanian wild chimpanzees were found to be infected by *C. hominis* (n = 6) and *C. suis* (n = 7) [50]. Remarkably, two of the six *C. hominis* isolates were identified as IfA12G2, the same subgenotype detected in humans living in the proximity of the Gombe National Park. Of note, members of the *gp60* genotype family If have been found to be prevalent in different African countries including Kenya, South Africa, and Tanzania [51]. Taken together, these data strongly suggest that the *C. hominis* IfA12G2 infections detected in chimpanzees at the Greater Gombe Ecosystem were of anthropogenic origin [50]. In that very same study, the presence of *C. suis* in wild chimpanzees was interpreted as the result of cross-species transmission from bush pigs to chimpanzees in Gombe, since domesticated pigs were locally absent [50]. In the present study, the two *Cryptosporidium*-positive isolates were confirmed as *C. hominis*, but failure to amplify them at the *gp60* locus precluded us from identifying the *gp60* genotype family involved in those infections. More research should be conducted to elucidate the origin of these infections, as captive and wild NHP including chimpanzees can be infected by a wide range of *gp60* genotype families including Ia, Ib, Id, Ie, If, Ii, Ik, Im, and In [52].

Entamoeba spp. infection has been associated with morbidity and even mortality in NHP [53,54], although at present it remains to be fully elucidated if the *Entamoeba* species carried by NHP include the pathogenic *E. histolytica*. Indeed, *E. histolytica* was detected by PCR in wild and semi-wild orangutans in Sumatra, Indonesia [6]. In that study, contact with humans was considered as an important risk factor for infection of wild primates with this protist parasite. Molecular-based epidemiological surveys conducted in African wild chimpanzee populations have generated different, even discrepant, results. Thus, *E. histolytica* was undetected in populations, including those surveyed here, from Senegal [55] and Tanzania [56]. However, a subsequent study conducted in the latter country found an *E. histolytica* prevalence of 34% in chimpanzees from the Greater Gombe Ecosystem [57], whereas the protists were also present at low rates in chimpanzees living in the Dja Faunal Reserve in Cameroon [58]. These surveys coincided in proposing that interventions targeting better sanitation and hygiene practices for humans living in proximity with wild chimpanzee populations may help in preventing *E. histolytica* infection in NHP, while also protecting the endangered species.

The epidemiology of *G. duodenalis* in African wildlife is poorly understood. The few studies focusing on NHP, including lowland *(G. g. gorilla)* and mountain gorillas *(G. b. beringei)*, chimpanzees *(Pan troglodytes* schweinfurthii), baboons *(Papio anubis)*, greater bamboo lemurs (*Prolemur simus*) and eastern rufous mouse lemurs (*Microcebus rufus*), have been conducted in wildlife areas in Gabon, Madagascar, Rwanda, South Africa, and Tanzania [51]. In wild chimpanzee's populations, *G. duodenalis* has been reported at infection rates of 3-9% by immunofluorescence microscopy in the Republic of Congo [47], of 6% by light microscopy in the Cantanhez National Park in Guinea Bissau [59], and of 1% by PCR in Senegal [49]. In the latter study, the assemblage/sub-assemblage of the only *G. duodenalis*-positive isolate detected could not be determined due to insufficient amount of starting parasitic DNA. This is also the case in the present study, where the five samples positive to the protists failed to be amplified at the three loci (*gdh, bg, tpi*) used for genotyping purposes. These results strongly suggest low *G. duodenalis* intensity infections and, very likely, the absence of clinical manifestations in the infected chimpanzees. In other

African wild NHP species, sub-assemblage AII has been identified in western lowland gorillas in the Central African Republic [60], assemblage A and sub-assemblage BIV in mountain gorillas in Rwanda [61] and Uganda [62], and sub-assemblages AII and BIV in colobus monkeys in Ghana [63] and Uganda [64]. Remarkably, several of these studies highlighted that anthropogenic habitat disturbance is enhancing interactions among people, livestock, and wildlife, and this could have negative consequences for wildlife conservation [61,63,64].

Experimental infections demonstrate that NHP, including chimpanzees, may serve as definitive hosts for zoonotic *Sarcocystis hominis* and *Sarcocystis suihominis* [65,66]. These coccidians are characterised by an obligatory prey–predator two-host life cycle. Definitive hosts become infected by ingesting extra-intestinal tissues (usually muscles) containing mature sarcocysts, while intermediate hosts become infected through ingestion of food or water contaminated with sporocysts [66]. In the present study, several different *Sarcocystis* genetic variants were confirmed in 11.5% (27/243) samples by nested PCR targeting *ssu* rRNA; obtained sequences displayed the highest genetic similarity with GenBank out-of-Africa available sequences of *Sarcocystis* spp. using Cervidae (absent in sub-Saharan Africa), Bovini (*Bos taurus*, present in the region) and Caprinae (*Capra* sp. Djallonké and *Ovis* sp. Djallonké, present in the region) as intermediate hosts [67–69]. Other antelopes in the region include *Cephalophus rufilatus* and *Tragelaphus scriptus*. There are three non-exclusive hypotheses for the presence of *Sarcocystis* spp. in chimpanzee faecal samples: (i) chimpanzees have been infected by consuming muscle tissue of some of these wild and domestic prey hosts, (ii) chimpanzees have acquired cysts from infected prey, but are not infected and do not represent definitive hosts for these parasites, and (iii) chimpanzees have acquired *Sarcocystis* spp. from the environment. Therefore, the results presented here do not serve as definitive evidence for these *Sarcocystis* spp. infecting chimpanzees, nor for locally novel prey items for chimpanzees. What is novel, to the best of our knowledge, is the presence of *Sarcocystis* genetic variants belonging to species considered to have Cervidae as intermediate hosts, mostly absent in the African continent [67].

Blastocystis sp. has been identified in more than 75 species of captive and wild NHP from 21 different countries. In free-living NHP, *Blastocystis* sp. infection/colonisation rates have been reported in the range of 22–100% globally [9]. Specifically, in African wild chimpanzee populations *Blastocystis* sp. has been documented at a prevalence of 22% by conventional microscopy in Cameroon [70], and of 41–71% by PCR in Senegal [49] and Tanzania [71]. The 6% prevalence rate reported here is much lower than those described above, but it should be noted that this figure is a conservative estimation as only isolates confirmed by Sanger sequencing were considered as true *Blastocystis*-positive samples. Regarding genetic diversity, ST1 is known to be by far the most prevalent (82–100%) *Blastocystis* subtype circulating in wild chimpanzees in Senegal and Tanzania, although in the former country few animals carried also ST2 and ST3 [49,71]. In line with these previous results, ST1 was also the only subtype detected in our chimpanzee population. Remarkably, sequence data revealed that all ST1 isolates detected involved allele 8 and, to a much lesser extent, alleles 1 and 7. In contrast, human infections/colonisation by ST1 are primarily due to allele 4 in both developed or developing countries [72,73]. Taken together, this data suggests that different host-adapted genetic variants of ST1 are circulating in human and NHP populations.

In this study, *E. bieneusi* was undetected in wild chimpanzees. Although this microsporidia has been reported in 32 species of NHP from 10 countries, most of these studies were conducted in captive animals. In wild NHP, *E. bieneusi* infection rates have been documented in the range of 3-28% worldwide [8]. Regarding the presence of *E. bieneusi* in African chimpanzees, there is no information available in wild populations, but this parasite has been reported in chimpanzees living in sanctuaries in Kenya (3.6%) and Cameroon (4.5%) [74]. In other African wild NHP species, *E. bieneusi* has been identified at prevalence rates of 4.0% in western lowland gorillas (*G. g. gorilla*) in Central African Republic, of 12.3% in olive baboons (*P. anubis*) in Kenya, and of 18.0% in mountain gorillas (*G. b. beringei*) in

Rwanda [8]. Further studies are needed to fully elucidate the role of wild chimpanzees in the transmission of *E. bieneusi*.

Balantioides coli presence has been reported in a wide range of host species, including chimpanzees [10]. It is considered a rare finding in wild populations; however, it is fairly common in captive animals. On the other hand, *T. abrassarti* is frequently reported in wild individuals but usually with low prevalence in captive ones [75–79]. While differences in diet have been suggested as an explanation for this pattern, the reasons behind it remain unclear [77,78]; other factors, such as the ground use by the chimpanzees and their interaction with human populations, could also explain differences in *B. coli* prevalence between captive and wild chimpanzees [75,80]. In the present work, the chimpanzee population under study was of wild animals; however, the study area is an anthropogenic landscape with chimpanzee–human-livestock interaction in the use of food and water resources [19,22]. The negative results obtained here can be explained by either absence of the parasite in human and livestock populations of the region, which should be confirmed by studying these host species, or by the fact these chimpanzees are wild, while more consistent and intense contact between humans and chimpanzees, such as what occurs in captivity, is required for zoonotic transmission. Equally surprising is the low prevalence of *T. abrassarti*, detected only in 5.6% of analysed samples. This ciliate does not form cysts and transmission is by ingestion of trophozoites contaminating food; they can be detected in faeces up to more than two days after defecation [81], which is twice the time after sampling collection in this survey. However, decomposition of trophozoites began immediately after defecation [81], so it is likely that DNA degradation under environmental conditions is responsible for the low PCR-based prevalence reported here. Despite *T. abrassarti* having a greater prevalence in wild chimpanzees, the colonisation density is apparently lower in wild chimpanzees than in captive ones [78]. This too has been considered a possible cause of negative results when searching for this ciliate in faecal samples [81]. Because of the above-mentioned limitations, data on the true prevalence of *T. abrassarti* should be interpreted with caution because the true prevalence of this ciliate is likely much higher. Moreover, we avoided analysing its seasonality, as it will only provide spurious results.

Faeces are not the optimal sample matrix for the detection of blood parasites such as *Plasmodium* or *Trypanosoma*, which are preferably identified in blood samples. However, several tests performed in humans [82] and NHP [83] have shown a good correlation between the diagnostic results obtained in dried blood spots and faecal samples, although large-scale field studies showed that the prevalence obtained was lower using faeces, at least in malaria [84]. In either case, the use of non-invasive, faecal-derived sampling methods, which can be implemented alongside routine monitoring and surveillance for other faecal parasites, gives it an added value, since otherwise these parasites would be difficult to identify in wild chimpanzee populations.

In this study, we detected five blood protist parasites: two *Plasmodium* spp. (probably both *P. malariae*) and three *T. brucei*. Several species of *Plasmodium* infect NHP exclusively, although they can sometimes cause sporadic human infections such as *P. cynomolgi* [85] or the most common *P. knowlesi* [86]. It is widely believed that human malaria parasites infect only people as a natural host, but many *Plasmodium* morphologically similar to those found in humans have been observed in NHP and, in certain cases, are even genetically identical. This is the case of *P. brasiliensis* in the Americas. In chimpanzees, *P. rodhaini* is morphologically identified as *P. malariae*-like parasite. Recent studies on African great apes have revealed the existence of a large diversity of *Plasmodium* parasites infecting chimpanzees and gorillas, some of them related to the deadliest human parasite *P. falciparum* (subgenus *Laverania*), others to the human parasites *P. malariae*, *P. ovale*, or *P. vivax* (subgenus *Plasmodium*) [87,88]. In this study, we identified a sample with *P. malariae*. A second sample could also be described as *P. malariae* according to phylogeny, as it fell within the same cluster of *P. malariae*/*P. brasiliensis*, although forming a different branch that could be considered as a *P. malariae*-like organism. Determining host specificity and host range of human malaria parasites is of great importance to understand the role of chimpanzees and other NHP in the epidemiology of the parasite, and to improve malaria control [89].

African trypanosomes of the *Trypanosoma brucei* group are the causative agents of sleeping sickness in humans and nagana in animals [90]. The group consists of seven species with the morphologically indistinguishable *T. b. rhodesiense* and *T. b. gambiense* causing the human African trypanosomiasis, and the rest being agents of infection in animals. In East Africa, animals are reservoirs for *T. b. rhodesiense*, whose infection produces nagana in them. However, in West Africa, where *T. b. gambiense* is primarily distributed, the role of animals as reservoirs of the parasite is not so clear [91]. In this study, the prevalence of *T. brucei* was low (1.3%). Of note, *T. brucei* has been previously reported in Western chimpanzees in Ivory Coast [14], but as in the case of our study, it was not possible to elucidate the subspecies present. In both cases, it was only possible to confirm their inclusion in the brucei group. As previous studies have shown, chimpanzees can become infected through the bites of the natural vector, tsetse fly (*Glossina* spp.), and complete the cycle, transmitting the disease to the vector [92]. Future studies determining the subspecies within the brucei group will be necessary to ascertain the frequency and directionality of zoonotic transmission events between chimpanzees and humans in endemic areas in West Africa. In this study, other trypanosomatidae, parasites of plants and insects, have been found in the faeces of the investigated chimpanzees, replicating the results obtained in earlier surveys [93]. These findings are possibly related to the chimpanzee diet or to the use that insects give to faeces to lay their eggs.

In this work, we have also detected DNA of filarial nematodes in chimpanzee faeces. Twenty-three samples (10%) were found to be positive, all of them characterized as *M. perstans* by phylogenetic analysis (Figure S3). Some of these sequences varied among them, forming a well-defined sub-cluster with a bootstrap support of 95%. In a recent study, *M. perstans* was found at a much lower frequency (2%) in chimpanzees from Cameroon and Gabon [15]. Furthermore, this survey provided sequences fitting within the same genus (*Mansonella*), but clustering in a different branch than that of *M. perstans*. In this case, the species could not be determined due to insufficient homology with previously described species. Unfortunately, in our study, we have not been able to compare the phylogeny of both groups of sequences as they correspond to different gene fragments. Some species of filariae, such as *Onchocerca volvulus* or *Wuchereria bancrofti*, have medical importance, although mansonellosis (the disease produced by *M. perstans*) is not considered a public health hazard [94]. The larvae of filariae (microfilariae) are found in the blood or the skin, while adults are found inside the host near the internal organs forming nodules or semi-protected remain confined in other areas. Therefore, the approach of attempting the detection of filarial DNA in faecal samples might seems unusual at first glimpse. In a seminal study, Gaillard et al. described for the first time the presence of *Mansonella* spp. DNA in faecal samples of infected chimpanzees [15]. These authors provided two potential explanations for this finding: (i) health issues associated with haematuria leading to the presence of blood parasites in the faeces of sick animals and (ii) free DNA or DNA included in exosomes secreted by the filarial parasite as mediators of the host–parasite interface. Consequently, the detection of these parasites in faeces is relevant to estimating prevalence rates without using invasive procedures to obtain other biological samples such as blood or tissues. The PCR-based method used here allowed for the detection of *M. perstans* (and possibly other filariae) present in chimpanzee faecal samples using pan-filarial primers regardless of species and hosts [38]. Additionally, this methodology can also provide relevant information on filariae abundance, pathogenesis, transmission and zoonotic potential in wild chimpanzees.

Data obtained here show that it is possible to detect *P. malariae*, *T. brucei*, and *M. perstans* in chimpanzee faecal material. These results suggest that stool samples are a promising, suitable matrix for the detection of blood parasites. However, we need to take into consideration that prevalence rates obtained in faecal samples do not necessarily reflect the ones we would identify in blood samples by conventional diagnostic methods. These three parasites normally infect humans and studying the role that chimpanzees can play as potential reservoirs would be important to optimize control programs for these diseases in areas where apes and humans live closely together.

Chimpanzee parasite seasonality has rarely been studied over more than a single dry or wet season [95–99]. Here, we aim to contrast results found in previous studies to our own. Beforehand, we must warn readers that independence of samples is not guaranteed in this study as individual chimpanzees could not be identified and results should, therefore, be interpreted with caution. *Blastocystis* was associated with rainfall, and undetected during the dry season in our study. One study in a degraded forest mosaic landscape in Uganda reports higher *Blastocystis* prevalence during the rainy season [95], and another report the same trend in a mixed evergreen and semi-deciduous forest habitat in Tanzania [96]. However, other studies report significant [98] and non-significant [98] associations between *Blastocystis* and the dry season. We found *Entamoeba dispar*'s prevalence to be higher during the dry season than the rainy season, and to be negatively associated with rainfall. However, previous studies reported a lack of association between *Entamoeba* spp. and seasonality [95–97]. Reports on *Giardia*'s seasonality do not find seasonal differences [96,98]. In them, however, prevalence was below 10%, as was the case in our study. Differences in prevalence rates according to season were also found for *M. perstans*, although absence of data in the literature precluded us from reaching strong conclusions. Further research, and in particular meta-analyses, are needed to clarify these trends, with particular focus on the factors that correlate with climate, rather than climate itself.

The results obtained and conclusions reached in this study may have been biased by some design and methodological constricts. For instance, collected faecal samples were stored in 70% ethanol at room temperature for several months before processing in the laboratory. This long storage period may have hampered the obtaining of sufficient amount of good quality parasitic DNA for PCR analyses. Low infection/colonisation rates for some of the parasite or commensal species investigated here may have negatively influenced the robustness of the statistical analyses conducted. Similarly, low parasitic densities (e.g., *G. duodenalis*, *C. hominis*) limited the performance of the PCR protocols used for genotyping and sub-genotyping purposes. Due to ethical and conservation concerns, no tissue or blood samples of animal origin were available for downstream molecular testing. In practical terms, this means that the occurrence of blood protist (*Plasmodium*, *Trypanosoma*) and filarial (*Mansonella*) species could only be indirectly assessed in faecal material, a sub-optimal matrix for the detection of these pathogens.

5. Conclusions

This molecular-based epidemiological study showed that potentially pathogenic protist/microsporidia species were present at low-medium infection rates (*Sarcocystis* spp., *G. duodenalis*, *C. hominis*) or apparently absent (*E. histolytica*, *E. bieneusi*) in a wild chimpanzee population in Dindefelo, Senegal. In contrast, protist of unclear pathogenic significance (*Blastocystis* sp.) and commensal species (*E. dispar*, *T. abrassarti*) were far more common. From a parasitological point of view, these data reflect an apparent healthy epidemiological situation characterised by low-to-moderate occurrence rates and parasitic burdens leading to absence of clinical manifestations in infected animals, although the extent of this claim should be confirmed in subsequent studies specifically devoted to assess the health status of these animals. The limited amount of parasitic DNA in faecal samples also explains the low amplification success rates accomplished by genotyping and sub-genotyping PCR protocols (e.g., for *G. duodenalis* and *C. hominis*, among others). Remarkably, this is the first survey reporting the presence of *Sarcocystis* spp. DNA in faeces from wild chimpanzees, raising the question of whether these findings are the result of true infections by this coccidian parasite or just spurious passage associated with diet or environmental contamination. Another major contribution of this survey is the demonstration that, even considering that faecal material is a suboptimal matrix for the detection of blood protist species, chimpanzees might be suitable hosts for *P. malariae* and *T. brucei*, two of the major contributors to the human disease burden by malaria and sleeping sickness in Africa. Overall, these data clearly indicate that wild chimpanzees play a more important role in the epidemiology and transmission of intestinal and blood protist parasites than initially anticipated. Taking into account that most of these agents are zoonotic, it is now clear that cross-transmission species is possible between wild chimpanzees and humans sharing habitats and natural resources. More research is needed to accurately assess the frequency and directionality of zoonotic events, and to estimate the proportion of chimpanzee infections of anthropogenic origin.

Supplementary Materials: The following are available online at https://www.mdpi.com/article/10.3390/ani11113291/s1, Figure S1: Phylogenetic relationships among *Plasmodium* sequences identified in free-living chimpanzees in the present study and homologous sequences retrieved from GenBank. Sequences accession numbers are indicated. The analysis was conducted by a neighbor-joining method of the *ssu* rRNA gene using Treecon software [100] after ClustalW alignment of the sequences [101]. Bootstrap values lower than 75% are not displayed. *Haemosporidia* was used as outgroup taxon to root the tree, Figure S2: Phylogenetic relationships among *Trypanosoma* sequences identified in free-living chimpanzees in the present study and homologous sequences retrieved from GenBank. Sequences accession numbers are indicated. The analysis was conducted by a neighbor-joining method of the *ssu* rRNA gene using Treecon software [100] after ClustalW alignment of the sequences [101]. Bootstrap values lower than 75% are not displayed. *Trypanosoma godfreyi* was used as outgroup taxon to root the tree, Figure S3: Phylogenetic relationships among *Mansonella perstans* sequences identified in free-living chimpanzees in the present study and homologous sequences retrieved from GenBank. Sequences accession numbers are indicated. The analysis was conducted by a neighbor-joining method of the *ssu* rRNA gene using Treecon software [100] after ClustalW alignment of the sequences [101]. Bootstrap values lower than 75% are not displayed. *Loa loa* was used as outgroup taxon to root the tree. Table S1: Oligonucleotides used for the molecular identification and/or characterization of the intestinal and blood protist species investigated in the present study. Table S2: Full dataset indicating main features of chimpanzee faecal samples at the time of collection and PCR and sequencing results for the intestinal and blood protist species investigated in the present study.

Author Contributions: Conceptualization, P.C.K., J.R.-H., L.D., P.P., D.G.-B., J.M.R., R.C.-B. and D.C.; methodology, J.R.-H., L.D., P.P., J.M.R., F.P.-G. and D.C.; software, P.C.K., J.R.-H. C.V.-I. and J.M.R.; valida-tion, J.R.-H., L.D., P.P., J.M.R. and D.C.; formal analysis, P.C.K., J.R.-H., L.D., P.P., D.S., J.M.R., C.V.-I. and D.C.; investigation, P.C.K., J.R.-H., L.D., P.P., D.S., A.D., B.B., M.L.(Manuel Llana), C.V.-I., A.J.-M., C.M.-G., A.S.M., M.L.(Marta Lanza) and J.M.R.; resources, J.R.-H., L.D., P.P., J.M.R., I.F. and D.C.; data curation, J.R.-H., P.P., D.G.-B., J.M.R., R.C.-B. and D.C.; writ-ing—original draft preparation, P.C.K., J.R.-H., L.D., P.P., D.G.-B., J.M.R., R.C.-B. and D.C.; writ-ing—review and editing, P.C.K., J.R.-H., L.D., P.P., A.D., D.G.-B., J.M.R., R.C.-B., F.P.-G. and D.C.; visualization, J.R.-H., P.P., J.M.R., R.C.-B. and D.C.; supervision, J.R.-H., P.P., J.M.R., R.C.-B. and D.C.; project administration, J.R.-H. and D.C.; funding acquisition, J.R.-H., L.D., P.P., J.M.R. and D.C. All authors have read and agreed to the published version of the manuscript.

Funding: This research was funded by the Health Institute Carlos III (ISCIII), Ministry of Economy and Competitiveness (Spain), under project PI16CIII/00024.

Institutional Review Board Statement: This study was carried out in accordance with Spanish legislation guidelines (RD 8/2003) and with the International Guiding Principles for Biomedical Research Involving Animals issued by the Council for International Organization of Medical Sciences and the International Council for Laboratory Animal Science (RD 53/2013). This study has been approved by Ethics Committee of the Health Institute Carlos III on 17 December 2018 under the reference number CEI PI 90_2018-v2. The samples for this study were not collected invasively. This research complied with the Guidelines of Best Practices for Field Primatology of the protocols of the International Primatological Society.

Data Availability Statement: All relevant data are within the article and its additional files. The sequences data were submitted to the GenBank database under the accession numbers MZ182323–MZ182324 (*C. hominis*), MZ182325–MZ182328 (*Blastocystis* sp.), MZ182329–MZ182352 (*Sarcocystis* spp.),

MZ224016 (*T. abrassarti*), MZ272002 (*P. malariae*), MZ272003 (*Plasmodium* sp.), MZ272004–MZ272006 (*T. brucei*), and MZ285880–MZ285897 (*M. perstans*).

Acknowledgments: David González-Barrio was recipient of a 'Sara Borrell' postdoctoral fellowship (CD19CIII/00011) funded by the Spanish Ministry of Science, Innovation and Universities. We thank the Direction des Eaux, Forets, Chasses et de la Conservation des Sols for permission to conduct research in Senegal. We also thank the field assistants of the Jane Goodall Institute Spain and Senegal for their help with sample collection in the field. The monitoring of the chimpanzees at Dindefelo during the sample collection period was supported by the Jane Goodall Institute USA, the Jane Goodall Institute Belgium, the Jane Goodall Institute Austria, and the Barcelona Zoo Foundation.

Conflicts of Interest: The authors declare no conflict of interest. The funders had no role in the design of the study; in the collection, analyses, or interpretation of data; in the writing of the manuscript, or in the decision to publish the results.

References

1. Kühl, H.S.; Sop, T.; Williamson, E.A.; Mundry, R.; Brugière, D.; Campbell, G.; Cohen, H.; Danquah, E.; Ginn, L.; Herbinger, I.; et al. The Critically Endangered western chimpanzee declines by 80%. *Am. J. Primatol.* **2017**, *79*, e22681. [CrossRef] [PubMed]
2. IUCN SSC Primate Specialist Group. *Regional Action Plan for the Conservation of Western Chimpanzees (Pan troglodytes verus) 2020–2030*; IUCN: Gland, Switzerland, 2020. [CrossRef]
3. Dunay, E.; Apakupakul, K.; Leard, S.; Palmer, J.L.; Deem, S.L. Pathogen transmission from humans to great apes is a growing threat to primate conservation. *Ecohealth* **2018**, *15*, 148–162. [CrossRef]
4. Ryan, U.; Fayer, R.; Xiao, L. *Cryptosporidium* species in humans and animals: Current understanding and research needs. *Parasitology* **2014**, *141*, 1667–1685. [CrossRef] [PubMed]
5. Dixon, B.R. *Giardia duodenalis* in humans and animals—Transmission and disease. *Res. Vet. Sci.* **2021**, *135*, 283–289. [CrossRef]
6. Stuart, P.; Yalcindag, E.; Ali, I.K.M.; Pecková, R.; Nurcahyo, W.; Morrogh-Bernard, H.; Foitová, I. *Entamoeba histolytica* infections in wild and semi-wild orangutans in Sumatra and Kalimantan. *Am. J. Primatol.* **2020**, *82*, e23124. [CrossRef]
7. Rosenthal, B.M. Zoonotic *Sarcocystis*. *Res. Vet. Sci.* **2021**, *136*, 51–157. [CrossRef]
8. Li, W.; Feng, Y.; Santin, M. Host specificity of *Enterocytozoon bieneusi* and public health implications. *Trends Parasitol.* **2019**, *35*, 436–451. [CrossRef] [PubMed]
9. Hublin, J.S.Y.; Maloney, J.G.; Santin, M. *Blastocystis* in domesticated and wild mammals and birds. *Res. Vet. Sci.* **2021**, *135*, 260–282. [CrossRef]
10. Ponce-Gordo, F.; García-Rodríguez, J.J. *Balantioides coli* . *Res. Vet. Sci.* **2021**, *135*, 424–431. [CrossRef]
11. Modrý, D.; Petrzelková, K.J.; Pomajbíková, K.; Tokiwa, T.; Krízek, J.; Imai, S.; Vallo, P.; Profousová, I.; Slapeta, J. The occurrence and ape-to-ape transmission of the entodiniomorphid ciliate *Troglodytella abrassarti* in captive gorillas. *J. Eukaryot. Microbiol.* **2009**, *56*, 83–87. [CrossRef]
12. Boundenga, L.; Ollomo, B.; Rougeron, V.; Mouele, L.Y.; Mve-Ondo, B.; Delicat-Loembet, L.M.; Moukodoum, N.D.; Okouga, A.P.; Arnathau, C.; Elguero, E.; et al. Diversity of malaria parasites in great apes in Gabon. *Malar. J.* **2015**, *14*, 111. [CrossRef] [PubMed]
13. Ngoubangoye, B.; Boundenga, L.; Arnathau, C.; Mombo, I.M.; Durand, P.; Tsoumbou, T.A.; Otoro, B.V.; Sana, R.; Okouga, A.P.; Moukodoum, N.; et al. The host specificity of ape malaria parasites can be broken in confined environments. *Int. J. Parasitol.* **2016**, *46*, 737–744. [CrossRef]
14. Jirků, M.; Votýpka, J.; Petrželková, K.J.; Jirků-Pomajbíková, K.; Kriegová, E.; Vodička, R.; Lankester, F.; Leendertz, S.A.; Wittig, R.M.; Boesch, C.; et al. Wild chimpanzees are infected by *Trypanosoma brucei*. *Int. J. Parasitol. Parasites Wildl.* **2015**, *4*, 277–282. [CrossRef]
15. Gaillard, C.M.; Pion, S.D.; Hamou, H.; Sirima, C.; Bizet, C.; Lemarcis, T.; Rodrigues, J.; Esteban, A.; Peeters, M.; Mpoudi Ngole, E.; et al. Detection of DNA of filariae closely related to *Mansonella perstans* in faecal samples from wild non-human primates from Cameroon and Gabon. *Parasites Vectors* **2020**, *13*, 313. [CrossRef] [PubMed]
16. Zahedi, A.; Bolland, S.J.; Oskam, C.L.; Ryan, U. *Cryptosporidium abrahamseni* n. sp. (Apicomplexa: Cryptosporidiiae) from red-eye tetra (*Moenkhausia sanctaefilomenae*). *Exp. Parasitol.* **2021**, in press. [CrossRef] [PubMed]
17. Ryan, U.; Cacciò, S.M. Zoonotic potential of *Giardia*. *Int. J. Parasitol.* **2013**, *43*, 943–956. [CrossRef] [PubMed]
18. Maloney, J.G.; Santin, M. Mind the gap: New full-length sequences of *Blastocystis* subtypes generated via Oxford Nanopore Minion sequencing allow for comparisons between full-length and partial sequences of the small subunit of the ribosomal RNA Gene. *Microorganisms* **2021**, *9*, 997. [CrossRef] [PubMed]
19. Pacheco, L.; Fraixedas, S.; Fernández-Llamazares, Á.; Estela, N.; Mominee, R.; Guallar, F. Perspectives on sustainable resource conservation in community nature reserves: A case study from Senegal. *Sustainability* **2012**, *4*, 3158–3179. [CrossRef]
20. Enthoven, D.; Pacheco, L.; Llana, M.; Reitan, T.; Hernandez-Aguilar, A. Nesting patterns of chimpanzees (*Pan troglodytes verus*) in a savannah habitat, Dindefelo, Senegal. *Folia Primatol.* **2017**, *88*, 114.
21. JGIS; APES Wiki Team. Dindefelo. 2021. Available online: https://apeswiki.eva.mpg.de/index.php/Dindefelo (accessed on 11 September 2021).

22. Ramon, M.; Llana, M.; Estela, N.; Pacheco, L.; Hockings, K.; Hill, C. The fruit of discord? *Saba senegalensis* use by chimpanzees (*Pan troglodytes verus*) and local people in the Dindefelo Community Nature Reserve (RNCD), Southeastern Senegal. *Folia Primatol.* **2017**, *88*, 167–168.
23. O'Donnell, L.J.; Virjee, J.; Heaton, K.W. Detection of pseudodiarrhoea by simple clinical assessment of intestinal transit rate. *BMJ* **1990**, *300*, 439–440. [CrossRef]
24. Huffman, G.J.; Stocker, E.F.; Bolvin, D.T.; Nelkin, E.J.; Tan, J. *GPM IMERG Final Precipitation L3 1 Day 0.1 Degree × 0.1 Degree V06*; Savtchenko, A., Ed.; Goddard Earth Sciences Data and Information Services Center (GES DISC): Greenbelt, MD, USA, 2019. [CrossRef]
25. AIRS Science Team/Joao Teixeira. *AIRS/Aqua L3 Daily Standard Physical Retrieval (AIRS-only) 1 Degree × 1 Degree V006*; Goddard Earth Sciences Data and Information Services Center (GES DISC): Greenbelt, MD, USA, 2013. [CrossRef]
26. Tiangtip, R.; Jongwutiwes, S. Molecular analysis of *Cryptosporidium* species isolated from HIV-infected patients in Thailand. *Trop. Med. Int. Health* **2002**, *7*, 357–364. [CrossRef]
27. Feltus, D.C.; Giddings, C.W.; Schneck, B.L.; Monson, T.; Warshauer, D.; McEvoy, J.M. Evidence supporting zoonotic transmission of *Cryptosporidium* spp. in Wisconsin. *J. Clin. Microbiol.* **2006**, *44*, 4303–4308. [CrossRef]
28. Gutiérrez-Cisneros, M.J.; Cogollos, R.; López-Vélez, R.; Martín-Rabadán, P.; Martínez-Ruiz, R.; Subirats, M.; Merino, F.J.; Fuentes, I. Application of real-time PCR for the differentiation of *Entamoeba histolytica* and *E. dispar* in cyst-positive faecal samples from 130 immigrants living in Spain. *Ann. Trop. Med. Parasitol.* **2010**, *104*, 145–149. [CrossRef]
29. Verweij, J.J.; Oostvogel, F.; Brienen, E.A.; Nang-Beifubah, A.; Ziem, J.; Polderman, A.M. Prevalence of *Entamoeba histolytica* and *Entamoeba dispar* in northern Ghana. *Trop. Med. Int. Health* **2003**, *8*, 1153–1156. [CrossRef]
30. Verweij, J.J.; Schinkel, J.; Laeijendecker, D.; van Rooyen, M.A.; van Lieshout, L.; Polderman, A.M. Real-time PCR for the detection of *Giardia lamblia*. *Mol. Cell. Probes* **2003**, *17*, 223–225. [CrossRef]
31. Read, C.M.; Monis, P.T.; Thompson, R.C. Discrimination of all genotypes of *Giardia duodenalis* at the glutamate dehydrogenase locus using PCR-RFLP. *Infect. Genet. Evol.* **2004**, *4*, 125–130. [CrossRef]
32. Lalle, M.; Pozio, E.; Capelli, G.; Bruschi, F.; Crotti, D.; Cacciò, S.M. Genetic heterogeneity at the beta-giardin locus among human and animal isolates of *Giardia duodenalis* and identification of potentially zoonotic subgenotypes. *Int. J. Parasitol.* **2005**, *35*, 207–213. [CrossRef]
33. Sulaiman, I.M.; Fayer, R.; Bern, C.; Gilman, R.H.; Trout, J.M.; Schantz, P.M.; Das, P.; Lal, A.A.; Xiao, L. Triosephosphate isomerase gene characterization and potential zoonotic transmission of *Giardia duodenalis*. *Emerg. Infect. Dis.* **2003**, *9*, 1444–1452. [CrossRef]
34. Scicluna, S.M.; Tawari, B.; Clark, C.G. DNA barcoding of *Blastocystis*. *Protist* **2006**, *157*, 77–85. [CrossRef]
35. Buckholt, M.A.; Lee, J.H.; Tzipori, S. Prevalence of *Enterocytozoon bieneusi* in swine: An 18-month survey at a slaughterhouse in Massachusetts. *Appl. Environ. Microbiol.* **2002**, *68*, 2595–2599. [CrossRef]
36. Ponce-Gordo, F.; Fonseca-Salamanca, F.; Martínez-Díaz, R.A. Genetic heterogeneity in internal transcribed spacer genes of *Balantidium coli* (Litostomatea, Ciliophora). *Protist* **2011**, *162*, 774–794. [CrossRef]
37. Vallo, P.; Petrželková, K.J.; Profousová, I.; Petrášová, J.; Pomajbíková, K.; Leendertz, F.; Hashimoto, C.; Simmons, N.; Babweteera, F.; Machanda, Z.; et al. Molecular diversity of entodiniomorphid ciliate *Troglodytella abrassarti* and its coevolution with chimpanzees. *Am. J. Phys. Anthropol.* **2012**, *148*, 525–533. [CrossRef]
38. Ta, T.-H.; Moya, L.; Nguema, J.; Aparicio, P.; Miguel-Oteo, M.; Cenzual, G.; Canorea, I.; Lanza, M.; Benito, A.; Crainey, J.L.; et al. Geographical distribution and species identification of human filariasis and onchocerciasis in Bioko Island, Equatorial Guinea. *Acta Trop.* **2018**, *180*, 12–17. [CrossRef]
39. Saitou, N.; Nei, M. The neighbor-joining method: A new method for reconstructing phylogenetic trees. *Mol. Biol. Evol.* **1987**, *4*, 406–425.
40. R Core Team. *R: A Language and Environment for Statistical Computing*; R Foundation for Statistical Computing: Vienna, Austria, 2019; Available online: https://www.R-project.org/ (accessed on 11 September 2021).
41. Best, D.J.; Roberts, D.E. Algorithm AS 89: The upper tail probabilities of Spearman's rho. *Appl. Stat.* **1975**, *24*, 377–379. [CrossRef]
42. Mehta, C.R.; Patel, N.R. A network algorithm for performing Fisher's exact test in r × c contingency tables. *J. Am. Stat. Assoc.* **1983**, *78*, 427–434. [CrossRef]
43. Benjamini, Y.; Hochberg, Y. Controlling the false discovery rate: A practical and powerful approach to multiple testing. *J. R. Stat. Soc. Ser. B* **1995**, *57*, 289–300. [CrossRef]
44. Claes, F.; Büscher, P.; Touratier, L.; Goddeeris, B.M. *Trypanosoma equiperdum*: Master of disguise or historical mistake? *Trends Parasitol.* **2005**, *21*, 316–321. [CrossRef]
45. Mapua, M.I.; Petrželková, K.J.; Burgunder, J.; Dadáková, E.; Brožová, K.; Hrazdilová, K.; Stewart, F.A.; Piel, A.K.; Vallo, P.; Fuehrer, H.P.; et al. A comparative molecular survey of malaria prevalence among Eastern chimpanzee populations in Issa Valley (Tanzania) and Kalinzu (Uganda). *Malar. J.* **2016**, *15*, 423. [CrossRef]
46. Wu, D.F.; Löhrich, T.; Sachse, A.; Mundry, R.; Wittig, R.M.; Calvignac-Spencer, S.; Deschner, T.; Leendertz, F.H. Seasonal and inter-annual variation of malaria parasite detection in wild chimpanzees. *Malar. J.* **2018**, *17*, 38. [CrossRef]
47. Gillespie, T.R.; Morgan, D.; Deutsch, J.C.; Kuhlenschmidt, M.S.; Salzer, J.S.; Cameron, K.; Reed, T.; Sanz, C. A legacy of low-impact logging does not elevate prevalence of potentially pathogenic protozoa in free-ranging gorillas and chimpanzees in the Republic of Congo: Logging and parasitism in African apes. *Ecohealth* **2009**, *6*, 557–564. [CrossRef]

48. Gonzalez-Moreno, O.; Hernandez-Aguilar, R.A.; Piel, A.K.; Stewart, F.A.; Gracenea, M.; Moore, J. Prevalence and climatic associated factors of *Cryptosporidium* sp. infections in savanna chimpanzees from Ugalla, Western Tanzania. *Parasitol. Res.* **2013**, *112*, 393–399. [CrossRef]
49. Renelies-Hamilton, J.; Noguera-Julian, M.; Parera, M.; Paredes, R.; Pacheco, L.; Dacal, E.; Saugar, J.M.; Rubio, J.M.; Poulsen, M.; Köster, P.C.; et al. Exploring interactions between *Blastocystis* sp., *Strongyloides* spp. and the gut microbiomes of wild chimpanzees in Senegal. *Infect. Genet. Evol.* **2019**, *74*, 104010. [CrossRef] [PubMed]
50. Parsons, M.B.; Travis, D.; Lonsdorf, E.V.; Lipende, I.; Roellig, D.M.; Collins, A.; Kamenya, S.; Zhang, H.; Xiao, L.; Gillespie, T.R. Epidemiology and molecular characterization of *Cryptosporidium* spp. in humans, wild primates, and domesticated animals in the Greater Gombe Ecosystem, Tanzania. *PLoS Negl. Trop. Dis.* **2015**, *9*, e0003529. [CrossRef]
51. Squire, S.A.; Ryan, U. *Cryptosporidium* and *Giardia* in Africa: Current and future challenges. *Parasites Vectors* **2017**, *10*, 195. [CrossRef]
52. Widmer, G.; Köster, P.C.; Carmena, D. *Cryptosporidium hominis* infections in non-human animal species: Revisiting the concept of host specificity. *Int. J. Parasitol.* **2020**, *50*, 253–262. [CrossRef]
53. Ragazzo, L.J.; Zohdy, S.; Velonabison, M.; Herrera, J.; Wright, P.C.; Gillespie, T.R. *Entamoeba histolytica* infection in wild lemurs associated with proximity to humans. *Vet. Parasitol.* **2018**, *249*, 98–101. [CrossRef]
54. Abdallah, M.D.; Wolf, R.F.; White, G.L.; Kosanke, S.D.; Carey, D.W.; Verweij, J.J.; El-Dessouky, Y.M.; Zhang, M.J.; Ravdin, J.I. Natural infection of baboons by *Entamoeba histolytica* elicits anti-gal-lectin heavy subunit IgA and IgG antibodies with shared epitope specificity to that of humans. *J. Egypt Soc. Parasitol.* **2013**, *43*, 723–735.
55. Howells, M.E.; Pruetz, J.; Gillespie, T.R. Patterns of gastro-intestinal parasites and commensals as an index of population and ecosystem health: The case of sympatric western chimpanzees (*Pan troglodytes verus*) and guinea baboons (*Papio hamadryas papio*) at Fongoli, Senegal. *Am. J. Primatol.* **2011**, *73*, 173–179. [CrossRef]
56. Jirků-Pomajbíková, K.; Čepička, I.; Kalousová, B.; Jirků, M.; Stewart, F.; Levecke, B.; Modrý, D.; Piel, A.K.; Petrželková, K.J. Molecular identification of *Entamoeba* species in savanna woodland chimpanzees (*Pan troglodytes schweinfurthii*). *Parasitology* **2016**, *143*, 741–748. [CrossRef]
57. Deere, J.R.; Parsons, M.B.; Lonsdorf, E.V.; Lipende, I.; Kamenya, S.; Collins, D.A.; Travis, D.A.; Gillespie, T.R. *Entamoeba Histolytica–1122*; Cambridge University Press: Cambridge, UK, 2018.
58. Vlčková, K.; Kreisinger, J.; Pafčo, B.; Čížková, D.; Tagg, N.; Hehl, A.B.; Modrý, D. Diversity of *Entamoeba* spp. in African great apes and humans: An insight from Illumina MiSeq high-throughput sequencing. *Int. J. Parasitol.* **2018**, *48*, 519–530. [CrossRef]
59. Sá, R.M.; Petrášová, J.; Pomajbíková, K.; Profousová, I.; Petrželková, K.J.; Sousa, C.; Cable, J.; Bruford, M.W.; Modrý, D. Gastrointestinal symbionts of chimpanzees in Cantanhez National Park, Guinea-Bissau with respect to habitat fragmentation. *Am. J. Primatol.* **2013**, *75*, 1032–1041. [CrossRef]
60. Sak, B.; Petrzelkova, K.J.; Kvetonova, D.; Mynarova, A.; Shutt, K.A.; Pomajbikova, K.; Kalousova, B.; Modry, D.; Benavides, J.; Todd, A.; et al. Long-term monitoring of microsporidia, *Cryptosporidium* and *Giardia* infections in western Lowland Gorillas (*Gorilla gorilla gorilla*) at different stages of habituation in Dzanga Sangha Protected Areas, Central African Republic. *PLoS ONE* **2013**, *8*, e71840. [CrossRef]
61. Hogan, J.N.; Miller, W.A.; Cranfield, M.R.; Ramer, J.; Hassell, J.; Noheri, J.B.; Conrad, P.A.; Gilardi, K.V. *Giardia* in mountain gorillas (*Gorilla beringei beringei*), forest buffalo (*Syncerus caffer*), and domestic cattle in Volcanoes National Park, Rwanda. *J. Wildl. Dis.* **2014**, *50*, 21–30. [CrossRef]
62. Graczyk, T.K.; Bosco-Nizeyi, J.; Ssebide, B.; Thompson, R.C.; Read, C.; Cranfield, M.R. Anthropozoonotic *Giardia duodenalis* genotype (assemblage) a infections in habitats of free-ranging human-habituated gorillas, Uganda. *J. Parasitol.* **2002**, *88*, 905–909. [CrossRef]
63. Teichroeb, J.A.; Kutz, S.J.; Parkar, U.; Thompson, R.C.; Sicotte, P. Ecology of the gastrointestinal parasites of Colobus vellerosus at Boabeng-Fiema, Ghana: Possible anthropozoonotic transmission. *Am. J. Phys. Anthropol.* **2009**, *140*, 498–507. [CrossRef] [PubMed]
64. Johnston, A.R.; Gillespie, T.R.; Rwego, I.B.; McLachlan, T.L.; Kent, A.D.; Goldberg, T.L. Molecular epidemiology of cross-species *Giardia duodenalis* transmission in western Uganda. *PLoS Negl. Trop. Dis.* **2010**, *4*, e683. [CrossRef]
65. Heydorn, A.O.; Gestrich, R.; Janitschke, K. Beiträge zum lebenszyklus der Sarkosporidien. VIII. Sporozysten von *Sarcocystis bovihominis* in den fäzes von rhesusaffen (*Macaca rhesus*) und pavianen (*Papio cynocephalus*). *Berl. Munch. Tierarztl. Wochenschr.* **1976**, *89*, 116–120.
66. Fayer, R.; Heydorn, A.O.; Johnson, A.J.; Leek, R.G. Transmission of *Sarcocystis suihominis* from humans to swine to nonhuman primates (*Pan troglodytes*, *Macaca mulatta*, *Macaca irus*). *Z. Parasitenkd.* **1979**, *59*, 15–20. [CrossRef]
67. Dubey, J.P.; Calero-Bernal, R.; Rosenthal, B.M.; Speer, C.A.; Fayer, R. *Sarcocystosis of Animals and Humans*, 2nd ed.; CRC Press: Boca Raton, FL, USA, 2016.
68. Gjerde, B. Molecular characterisation of *Sarcocystis bovifelis*, *Sarcocystis bovini* n. sp., *Sarcocystis hirsuta* and *Sarcocystis cruzi* from Cattle (*Bos taurus*) and *Sarcocystis sinensis* from Water Buffaloes (*Bubalus bubalis*). *Parasitol. Res.* **2016**, *115*, 1473–1492. [CrossRef]
69. Abe, N.; Matsuo, K.; Moribe, J.; Takashima, Y.; Irie, T.; Baba, T.; Gjerde, B. Morphological and molecular characteristics of seven *Sarcocystis* species from sika deer (*Cervus nippon centralis*) in Japan, including three new species. *Int. J. Parasitol. Parasites Wildl.* **2019**, *10*, 252–262. [CrossRef]

70. Drakulovski, P.; Bertout, S.; Locatelli, S.; Butel, C.; Pion, S.; Mpoudi-Ngole, E.; Delaporte, E.; Peeters, M.; Mallié, M. Assessment of gastrointestinal parasites in wild chimpanzees (*Pan troglodytes troglodytes*) in southeast Cameroon. *Parasitol. Res.* **2014**, *113*, 2541–2550. [CrossRef]
71. Petrášová, J.; Uzlíková, M.; Kostka, M.; Petrželková, K.J.; Huffman, M.A.; Modrý, D. Diversity and host specificity of *Blastocystis* in syntopic primates on Rubondo Island, Tanzania. *Int. J. Parasitol.* **2011**, *41*, 1113–1120. [CrossRef]
72. Muadica, A.S.; Köster, P.C.; Dashti, A.; Bailo, B.; Hernández-de-Mingo, M.; Reh, L.; Balasegaram, S.; Verlander, N.Q.; Ruiz Chércoles, E.; Carmena, D. Molecular diversity of *Giardia duodenalis*, *Cryptosporidium* spp. and *Blastocystis* sp. in asymptomatic school children in Leganés, Madrid (Spain). *Microorganisms* **2020**, *8*, 466. [CrossRef] [PubMed]
73. Muadica, A.S.; Köster, P.C.; Dashti, A.; Bailo, B.; Hernández-de-Mingo, M.; Balasegaram, S.; Carmena, D. Molecular diversity of *Giardia duodenalis*, *Cryptosporidium* spp., and *Blastocystis* sp. in symptomatic and asymptomatic schoolchildren in Zambézia Province (Mozambique). *Pathogens* **2021**, *10*, 255. [CrossRef] [PubMed]
74. Sak, B.; Kvác, M.; Petrzelková, K.; Kvetonová, D.; Pomajbíková, K.; Mulama, M.; Kiyang, J.; Modrý, D. Diversity of microsporidia (Fungi: Microsporidia) among captive great apes in European zoos and African sanctuaries: Evidence for zoonotic transmission? *Folia Parasitol.* **2011**, *58*, 81–86. [CrossRef]
75. Boudenga, L.; Ngoubangoye, B.; Moukodoum, N.; Dibakou, S.-E.; Moussadji, C.; Hugot, J.P. Diversity of parasites in two captive chimpanzee populations in souther Gabon. *Infect. Genet. Evol.* **2021**, *91*, 104807. [CrossRef]
76. Bakuza, J.S.; Nkwengulila, G. Variation over time in parasite prevalence among free-ranging chimpanzees at Gombe National Park, Tanzania. *Int. J. Primatol.* **2009**, *30*, 43–53. [CrossRef]
77. Pomajbíková, K.; Petrželková, K.J.; Profousová, I.; Petrášová, J.; Modrý, D. Discrepancies in the occurrence of *Balantidium coli* between wild and captive African great apes. *J. Parasitol.* **2010**, *96*, 1139–1144. [CrossRef] [PubMed]
78. Pomajbíková, K.; Petrželková, K.J.; Profousová, I.; Petrášova, J.; Kišidayová, S.; Varádyová, Z.; Modrý, D. A survey of entodiniomorphid ciliates in chimpanzees and bonobos. *Am. J. Phys. Anthropol.* **2010**, *142*, 42–48. [CrossRef] [PubMed]
79. Lilly, A.A.; Mehlman, P.T.; Doran, D. Intestinal parasites in gorillas, chimpanzees, and humans at Mondika Research Site, Dzanga-Ndoki National Park, Central African Republic. *Int. J. Primatol.* **2002**, *23*, 555–573. [CrossRef]
80. Zommers, Z.; Macdonald, D.W.; Johnson, P.J.; Gillespie, T.R. Impact of human activities on chimpanzee ground use and parasitism (*Pan troglodytes*). *Conserv. Lett.* **2013**, *6*, 264–273. [CrossRef]
81. Profousová, I.; Petrželková, K.J.; Pomajbíková, K.; Modrý, D. Survival and morphological changes of entodiniomorphid ciliate *Troglodytella abrassarti* in chimpanzee feces. *Zoo Wild Med.* **2011**, *42*, 69–74. [CrossRef] [PubMed]
82. Jirků, M.; Pomajbíková, K.; Petrželková, K.J.; Hůzová, Z.; Modrý, D.; Lukeš, J. Detection of *Plasmodium* spp. in human feces. *Emerg. Infect. Dis.* **2012**, *18*, 634–636. [CrossRef]
83. Assis, G.M.; Alvarenga, D.A.; Costa, D.C.; Souza, J.C., Jr.; Hirano, Z.M.; Kano, F.S.; Sousa, T.N.; Brito, C.F. Detection of *Plasmodium* in faeces of the New World primate *Alouatta clamitans*. *Mem. Inst. Oswaldo Cruz* **2016**, *111*, 570–576. [CrossRef] [PubMed]
84. Al-Shehri, H.; Power, B.J.; Archer, J.; Cousins, A.; Atuhaire, A.; Adriko, M.; Arinaitwe, M.; Alanazi, A.D.; LaCourse, E.J.; Kabatereine, N.B.; et al. Non-invasive surveillance of *Plasmodium* infection by real-time PCR analysis of ethanol preserved faeces from Ugandan school children with intestinal schistosomiasis. *Malar. J.* **2019**, *18*, 109. [CrossRef]
85. Ta, T.H.; Hisam, S.; Lanza, M.; Jiram, A.I.; Ismail, N.; Rubio, J.M. First case of a naturally acquired human infection with *Plasmodium cynomolgi*. *Malar. J.* **2014**, *13*, 68. [CrossRef]
86. Ta, T.T.; Salas, A.; Ali-Tammam, M.; Martínez, M.D.C.; Lanza, M.; Arroyo, E.; Rubio, J.M. First case of detection of *Plasmodium knowlesi* in Spain by Real Time PCR in a traveller from Southeast Asia. *Malar. J.* **2010**, *27*, 219.
87. Coatney, G.R.; Collins, W.E.; Warren, M.; Contacos, P.G. *CD-ROM. The Primate Malarias*; Original Book Published, 1971; Division of Parasitic Disease, Producers; Version 1.0; CDC: Atlanta, GA, USA, 2003.
88. Hayakawa, T.; Arisue, N.; Udono, T.; Hirai, H.; Sattabongkot, J.; Toyama, T.; Tsuboi, T.; Horii, T.; Tanabe, K. Identification of *Plasmodium malariae*, a human malaria parasite, in imported chimpanzees. *PLoS ONE* **2009**, *4*, e7412. [CrossRef]
89. Makanga, B.; Yangari, P.; Rahola, N.; Rougeron, V.; Elguero, E.; Boundenga, L.; Moukodoum, N.D.; Okouga, A.P.; Arnathau, C.; Durand, P.; et al. Ape malaria transmission and potential for ape-to-human transfers in Africa. *Proc. Natl. Acad. Sci. USA* **2016**, *113*, 5329–5334. [CrossRef]
90. Simarro, P.P.; Diarra, A.; Postigo, J.A.R.; Franco, J.R.; Jannin, J.G. The human African trypanosomiasis control and surveillance programme of the World Health Organization 2000–2009: The way forward. *PLoS Negl. Trop. Dis.* **2011**, *5*, e1007. [CrossRef]
91. Cordon-Obras, C.; Berzosa, P.; Ndong-Mabale, N.; Bobuakasi, L.; Buatiche, J.N.; Ndongo-Asumu, P.; Benito, A.; Cano, J. *Trypanosoma brucei gambiense* in domestic livestock of Kogo and Mbini foci (Equatorial Guinea). *Trop. Med. Int. Health* **2009**, *14*, 535–541. [CrossRef]
92. Godfrey, D.G.; Killick-Kendrick, R. Cyclically transmitted infections of *Trypanosoma brucei*, *T. rhodesiense* and *T. gambiense* in chimpanzees. *Trans. R. Soc. Trop. Med. Hyg.* **1967**, *61*, 781–791. [CrossRef]
93. Votýpka, J.; Pafčo, B.; Modrý, D.; Mbohli, D.; Tagg, N.; Petrželková, K.J. An unexpected diversity of trypanosomatids in fecal samples of great apes. *Int. J. Parasitol. Parasites Wildl.* **2018**, *7*, 322–325. [CrossRef] [PubMed]
94. Ta-Tang, T.H.; Luz, S.L.; Crainey, J.L.; Rubio, J.M. An overview of the management of Mansonellosis. *Res. Rep. Trop. Med.* **2021**, *12*, 93–105. [CrossRef]

95. McLennan, M.R.; Hasegawa, H.; Bardi, M.; Huffman, M.A. Gastrointestinal parasite infections and self-medication in wild chimpanzees surviving in degraded forest fragments within an agricultural landscape mosaic in Uganda. *PLoS ONE* **2017**, *12*, e0180431. [CrossRef]
96. Petrášová, J.; Modrý, D.; Huffman, M.A.; Mapua, M.I.; Bobáková, L.; Mazoch, V.; Singh, J.; Kaur, T.; Petrželková, K.J. Gastrointestinal parasites of indigenous and introduced primate species of Rubondo Island National Park, Tanzania. *Int. J. Primatol.* **2010**, *31*, 920–936. [CrossRef]
97. Kalousová, B.; Piel, A.K.; Pomajbíková, K.; Modrý, D.; Stewart, F.A.; Petrželková, K.J. Gastrointestinal parasites of savanna chimpanzees (*Pan troglodytes schweinfurthii*) in Ugalla, Tanzania. *Int. J. Primatol.* **2014**, *35*, 463–475. [CrossRef]
98. Ashford, R.W.; Reid, G.D.; Wrangham, R.W. Intestinal parasites of the chimpanzee *Pan troglodytes* in Kibale Forest, Uganda. *Ann. Trop. Med. Parasitol.* **2000**, *94*, 173–179. [CrossRef]
99. Masi, S.; Chauffour, S.; Bain, O.; Todd, A.; Guillot, J.; Krief, S. Seasonal effects on great ape health: A case study of wild chimpanzees and Western gorillas. *PLoS ONE* **2012**, *7*, e49805. [CrossRef] [PubMed]
100. Van de Peer, Y.; De Wachter, R. TREECON for Windows: A software package for the construction and drawing of evolutionary trees for the Microsoft Windows environment. *Comput. Appl. BioSci.* **1994**, *10*, 569–570. [CrossRef] [PubMed]
101. Thompson, J.D.; Higgins, D.G.; Gibson, T.J. CLUSTAL W: Improving the sensitivity of progressive multiple sequence alignment through sequence weighting, position-specific gap penalties and weight matrix choice. *Nucleic Acids Res.* **1994**, *22*, 4673–4680. [CrossRef] [PubMed]

animals

MDPI

Article

Babesia microti in Rodents from Different Habitats of Lithuania

Dalytė Mardosaitė-Busaitienė [1], Jana Radzijevskaja [1,*], Linas Balčiauskas [2] and Algimantas Paulauskas [1]

[1] Faculty of Natural Sciences, Vytautas Magnus University, LT-44243 Kaunas, Lithuania; dalyte.mardosaite-busaitiene@vdu.lt (D.M.-B.); algimantas.paulauskas@vdu.lt (A.P.)
[2] Laboratory of Mammalian Ecology, Nature Research Centre, LT-08412 Vilnius, Lithuania; linas.balciauskas@gamtc.lt
* Correspondence: jana.radzijevskaja@vdu.lt; Tel.: +370-6744-8834

Simple Summary: *Babesia microti*, the causative agent of human babesiosis, is an intraerythrocytic protozoan parasite, that circulates among small rodents and ixodid ticks in many countries worldwide. Zoonotic and non-zoonotic *B. microti* strains have been identified in rodent populations in Europe. Analyzing eight species of small rodents collected from different habitats (meadows, forests and their ecotones) in Lithuania, we checked for the presence of *B. microti* and found the highest infection prevalence to be in *Microtus oeconomus* and *Microtus agrestis* rodents. Of note, this study also detected the first reported cases of *Babesia* parasites in *Micromys minutus* mice. In term of habitat, the highest prevalence of *Babesia* parasites was detected in rodents trapped in meadows. Our results demonstrate that rodents, especially *Microtus* voles, can play an important role in the circulation of the zoonotic *B. microti* 'Jena/Germany' strain in Lithuania.

Abstract: *Babesia microti* (Aconoidasida: Piroplasmida) (Franca, 1910) is an emerging tick-borne parasite with rodents serving as the considered reservoir host. However, the distribution of *B. microti* in Europe is insufficiently characterized. Based on the sample of 1180 rodents from 19 study sites in Lithuania, the objectives of this study were: (1) to investigate the presence of *Babesia* parasites in eight species of rodents, (2) to determine the prevalence of *Babesia* parasites in rodents from different habitats, and (3) to characterize the detected *Babesia* strains using partial sequencing of the 18S rRNR gene. *Babesia* DNA was detected in 2.8% rodents. The highest prevalence of *Babesia* was found in *Microtus oeconomus* (14.5%) and *Microtus agrestis* (7.1%) followed by *Clethrionomys glareolus* (2.3%), *Apodemus flavicollis* (2.2%) and *Micromys minutus* (1.3%). In *M. minutus*, *Babesia* was identified for the first time. The prevalence of *Babesia*-infected rodents was higher in the meadow (5.67%) than in the ecotone (1.69%) and forest (0.31%) habitats. The sequence analysis of the partial 18S rRNA gene reveals that *Babesia* isolates derived from rodents were 99–100% identical to human pathogenic *B. microti* 'Jena/Germany' strain.

Keywords: 18S rRNA; *Babesia*; rodents; voles; mice; Lithuania

Citation: Mardosaitė-Busaitienė, D.; Radzijevskaja, J.; Balčiauskas, L.; Paulauskas, A. *Babesia microti* in Rodents from Different Habitats of Lithuania. *Animals* **2021**, *11*, 1707. https://doi.org/10.3390/ani11061707

Academic Editors: Rafael Calero-Bernal and Ignacio Garcia-Bocanegra

Received: 14 May 2021
Accepted: 3 June 2021
Published: 7 June 2021

1. Introduction

Babesiae are emerging tick-borne protozoan parasites circulating in many countries worldwide in vertebrate hosts and vectors. The *Babesia* species including *Babesia microti*, *Babesia divergens*, *B. divergens*-like, *Babesia venatorum* and *Babesia duncani* are known to cause infection in humans. In Europe, Asia and North America respectively, the main vectors of zoonotic *Babesia* species are *Ixodes ricinus*, *Ixodes persulcatus* and *Ixodes scapularis* ticks. [1]. *B. microti* is the main causative agent of human babesiosis, especially in North America [2]. In Europe however, human babesiosis cases are less frequently reported and mostly related to *B. divergens*, *B. divergens*-like and *B. venatorum* [3]. However, a few cases of human babesiosis resulting from *B. microti* have also been reported in Europe [4–6]. To the best of our knowledge, no cases of human babesiosis have been documented in Lithuania.

The common vole (*Microtis arvalis*), field vole (*Microtus agrestis*) and root vole (*Microtus oeconomus*) are microtine rodents that play important roles in the circulation of *B. microti* in Europe [7,8]. *B. microti* infection was also detected in yellow-necked mouse (*Apodemus flavicollis*), striped field mouse (*Apodemus agrarius*), wood mouse (*Apodemus sylvaticus*) and bank vole (*Clethrionomys glareolus*), which are the main hosts for the immature stages of *Ixodes* ticks [7,9–11]. In general, *Ixodes trianguliceps* (with all three developmental stages feeding on rodents and does not bite humans) is the main vector of *B. microti* [12,13], while *I. ricinus* (with larvae and nymphs feeding on rodents) would only serve as a bridge vector of *B. microti* [9,14,15].

Molecular phylogenetic analysis demonstrated that *B. microti* consisting of genetically diverse isolates that belong to different clades [16]. *B. microti* isolates from rodents are subdivided within these clades into the non-zoonotic and zoonotic strains [17,18]. Different strains of *B. microti* have been reported in rodents in Slovenia, Croatia, Poland, Finland, Germany, Slovakia and France [7,9,12,19–22]. Various *B. microti* strains may circulate in rodent community at the same time [7]. However, distributions of *B. microti* strains in Europe are still insufficiently characterized.

The aims of the present study were: (1) to investigate the presence of *Babesia* parasites in eight species of Lithuanian rodents, (2) to determine the prevalence of *Babesia* parasites in rodents from meadows, forests and their ecotones, and (3) to characterize the detected *B. microti* strains using partial sequencing of 18S rRNR gene.

2. Materials and Methods

2.1. Study Sites

Rodents were trapped in 19 locations of different habitats in western (Curonian Spit; sites 1–8 and Nemunas River Delta; sites 9–10) and eastern (sites 11–19) parts of Lithuania during 2013–2017 (Figure 1). Rodents were captured in meadows, forests and their ecotones. Rodent sampling in the Curonian Spit was conducted in the coastal meadows (sites 1, 2, 5–8), mixed forests (site 4) and meadow-mixed forest ecotone (site 3). In the Nemunas River Delta, the trapping was conducted in two habitats: in a flooded meadow (site 9) and in a spring-flooded black alder stands forest (site 10).

Figure 1. Rodent trapping sites in Lithuania, 2013–2017. Δ/▲—forests; ○/●—meadows; □/■—ecotones; Δ/○/□—rodents negative for *Babesia* parasites; ▲/●/■—rodents infected with *Babesia* parasites. Map was created using Open street map data (open source https://www.openstreetmap.org, accessed on 22 April 2020).

Eastern Lithuania was represented by different habitats—mixed forests (sites 11,13, 14, 16,19), mixed forest-meadow ecotones (sites 15, 17), and deciduous forests, one of them in peninsula of Lukštas lake (site 12) and the other on an island in an artificial water body, Elektrėnai Reservoir (site 18) (Table 1).

Table 1. Prevalence of *Babesia* parasites in Lithuanian rodents, 2013–2017 (presented as n/N, %) [1].

No	Habitat	*A. flavicollis*	*A. agrarius*	*M. musculus*	*M. minutus*	*C. glareolus*	*M. oeconomus*	*M. agrestis*	*M. arvalis*	Total
1	coastal meadow	4/59 (6.8)			1/40 (2.5)	1/9	1/4		0/1	7/113 (6.2)
2	coastal meadow							2/2		2/2
3	forest-meadow ecotone	2/192 (1.1)			0/2	0/36	1/2	0/1		3/233 (1.3)
4	mixed forest	0/29								0/29
5	coastal meadow	0/33			0/26	1/5	0/2			1/66 (1.5)
6	coastal meadow	2/54 (3.7)			0/1	1/3	8/18 (44.5)		0/1	11/77 (14.3)
7	coastal meadow	3/37 (8.1)			0/4	0/8			0/2	3/51 (5.9)
8	coastal meadow	0/5				1/1				1/6
9	flooded meadows		0/52		0/3	0/19	0/40	0/12		0/126
10	flooded forest	0/5	0/7		0/1	1/14 (2.5)	0/2	0/9		1/38 (2.6)
11	mixed forest						0/1		0/13	0/14
12	deciduous forest	0/12	0/6			0/52				0/70
13	mixed forest	0/3				0/7				0/10
14	mixed forest	0/3				0/11				0/14
15	forest-meadow ecotone	0/20				4/117 (3.5)				4/137 (2.9)
16	mixed forest	0/6				0/14				0/20
17	forest-meadow ecotone	0/17	0/10	0/12				0/4		0/43
18	deciduous forest	0/2	0/7			0/88				0/97
19	mixed forest	0/22				0/12				0/34
	Total	11/499 (2.2)	0/82	0/12	1/77 (1.3)	9/396 (2.3)	10/69 (14.5)	2/28 (7.1)	0/17	33/1180 (2.8)

[1] n, number of individuals infected; N, number of individuals tested; No, site number.

2.2. Rodent Trapping

Rodents were trapped by using live or snap traps baited with bread immersed in unrefined sunflower oil. One trapping session consist of three days. The traps were checked two times per day [23]. All rodents were identified to species level and gender morphologically and under dissection, with specimens of *Microtus* voles identified by their teeth [24].

2.3. Molecular Analyses

DNA from rodent spleen was extracted using Genomic DNA Purification Kit (Thermo Fisher Scientific, Vilnius, Lithuania), according to the manufacturer's protocol. The presence of *Babesia* pathogens were conducted through the amplification of the 330 bp fragment of the 18S rRNA gene in nested PCR using two primer sets BS1/BS2 and PiroA/PiroC as described by Rar et al. [25,26]. The primary PCR reaction was carried out in a 20 μL final volume containing: 1 × PCR buffer, 2 mM $MgCl_2$, 0.2 mM dNTPs, 10 pmol of each primer, 2 U Taq DNA polymerase (Thermo Fisher Scientific, Lithuania), double-distilled water and 2 μL of DNA template. Reaction was performed according to the conditions: initial denaturation at 94 °C for 3 min, 35 cycles: denaturation at 94 °C for 60 s, annealing at 58 °C for 60 s and extension at 72 °C for 90 s, and final extension step at 72 °C for 3 min. In the second PCR, the reaction mix was similarly prepared as it was in the first step, with exception that instead of the DNA, 1 μL of the PCR product was added. The PCR conditions were: initial denaturation at 94 °C for 3 min, followed by 35 cycles: denaturation at 94 °C for 60 s, annealing at 64 °C for 60 s, and extension at 72 °C for 90 s. The final extending was at 72 °C for 3 min. In each PCR run negative (double-distilled water) and positive (DNA of *Babesia* positive ticks, infection confirmed by sequencing) controls were used. The PCR products were analyzed by horizontal electrophoresis in 1.5% agarose gel and visualized with ethidium bromide solution (20 ng/μL) using ultra-violet transilluminator UVP GelDoc-It 310 model (Ultra-Violet Products Ltd., Cambridge, UK). The good quality PCR products of *Babesia*-positive samples were extracted from agarose gel. GenJet PCR purification kit (Thermo Fisher Scientific, Lithuania) was used for purification and, after preparation, samples were sent for direct sequencing by Sanger method to Macrogen Europe company (Amsterdam, The Netherlands).

The partial 18S rRNA sequences were analyzed using MEGA X software package, version 10.0.5. [27] and compared with the sequence data available in NCBI GenBank database using the NCBI BLAST® software (http://blast.ncbi.nlm.nih.gov, accessed on

6 June 2020). A phylogenetic tree was constructed by applying maximum-likelihood (ML) method implemented with Tamura-Nei model. Partial 18S rRNA sequences for representative samples were submitted to GenBank under the accession numbers: MT745579 to MT745583.

2.4. Statistical Analysis

The between-species and between-location differences in the prevalence of *Babesia* infection were tested. For these differences, we used Fisher's exact test and the Mantel–Haenszel common odds ratio estimate. Calculations were performed in SPSS software version 22 (IBM SPSS, Chicago, IL, USA), using 95% confidence intervals. We assessed the prevalence of *Babesia* in all investigated rodent species; calculations were performed in OpenEpi software [28]. 95% CI for prevalence was calculated according to the Wilson method [29]. We tested the significance of differences in the prevalence between species and between habitats. These calculations were performed in WinPepi, ver. 11.39. We used the chi-squared test with Upton's approximation for small and medium sample sizes. To express the effect size, we used adjusted Cohen's w [30]. In all tests, $p < 0.05$ was considered significant.

3. Results

We analyzed 1180 rodent individuals, belonging to eight species, best represented by *A. flavicollis* and *C. glareolus* (Table 1). *Babesia* infected rodents were trapped in nine out of 19 sampling locations (Figure 1). *A. flavicollis* was the dominant trapped rodent species in Curonian Spit (70.9%; 409/577) with the prevalence of infection ranging in four locations (where the infected rodents were captured) from 1.1% to 8.1%. *Babesia* infected *C. glareolus* were found in six of the fifteen sampling locations: with the overall prevalence of infection estimated at 6.5% on the Curonian Spit, 3% in the Nemunas River Delta and 1.3% in the eastern part of the country. *Babesia* infected *M. oeconomus* and *M. agrestis* were found in three and one sampling locations in the Curonian Spit, respectively. One *Babesia* infected harvest mouse (*Micromys minutus*) specimen was found in one location in the Curonian Spit (site 1) (Table 1). All trapped house mice (*Mus musculus*), *A. agrarius* and *M. arvalis* were not infected.

3.1. Prevalence of Babesia Parasites in Various Rodent Species

A total of 33 (2.8%, CI = 1.74–4.23%) out of 1180 DNA samples of rodents were positive for *Babesia* DNA. The species-based differences of prevalence of *Babesia* parasites were significant (14.5%; OR, 3.6; 95% CI, 1.330–9.625; $p < 0.012$) and are presented in Figure 2. The highest prevalence of *Babesia* was found in *M. oeconomus* and *M. agrestis* (14.5% vs. 7.1%, $\chi 2 = 0.98$, NS; Cohen's w = 0.101, small effect size). The prevalence in *M. oeconomus* was significantly higher than that in *C. glareolus* ($\chi 2 = 23.1$, w = 0.221), *A. flavicollis* ($\chi 2 = 25.7$; w = 0.213) and *M. minutus* ($\chi 2 = 9.0$; w = 0.250). All differences are significant at $p < 0.001$, effect sizes medium.

3.2. Habitat-Based Differences

In general, the highest prevalence of *Babesia* parasites was characteristic to rodents, trapped in meadows (5.67%, CI = 3.87–8.23%), exceeding that in forests (0.31%, CI = 0.05–1.72%), with intermediate prevalence values observed in ecotones (1.69%, CI = 0.82–3.46%). The differences between meadow-forest ($\chi 2 = 16.4$, $p < 0.001$) and meadow-ecotone ($\chi 2 = 9.3$, $p = 0.002$) were highly significant with medium effect size, while that of forest-ecotone ($\chi 2 = 3.3$, $p = 0.07$) had only a trend without effect (w = 0.067).

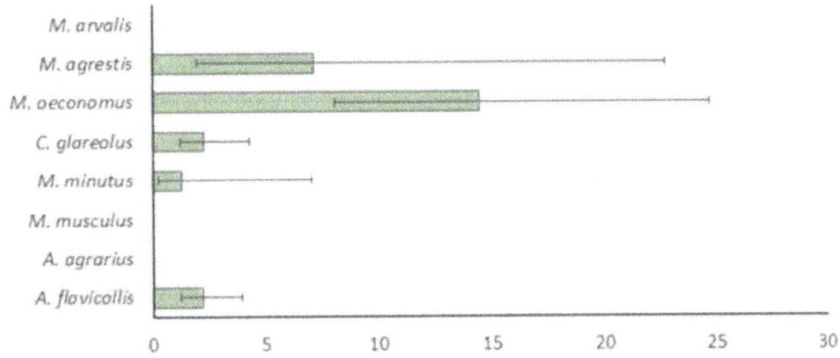

Figure 2. Prevalence (in %, bars represent 95% CI) of *Babesia* in the eight rodent species, irrespective to the habitat.

In the forests, only *C. glareolus* was infected by *Babesia* with low prevalence (Figure 3a). In the meadows, minimum observed prevalence of *Babesia* in *M. minutus* (Figure 3b) was significantly exceeded by prevalence in *M. oeconomus* ($\chi2 = 8.2$, $p < 0.01$; w = 0.245, medium effect size), *M. agrestis* ($\chi2 = 5.9$, $p = 0.015$; w = 0.261, medium effect) and *C. glareolus* ($\chi2 = 3.9$, $p < 0.05$; w = 0.262, small effect size). Other differences of *Babesia* prevalence between rodent species in meadows were not significant. In the forest-meadow ecotone, prevalence of *Babesia* in *M. oeconomus* was higher than in other species, despite minimum sample size (Figure 3c).

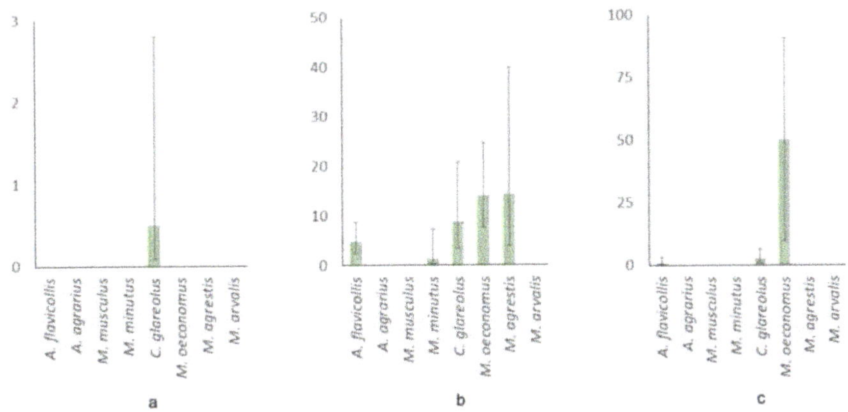

Figure 3. Habitat-based differences of prevalence (in %, bars represent 95% CI) of *Babesia* in the eight rodent species: (**a**) forests, (**b**) meadows, (**c**) ecotones.

3.3. Molecular Characterization of Babesia Isolates

A total of 19 18S rRNA sequences derived from four rodent species *A. flavicollis* (n = 7), *C. glareolus* (n = 3), *M. oeconomus* (n = 7) and *M. agrestis* (n = 2) were analyzed. The sequence analysis of the partial 18S rRNA gene revealed that *Babesia* isolates derived from rodents were 99–100% identical to *B. microti* 'Jena/Germany' strain (GenBank: KC470047; EF413181). Two genotypes with one nucleotide difference were detected in *M. oeconomus* (Figure 4).

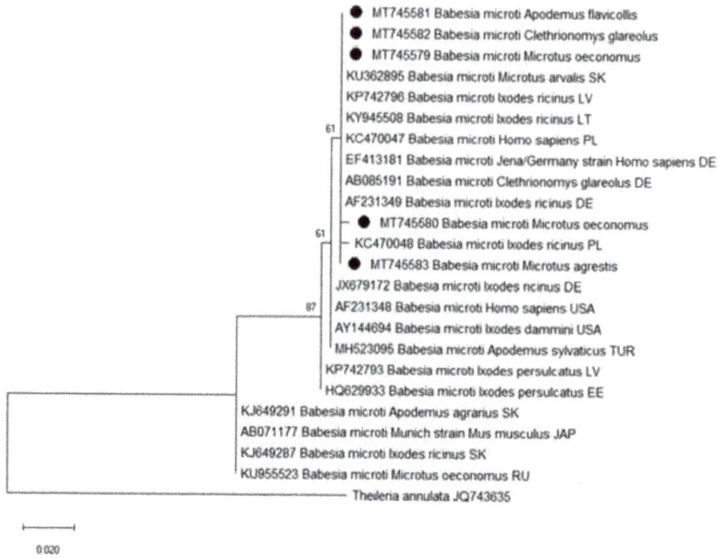

Figure 4. Phylogenetic tree of the partial 18S rRNA gene of *Babesia microti* inferred by ML method, the Tamura–Nei model and bootstrap analysis of 1000 replicates. Marked with dark circle are samples sequenced in the present study. Sequences MT745579 and MT745581 are representative of six and five other samples obtained in this study (from *A. flavicollis* and *M. oeconomus*), respectively. Sequences MT745582 and MT745583 are representative of two and one other samples sequenced in the present study (from *C. glareolus* and *M. agrestis*), respectively.

4. Discussion

In this study, *Babesia* DNA was detected in 33 of 1180 (2.8%) spleen tissue samples of five small rodent species. The overall prevalence of *Babesia* varied among rodent species with the highest prevalence detected in voles *M. oeconomus* (14.5%) and *M. agrestis* (7.1%) (Table 1). *Babesia* infected *M. oeconomus* and *M. agrestis* have been found in north-eastern Poland with the 39.5% (30/76) and 17.7% (3/17) prevalence of infection, respectively [8]. These figures are almost three times higher compered than that obtained in this study. A high prevalence of *Babesia* spp. in *M. agrestis* has been reported in Austria 30.4% (14/46) [31] and the United Kingdom 27.9% (671/2402) [14], while Šebek [32] reported a much lower 0.5% (1/218) prevalence of infection in this rodent species in the former Czechoslovakia. The low overall prevalence of *Babesia* in this study was detected in *C. glareolus* (2.3%). In other European countries, the prevalence of *Babesia* infection in *C. glareolus* varied: 39.7% (60/151) reported in Finland [21], 15.9% (60/151) in Slovenia [12], 11.9% (59/495) in north-eastern Poland [33], 6.1% (3/49) in Croatia [19,34], 6% (25/405) in the Netherlands [11], 0.8% (4/498) in Slovakia [7,35], 0.68% (1/147) in France [22], 0.03% (11/396) in Germany [9].

Babesia parasites were found with low prevalence in mice *A. flavicollis* (2.2%) and *M. minutus* (1.3%) irrespective of the study sites. In line with our results, a low prevalence of *Babesia* in *A. flavicollis* has been reported in Slovakia (1.7%; 12/706; [7,35]) and in Germany (0.01%; 1/178; [9]). However, higher *Babesia* infection rate detected in *A. flavicollis* was documented in Croatia 16.9% (11/65) [19,34], in north-eastern Poland 13.1% (8/61) [20] and in Slovenia 11.8% (15/127) [12]. To the best of our knowledge, our study is the first report of *Babesia* infection in *M. minutus*.

Although, *Babesia* parasites were not found in *A. agrarius* and *M. arvalis* in this study, it was detected in these rodent species trapped in other locations in Lithuania, with a prevalence of 2.1% and 9.1%, respectively [36].

In the present study, a significantly higher overall prevalence of *Babesia* among investigated areas was found on the Curonian Spit (4.9%, 28/577; OR, 3.3; 95% CI, 2.346–4.720; $p < 0.000$) with the highest prevalence of infection (among all locations examined) detected in coastal meadow, site 6 (14.3%, OR, 0.35; 95% CI, 0.173–0.708; $p < 0.004$). In line with this, five various rodent species have been found positive with *Babesia* in seven out of eight locations on the Curonian Spit (Table 1). The detected differences in the prevalence of *Babesia* parasites in the investigated locations might be explained by habitat factors: the highest infection prevalence was detected in the coastal meadows habitat (which on the Curonian Spit most frequent), and additionally by the fact that examined rodents trapped in the Curonian Spit were frequently infested with immature *I. ricinus* (mostly larvae) [37]. The overall prevalence of infestation with immature *I. ricinus* varied between rodent hosts and was highest for *A. flavicollis* (56%). The mean intensity of infestation with *I. ricinus* was 5.2 per rodent hosts (5.6 in *A. flavicollis*, 3.3 in *M. minutus*, 3.0 in *M. oeconomus*, 3.0 in *M. arvalis* and 2.0 in *C. glareolus*) (personal authors data).

Worldwide, seven *B. microti* strains—'USA', 'Hobetsu' ('Otsu'), 'Nagano', 'Kobe', 'Jena/Germany', 'Munich' and 'Baltic' were identified [18,20,38–40]. In Europe, from those, four *B. microti* strains have been detected: the zoonotic 'Jena/Germany' and 'USA' strains, and non-zoonotic 'Munich' strain reported in *Ixodes* ticks and rodents [7–9,20,22,33] and the 'Baltic' strain detected in *I. persulcatus* collected from Estonia and Latvia [39,41] which pathogenicity for human is not known.

The zoonotic 'Jena/Germany' strain has been detected in *A. flavicollis*, *A. agrarius*, *M. arvalis* and *C. glareolus* from Slovakia [7,35], in *M. oeconomus*, *M. arvalis* and *M. agrestis* from Poland [8], in *A. flavicollis*, *C. glareolus* and *M. arvalis* from Germany [9] and in *A. flavicollis* and *C. glareolus* from central Croatia [19], while zoonotic *B. microti* 'USA' strain was detected in microtine rodents from north-eastern Poland [33]. In this study, the *B. microti* 'Jena/Germany' strains were detected in *A. flavicollis* mice and three voles species—*M. oeconomus*, *M. agrestis* and *C. glareolus*. In previous studies, the zoonotic 'Jena/Germany' strains were detected in *I. ricinus* ticks in Europe, including Baltic countries [39,41,42]. Autochthonous cases of human babesiosis due to the *B. microti* 'Jena/Germany' strain have been reported in Germany [4] and Poland [5]. Worldwide, most of the human babesiosis cases have been related to the zoonotic *B. microti* 'USA' strain. A lower virulence of European *B. microti* strains compared to those circulating in North America may be the reason of a lack of recognized human cases associated with European *B. microti* strains, despite human exposure to infectious tick bites in this continent. [43].

The non-zoonotic *B. microti* 'Munich' strain has been found in *C. glareolus* from Slovakia [7], in *A. flavicollis* and *C. glareolus* from central Croatia [19] and in *C. glareolus* from Finland [21] and France [22]. As a general rule, the *B. microti* 'Munich' strain was not found outside of the *I. trianguliceps* distribution area (from Great Britain to Baikal). It is thought that *I. trianguliceps* ticks play an important role for the maintenance of the non-zoonotic *B. microti* 'Munich' strain to mammalian hosts [17]. In Lithuania, *I. trianguliceps* ticks are present and previously were found on *A. flavicollis* and *C. glareolus* rodents [44]. However, in the present study, trapped rodents were infested only with immature *I. ricinus*.

5. Conclusions

Our findings suggest that rodents, especially *Microtus* voles, play an important role in the circulation of the zoonotic *B. microti* 'Jena/Germany' strain in Lithuania. The highest prevalence of *Babesia* parasites was detected in rodents trapped in coastal meadows. This study also detected *Babesia* infection in *M. minutus*, the first recorded infection in this species to the best of the authors' knowledge.

Author Contributions: Conceptualization, A.P., J.R. and D.M.-B.; methodology, L.B., J.R. and D.M.-B.; software, L.B., J.R. and D.M.-B.; formal analysis, L.B., J.R. and D.M.-B.; investigation, J.R. and D.M.-B.; writing—original draft preparation, A.P., J.R., L.B. and D.M.-B.; writing—review and editing, A.P., J.R., L.B. and D.M.-B.; funding acquisition, A.P. and J.R. All authors have read and agreed to the published version of the manuscript.

Funding: Part of this research is funded by the European Social Fund under the Nr. 09.3.3-LMT-K-712-09-0270 "Development of Competences of Scientists, other Researchers and Students through Practical Research Activities" measure.

Institutional Review Board Statement: Rodent sampling was conducted with permission from the Environmental Protection Agency (EPA) and approved by the Ministry of Environment of the Republic of Lithuania, licenses (No. 1 (10 April 2013), No. 15 (31 March 2014), No. 22 (10 April 2015), No. 12 (30 March 2016) and No. 13 (22 March 2017)) in accordance with Lithuanian (the Republic of Lithuania Law on Welfare and Protection of Animals No. XI-2271) and European legislation (Directive 2010/63/EU) on the protection of animals.

Data Availability Statement: Partial 18S rRNA sequences for representative samples were submitted to GenBank under the accession numbers: MT745579 to MT745583.

Acknowledgments: The authors are grateful to Vaclovas Gedminas for collection of rodents on the Curonian Spit.

Conflicts of Interest: The authors declare no conflict of interest.

References

1. Young, K.M.; Corrin, T.; Wilhelm, B.; Uhland, C.; Greig, J.; Mascarenhas, M.; Waddell, L.A. Zoonotic *Babesia*: A scoping review of the global evidence. *PLoS ONE* **2019**, *14*, e0226781. [CrossRef]
2. Westblade, L.F.; Simon, M.S.; Mathison, B.A.; Kirkman, L.A. *Babesia microti*: From mice to ticks to an increasing number of highly susceptible humans. *J. Clin. Microbiol.* **2017**, *55*, 2903–2912. [CrossRef]
3. Hildebrandt, A.; Gray, J.S.; Hunfeld, K.P. Human Babesiosis in Europe: What clinicians need to know. *Infection* **2013**, *41*, 1057–1072. [CrossRef]
4. Hildebrandt, A.; Hunfeld, K.P.; Baier, M.; Krumbholz, A.; Sachse, S.; Lorenzen, T.; Kiehntopf, M.; Fricke, H.-J.; Straube, E. First confirmed autochthonous case of human *Babesia microti* infection in Europe. *Eur. J. Clin. Microbiol. Infect. Dis.* **2007**, *26*, 595–601. [CrossRef] [PubMed]
5. Welc-Falęciak, R.; Pawełczyk, A.; Radkowski, M.; Pancewicz, S.A.; Zajkowska, J.; Siński, E. First report of two asymptomatic cases of human infection with *Babesia microti* (Franca, 1910) in Poland. *Ann. Agric. Environ. Med.* **2015**, *22*, 51–54. [CrossRef]
6. Moniuszko-Malinowska, A.; Swiecicka, I.; Dunaj, J.; Zajkowska, J.; Czupryna, P.; Zambrowski, G.; Czupryna, P.; Garkowski, A. Infection with *Babesia microti* in humans with non-specific symptoms in North East Poland. *Infect. Dis.* **2016**, *48*, 537–543. [CrossRef]
7. Blaňarová, L.; Stanko, M.; Miklisová, D.; Víchová, B.; Mošanský, L.; Kraljik, J.; Bona, M.; Derdáková, M. Presence of *Candidatus Neoehrlichia mikurensis* and *Babesia microti* in rodents and two tick species (*Ixodes ricinus* and *Ixodes trianguliceps*) in Slovakia. *Ticks Tick Borne Dis.* **2016**, *7*, 319–326. [CrossRef]
8. Tołkacz, K.; Bednarska, M.; Alsarraf, M.; Dwużnik, D.; Grzybek, M.; Welc-Falęciak, R.; Behnke, J.M.; Bajer, A. Prevalence, genetic identity and vertical transmission of *Babesia microti* in three naturally infected species of vole, *Microtus* spp. (Cricetidae). *Parasites Vectors* **2017**, *10*, 66. [CrossRef] [PubMed]
9. Obiegala, A.; Pfeffer, M.; Pfister, K.; Karnath, C.; Silaghi, C. Molecular examinations of *Babesia microti* in rodents and rodent-attached ticks from urban and sylvatic habitats in Germany. *Ticks Tick Borne Dis.* **2015**, *6*, 445–449. [CrossRef]
10. Krawczyk, A.I.; van Duijvendijk, G.L.A.; Swart, A.; Heylen, D.; Jaarsma, R.I.; Jacobs, F.H.H.; Fonville, M.; Sprong, H.; Takken, W. Effect of rodent density on tick and tick-borne pathogen populations: Consequences for infectious disease risk. *Parasites Vectors* **2020**, *13*, 34. [CrossRef] [PubMed]
11. Azagi, T.; Jaarsma, R.I.; Docters van Leeuwen, A.; Fonville, M.; Maas, M.; Franssen, F.F.J.; Kik, M.; Rijks, J.M.; Montizaan, M.G.; Groenevelt, M.; et al. Circulation of *Babesia* Species and Their Exposure to Humans through *Ixodes ricinus*. *Pathogens* **2021**, *10*, 386. [CrossRef]
12. Duh, D.; Petrovec, M.; Trilar, T.; Avsic-Zupanc, T. The molecular evidence of *Babesia microti* infection in small mammals collected in Slovenia. *Parasitology* **2003**, *126*, 113–117. [CrossRef]
13. Nefedova, V.V.; Korenberg, E.I.; Kovalevskii, Y.V.; Samokhvalov, M.V.; Gorelova, N.B. The role of *Ixodes trianguliceps* tick larvae in circulation of *Babesia microti* in the Middle Urals. *Entomol. Rev.* **2013**, *93*, 258–266. [CrossRef]
14. Bown, K.J.; Lambin, X.; Telford, G.R.; Ogden, N.; Telfer, S.; Woldehiwet, Z.; Birtles, R. Relative importance of *Ixodes ricinus* and *Ixodes trianguliceps* as vectors for *Anaplasma phagocytophilum* and *Babesia microti* in field vole (*Microtus agrestis*) populations. *Appl. Environ. Microbiol.* **2008**, *74*, 7118–7125. [CrossRef]
15. Cayol, C.; Jääskeläinen, A.; Koskela, E.; Kyröläinen, S.; Mappes, T.; Siukkola, A.; Kallio, E.R. Sympatric *Ixodes*-tick species: Pattern of distribution and pathogen transmission within wild rodent populations. *Sci. Rep.* **2018**, *8*, 16660. [CrossRef] [PubMed]
16. Baneth, G.; Florin-Christensen, M.; Cardoso, L.; Schnittger, L. Reclassification of *Theileria annae* as *Babesia* vulpes sp. nov. *Parasites Vectors* **2015**, *8*, 207. [CrossRef]

17. Rar, V.; Yakimenko, V.; Makenov, M.; Tikunov, A.; Epikhina, T.; Tancev, A.; Bobrova, O.; Tikunova, N. High prevalence of *Babesia microti* 'Munich' type in small mammals from an *Ixodes persulcatus/Ixodes trianguliceps* sympatric area in the Omsk region, Russia. *Parasitol. Res.* **2016**, *115*, 3619–3629. [CrossRef]
18. Usluca, S.; Celebi, B.; Karasartova, D.; Gureser, A.S.; Matur, F.; Oktem, M.A.; Sozen, M.; Karatas, A.; Babur, C.; Mumcuoglu, K.Y.; et al. Molecular Survey of *Babesia microti* (Aconoidasida: Piroplasmida) in Wild Rodents in Turkey. *J. Med. Entomol.* **2019**, *56*, 1605–1609. [CrossRef]
19. Beck, R.; Vojta, L.; Ćurković, S.; Mrljak, V.; Margaletić, J.; Habrun, B. Molecular survey of *Babesia microti* in wild rodents in central Croatia. *Vector Borne Zoonotic Dis.* **2011**, *11*, 81–83. [CrossRef]
20. Welc-Falęciak, R.; Bajer, A.; Paziewska-Harris, A.; Baumann-Popczyk, A.; Siński, E. Diversity of *Babesia* in *Ixodes ricinus* ticks in Poland. *Adv. Med. Sci.* **2012**, *57*, 364–369. [CrossRef]
21. Kallio, E.R.; Begon, M.; Birtles, R.J.; Bown, K.J.; Koskela, E.; Mappes, T.; Watts, P.C. First report of *Anaplasma phagocytophilum* and *Babesia microti* in rodents in Finland. *Vector Borne Zoonotic Dis.* **2014**, *14*, 389–393. [CrossRef]
22. Jouglin, M.; Perez, G.; Butet, A.; Malandrin, L.; Bastian, S. Low prevalence of zoonotic *Babesia* in small mammals and *Ixodes ricinus* in Brittany, France. *Vet. Parasitol.* **2017**, *238*, 58–60. [CrossRef]
23. Balčiauskas, L. *Methods of Investigation of Terrestrial Ecosystems. Part I. Animal Surveys*; Vilnius University: Vilnius, Lithuania, 2004; p. 183.
24. Prūsaitė, J.; Mažeikytė, R.; Pauža, D.; Paužienė, N.; Baleišis, R.; Juškaitis, R.; Mickus, A.; Grušas, A.; Skeiveris, R.; Bluzma, P.M. *Fauna of Lithuania. Mammals*; Mokslas: Vilnius, Lithuania, 1988; p. 295.
25. Rar, V.A.; Fomenko, N.V.; Dobrotvorsky, A.K.; Livanova, N.N.; Rudakova, S.A.; Fedorov, E.G.; Astanin, V.B.; Morozova, O.V. Tick-borne pathogen detection, western Siberia, Russia. *Emerg. Infect. Dis.* **2005**, *11*, 1708. [CrossRef]
26. Rar, V.A.; Epikhina, T.I.; Livanova, N.N.; Panov, V.V. Genetic diversity of *Babesia* in *Ixodes persulcatus* and small mammals from North Ural and West Siberia, Russia. *Parasitology* **2011**, *138*, 175–182. [CrossRef]
27. Kumar, S.; Stecher, G.; Li, M.; Knyaz, C.; Tamura, K. MEGA X: Molecular evolutionary genetics analysis across computing platforms. *Mol. Biol. Evol.* **2018**, *35*, 1547–1549. [CrossRef]
28. Brown, L.D.; Cat, T.T.; DasGupta, A. Interval Estimation for a Proportion. *Stat. Sci.* **2001**, *16*, 101–133. [CrossRef]
29. Abramson, J.H. WINPEPI Updated: Computer Programs for Epidemiologists, and their Teaching Potential. *Epidemiol. Perspect. Innov.* **2011**, *8*, 1. [CrossRef]
30. Thomas, J.R.; Salazar, W.; Landers, D.M. What is Missing in $p < 0.05$? Effect Size. *Res. Q. Exerc. Sport* **1991**, *62*, 344–348. [CrossRef]
31. Šebek, Z.; Sixl, W.; Stünzner, D.; Valová, M.; Hubálek, Z.; Troger, H. Zur Kenntnis der Blutparasiten wildlebender Kleinsäuger in der Steiermark und im Burgenland [Blood parasites of small wild mammals in Steiermark and Burgenland]. *Folia Parasitol.* **1980**, *27*, 295–301. (In German)
32. Šebek, Z. Blutparasiten der wildlebenden Kleinsäuger in der Tschechoslowakei [Blood parasites of small wild mammals in Czechoslovakia]. *Folia Parasitol.* **1975**, *22*, 11–20. (In German)
33. Welc-Falęciak, R.; Bajer, A.; Behnke, J.M.; Siński, E. Effects of host diversity and the community composition of hard ticks (Ixodidae) on *Babesia microti* infection. *Int. J. Med. Microbiol.* **2008**, *298*, 235–242. [CrossRef]
34. Tadin, A.; Turk, N.; Korva, M.; Margaletić, J.; Beck, R.; Vucelja, M.; Habuš, J.; Svoboda, P.; Županc, T.A.; Henttonen, H.; et al. Multiple co-infections of rodents with hantaviruses, *Leptospira*, and *Babesia* in Croatia. *Vector Borne Zoonotic Dis.* **2012**, *12*, 388–392. [CrossRef]
35. Hamšíková, Z.; Kazimírová, M.; Haruštiaková, D.; Mahríková, L.; Slovák, M.; Berthová, L.; Kocianová, E.; Schnittger, L. *Babesia* spp. in ticks and wildlife in different habitat types of Slovakia. *Parasites Vectors* **2016**, *9*, 292. [CrossRef]
36. Baltrūnaitė, L.; Kitrytė, N.; Križanauskienė, A. Blood parasites (*Babesia, Hepatozoon* and *Trypanosoma*) of rodents, Lithuania: Part I. Molecular and traditional microscopy approach. *Parasitol. Res.* **2020**, *119*, 687–694. [CrossRef] [PubMed]
37. Radzijevskaja, J.; Kaminskienė, E.; Lipatova, I.; Mardosaitė-Busaitienė, D.; Balčiauskas, L.; Stanko, M.; Paulauskas, A. Prevalence and diversity of *Rickettsia* species in ectoparasites collected from small rodents in Lithuania. *Parasites Vectors* **2018**, *11*, 375. [CrossRef] [PubMed]
38. Saito-Ito, A.; Kawai, A.; Ohmori, S.; Nagano-Fujii, M. Continuous in vivo culture and indirect fluorescent antibody test for zoonotic protozoa of *Babesia microti*. *Parasitol. Int.* **2016**, *65*, 526–531. [CrossRef] [PubMed]
39. Capligina, V.; Berzina, I.; Bormane, A.; Salmane, I.; Vilks, K.; Kazarina, A.; Bandere, D.; Baumanis, V.; Ranka, R. Prevalence and phylogenetic analysis of *Babesia* spp. in *Ixodes ricinus* and *Ixodes persulcatus* ticks in Latvia. *Exp. Appl. Acarol.* **2016**, *68*, 325–336. [CrossRef] [PubMed]
40. Wei, C.Y.; Wang, X.M.; Wang, Z.S.; Wang, Z.H.; Guan, Z.Z.; Zhang, L.H.; Dou, X.F.; Wang, H. High prevalence of *Babesia microti* in small mammals in Beijing. *Infect. Dis. Poverty* **2020**, *9*, 1–11. [CrossRef] [PubMed]
41. Katargina, O.; Geller, J.; Vasilenko, V.; Kuznetsova, T.; Järvekülg, L.; Vene, S.; Lundkvist, Å.; Golovljova, I. Detection and characterization of *Babesia* species in *Ixodes* ticks in Estonia. *Vector-Borne Zoonotic Dis.* **2011**, *11*, 923–928. [CrossRef]
42. Radzijevskaja, J.; Mardosaitė-Busaitienė, D.; Aleksandravičienė, A.; Paulauskas, A. Investigation of *Babesia* spp. in sympatric populations of *Dermacentor reticulatus* and *Ixodes ricinus* ticks in Lithuania and Latvia. *Ticks Tick Borne Dis.* **2018**, *9*, 270–274. [CrossRef]

43. Rudolf, I.; Golovchenko, M.; Sikutová, S.; Rudenko, N. *Babesia microti* (Piroplasmida: Babesiidae) in nymphal *Ixodes ricinus* (Acari: Ixodidae) in the Czech Republic. *Folia Parasitol.* **2005**, *52*, 274. [CrossRef] [PubMed]
44. Paulauskas, A.; Radzijevskaja, J.; Turčinavičienė, J.; Ambrasienė, D.; Galdikaitė, E. Data on some Ixodid tick species (Acari, Ixodidae) in the Baltic countries. *New Rare Lith. Insect Species* **2010**, *22*, 43–51.

animals

MDPI

Article

Evidence for Unknown *Sarcocystis*-Like Infection in Stranded Striped Dolphins (*Stenella coeruleoalba*) from the Ligurian Sea, Italy

Federica Giorda [1,2,†], Umberto Romani-Cremaschi [3,†], Antoinette E. Marsh [4], Carla Grattarola [1], Barbara Iulini [1], Alessandra Pautasso [5], Katia Varello [1], Enrica Berio [1], Paola Gazzuola [1], Letizia Marsili [6], Cristina E. Di Francesco [7], Maria Goria [1], Federica Verna [8], Tania Audino [1], Simone Peletto [1], Maria Caramelli [1], Mercedes Fernández-Escobar [9], Eva Sierra [2], Antonio Fernández [2], Rafael Calero-Bernal [9] and Cristina Casalone [1,*]

1 Istituto Zooprofilattico Sperimentale del Piemonte, Liguria e Valle d'Aosta, 10154 Torino, Italy; federica.giorda@izsto.it (F.G.); carla.grattarola@izsto.it (C.G.); barbara.iulini@izsto.it (B.I.); katia.varello@izsto.it (K.V.); enrica.berio@izsto.it (E.B.); paola.gazzuola@izsto.it (P.G.); maria.goria@izsto.it (M.G.); tania.audino@izsto.it (T.A.); simone.peletto@izsto.it (S.P.); maria.caramelli@izsto.it (M.C.)

2 Institute for Animal Health and Food Safety (IUSA), Veterinary School, University of Las Palmas de Gran Canaria, Las Palmas de Gran Canaria, 35416 Canary Islands, Spain; eva.sierra@ulpgc.es (E.S.); antonio.fernandez@ulpgc.es (A.F.)

3 Veterinary Department, Mundomar, Calle Sierra Helada s/n, 03503 Benidorm, Spain; umberto.romanicremaschi@gmail.com

4 Department of Veterinary Preventive Medicine, College of Veterinary Medicine, The Ohio State University, 1920 Coffey Road, Columbus, OH 43210, USA; marsh.2061@osu.edu

5 Department of Prevention, Local Veterinary Services (ASL1 Imperiese), Via Aurelia Ponente 97, Bussana di Sanremo, 18038 Imperia, Italy; a.pautasso@asl1.liguria.it

6 Dipartimento di Scienze Fisiche, della Terra e dell'Ambiente, University of Siena, Via Mattioli 4, 53100 Siena, Italy; letizia.marsili@unisi.it

7 Faculty of Veterinary Medicine, University of Teramo, Strada Provinciale 18 Località Piano d'Accio, 64100 Teramo, Italy; cedifrancesco@unite.it

8 Department of Prevention, Local Veterinary Services, Via Conte Verde 125, 35040 Asti, Italy; fverna@asl.at.it

9 SALUVET, Department of Animal Health, Faculty of Veterinary, Complutense University of Madrid, 28040 Madrid, Spain; merfer02@ucm.es (M.F.-E.); r.calero@ucm.es (R.C.-B.)

* Correspondence: cristina.casalone@izsto.it

† These authors have equally contributed to this work.

Citation: Giorda, F.; Romani-Cremaschi, U.; Marsh, A.E.; Grattarola, C.; Iulini, B.; Pautasso, A.; Varello, K.; Berio, E.; Gazzuola, P.; Marsili, L.; et al. Evidence for Unknown *Sarcocystis*-Like Infection in Stranded Striped Dolphins (*Stenella coeruleoalba*) from the Ligurian Sea, Italy. *Animals* **2021**, *11*, 1201. https://doi.org/10.3390/ani11051201

Academic Editor: Pablo Diaz

Received: 20 March 2021
Accepted: 20 April 2021
Published: 22 April 2021

Simple Summary: Two stranded striped dolphins presented meningoenchepalitic lesions associated with the presence of unknown protozoan tissue cysts. The present study aimed at fully characterizing these previously undescribed parasites. Light microscopy re-examination of affected CNS areas showed high numbers of tissue cysts with morphological features resembling those of *Sarcocystis* species. Tissue cyst bradyzoites positively stained when labeled with polyclonal antisera but cross-reactivity could not be precluded. *Sarcocystis* sp. sequences with high homology to species infecting livestock were amplified by means of PCR from myocardial and muscle tissues. This is the first report of *Sarcocystis*-like tissue cysts in the cerebral tissue of stranded cetaceans with muscular sarcocystosis in Mediterranean dolphins. The obtained results may suggest a land-to-sea cycling of Apicomplexan parasites in this region and the need for further investigations in order to foster marine mammal conservation.

Abstract: Two striped dolphins (SD1, SD2), stranded along the Ligurian coast of Italy, were diagnosed with a nonsuppurative meningoencephalitis associated with previously undescribed protozoan tissue cysts. As tissue cysts were morphologically different from those of *Toxoplasma gondii*, additional histopathological, immunohistochemical, ultrastructural, and biomolecular investigations were performed, aiming to fully characterize the organism. Histopathology revealed the presence of large *Sarcocystis*-like tissue cysts, associated with limited inflammatory lesions in all CNS areas studied. IHC was inconclusive, as positive staining with polyclonal antisera did not preclude cross-reaction with other Sarcocystidae coccidia. Applied to each animal, 11 different PCR protocols precluded a neural infection by *Sarcocystis neurona, Sarcocystis falcatula, Hammondia hammondi*, and *Neospora caninum*. *T. gondii* coinfection was confirmed only in dolphin SD2. *Sarcocystis* sp. sequences, showing the highest homology to species infecting the Bovidae family, were amplified from SD1 myocardium and SD2 skeletal muscle. The present study represents the first report of *Sarcocystis*-like tissue cysts in the brain of stranded cetaceans along with the first description of *Sarcocystis* sp. infection in muscle tissue of dolphins from the Mediterranean basin.

Keywords: striped dolphin; tissue cysts; neuropathology; *Toxoplasma gondii*; *Sarcocystis*-like; genotype

1. Introduction

Tissue cyst-forming coccidia from the genera *Toxoplasma*, *Sarcocystis*, and *Neospora* (Apicomplexa) are capable of infecting several species of marine mammals and are responsible for either chronic diseases or acute mortality [1]. While *Neospora caninum* does not seem to pose a major threat to marine wildlife, *Toxoplasma gondii* and *Sarcocystis neurona* are the two coccidian parasites most widely reported in North American marine mammals [1], especially in coastal species that are more likely to be overexposed to immunosuppressant chemical pollutants and to high concentrations of land-derived oocysts [2–4]. The most probable exposure route in these animals is through ingestion of environmentally resistant oocysts or sporocysts shed on land by the definitive hosts and passed into the sea through freshwater runoffs or the release of contaminated ship waters. Eventually, protozoal infectious stages may then accumulate in marine invertebrates, bivalve mollusks, or fish on which intermediate hosts prey [5]. In the Mediterranean basin, *T. gondii* is a frequent finding in stranded odontocetes, and it is often associated with protozoal meningoencephalitis [6,7].

On the other hand, neither muscular nor neural sarcocystosis has ever been officially reported in this geographical area, to the authors' best knowledge. However, a fatal case of hepatic sarcocystosis [8], caused by an unknown species, is the only account of a *Sarcocystis* sp. infection in a Mediterranean cetacean.

In this study, we provide evidence of an infection sustained by a *Sarcocystis*-like organism in two striped dolphins (*Stenella coeruleoalba*) stranded along the Ligurian coast of Italy, in the marine protected area of the Pelagos Sanctuary, in 2011 and 2017.

2. Materials and Methods

2.1. Naturally Infected Dolphins

The cases, SD1 and SD2, included in the present study were diagnosed during routine pathological and cause-of-death assessment in stranded cetaceans at the Istituto Zooprofilattico Sperimentale del Piemonte, Liguria e Valle d'Aosta. The two striped dolphins (*Stenella coeruleoalba*), stranded along the Ligurian Sea coast in 2011 (SD1) and 2017 (SD2) (Figure 1), were submitted for a complete postmortem examination, according to standard protocols [9]. Dolphin SD1 was a 206 cm (total length, TL) adult male, in a poor nutritional status and in a postmortem condition code 3 (moderately decomposed). Dolphin SD2 was a 177 cm (TL) juvenile male, in a moderate nutritional status and in postmortem condition code 2 (fresh). Neither animals displayed any evidence of interaction with fishing activities, and the stomach chambers were devoid of intake. Initially, postmortem findings from SD1 and SD2 were previously published [10,11] and, for the present study, these two cases were re-examined with the focus on neurotrophic causes of inflammation, CNS, and tissue protozoal cysts, parasite identification, and distribution.

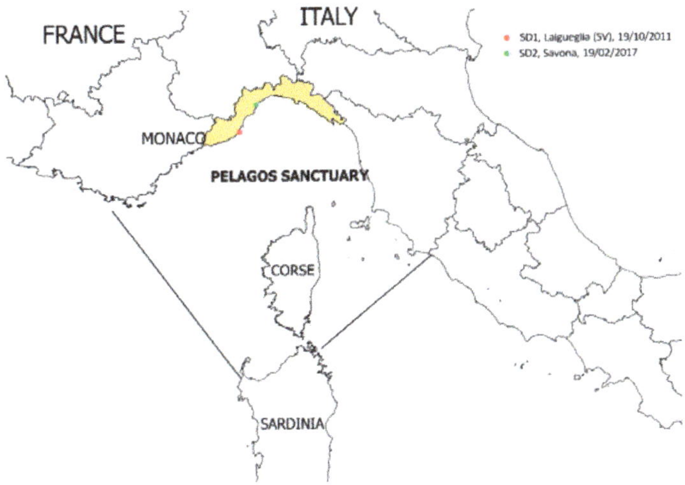

Figure 1. Map of the study area (Ligurian coastline), displaying the stranding locations (red and green dots) of the two striped dolphins infected by Sarcocystidae organisms. The map was created by A.P. with QGIS (QGIS Development Team (2018). QGIS Geographic Information System. Open Source Geospatial Foundation Project. http://qgis.osgeo.org, accessed on 17 August 20).

During necropsy, the tissue samples of all the major organs and lesions were collected and split into aliquots for subsequent analyses: one was kept frozen at −20 °C for microbiological and toxicological investigations, one was kept frozen at −80 °C for biomolecular analyses, and the other was preserved in 10% buffered formalin for histological and immunohistochemical (IHC) investigations. Blood serum, aqueous humor, and cerebrospinal fluid (CSF) were collected, when available, and kept frozen at −20 °C for serological investigations.

2.2. Histology and Immunohistochemistry

Representative tissues from SD1 (brain, lung, heart, liver, spleen, kidney, prescapular lymph node, urinary bladder, and reproductive system) and from SD2 (brain, tonsils, lung, prescapular and tracheobronchial lymph nodes, heart, liver, spleen, pancreas, intestine, skeletal muscle, skin, kidney, urinary bladder, adrenal gland, tongue lesion, and reproductive system) were collected and fixed in 10% neutral buffered formalin, embedded in paraffin, sectioned at 4 ± 2 µm, stained with hematoxylin and eosin (H&E), and examined through a light microscope.

Nine different areas from the brain were sampled and examined, including basal nuclei, thalamus, mesencephalon, pons, obex, and frontal, parietal, occipital, and cerebellar cortex. Immunohistochemistry (IHC) for *Morbillivirus* was performed on tissue sections from SD1 (brain) and from SD2 (brain, tonsils, lung, prescapular and pulmonary lymph nodes, spleen, kidney, urinary bladder, liver, skin, and muscle), using a monoclonal anti-*Canine distemper virus* (CDV) antibody (VMRD, Pullman, WA, USA) [6]. *Toxoplasma gondii* IHC was carried out on the nine aforementioned brain tissues of each case, using a polyclonal serum of caprine origin (VMRD, Pullman, WA, USA) [6].

2.3. PCR and Sequence Analysis

Molecular detection of *Dolphin morbillivirus* (DMV) [12], *Herpesvirus* (HV) [13], *T. gondii* [14], and *Brucella* spp. [15] was routinely achieved on target tissues available from each case, consisting of brain, lung, tonsils, lymph nodes, liver, spleen, kidney, bladder, and blood for DMV, brain, lung, lymph nodes, spleen, and kidney for HV, brain, lymph nodes, liver, spleen, heart, and muscle for *T. gondii*, and brain, lung, tonsils, lymph nodes, liver, spleen, kidney, and blood for *Brucella* spp.

For DMV assays, amplicons were directly sequenced using PCR primers on a 3130XL Genetic Analyzer (Thermo Fisher Scientific Inc., Waltham, MA, USA). Sequences were aligned using the SeqMan software (Lasergene package. DNASTAR Inc., Madison, WI, USA) to obtain a consensus sequence and compared with available sequences retrieved from the National Center for Biotechnology Information (NCBI) database through the BLAST tool (http://blast.ncbi.nlm.nih.gov/Blast.cgi, accessed on 20 February 2021).

2.4. Serological, Toxicological, and Microbiological Analyses

Serological investigations to screen for the presence of specific antibodies against DMV and *T. gondii* were performed [6] on serum, CSF, and aqueous humor, when these samples were available from SD1 and SD2. These same samples were also tested by rapid serum agglutination (Rose Bengal plate test, RBT) using RBT antigen produced from *B. abortus* strain S99 [6,16] to detect anti-smooth *Brucella* spp. antibodies.

Toxicological investigations were carried out only in SD2. The toxicological analysis did not include tissues from SD1 as they were collected before implementing a sampling protocol that included contaminant analyses. Polychlorinated biphenyls (PCBs), hexachlorobenzene (HCB), and dichlorodiphenyltrichloroethanes (DDTs) were measured in blubber. Measurements were made according to Environmental Protection Agency method 8081/8082, with modifications [17], and toxicological stress was evaluated using a theoretical model [18].

Tissue samples including brain, lung, lymph nodes, liver, spleen, and kidney (SD1 and SD2) were processed for standard aerobic, anaerobic, and microaerobic (5% CO_2) bacterial culture and identification, using biochemical and/or molecular analyses. Following international recommendations [19], samples from target tissues underwent specific bacteriological procedures to screen for *Salmonella* spp., *Listeria* spp., and *Brucella* spp.

2.5. Light Microscopy Re-Examination for Parasite Characterization: Histology and Immunohistochemistry

CNS and heart sections from both SD1 and SD2 and skeletal muscle tissue only from SD2 were hematoxylin and eosin (H&E)-stained for light microscopy re-examination. For each organ and the aforementioned CNS areas studied, nine additional 5 μm thick sections were cut in series from stored paraffin blocks. Sections n° 1, 4, and 7 were stained with H&E for light microscopy examination. For IHC, only sections adjacent to the H&E slides presenting tissue cysts which were clearly distinct from *T. gondii* tissue cysts were used (well-defined cyst wall and >8 μm long bradyzoites). The IHC protocols included *S. neurona* polyclonal antibodies (PoAb Rabbit 1 R81, [20]), *S. falcatula* polyclonal antibodies (PoAb Rabbit 2 R-anti SF, [20]), and *S. neurona* monoclonal antibody (MAb 2G5, [21]). The IHC analyses were performed as previously described [21,22]. *Sarcocystis neurona*-infected and noninfected murine brains from interferon gamma (IFN-γ) knockout B6.129S7-Ifngtm1Ts (*Mus musculus*) (Jackson Laboratories, Bar Harbor, ME, USA) mice [23] were used, respectively, as positive and negative controls for the polyclonal antibodies, whereas *S. neurona*-infected opossum (*Didelphis virginiana*) intestine tissues and the brain of a bottlenose dolphin (*Tursiops truncatus*) foetus born under human care were used, respectively, as positive and negative control for the monoclonal antibody.

2.6. Electron Microscopy Examination

Transmission electron microscopy (TEM) was performed on CNS samples of both specimens at the Spanish National Center for Electron Microscopy (Complutense University of Madrid). For each animal, two cysts were excised from the paraffin block and samples were prepared for TEM as previously described [24]. Ultra-thin sections were cut on a Leica UC6 ultramicrotome (Leica Microsystems GmbH, Wetzlar, Germany), mounted onto TEM grids, and stained with 6% saturated uranyl acetate and 3% lead citrate. Sections were examined with a JEOL JEM 1400 Plus (JEOL USA Inc., Peabody, MA, USA) transmission electron microscope operated at 80 kV.

2.7. Molecular Analyses: Parasite Detection, Identification, and Characterization

Genomic DNA was extracted from CNS, myocardium, and skeletal muscle tissue samples from SD1 and SD2 and screened for the presence of tissue-cyst forming coccidia (*Sarcocystis* spp., *T. gondii* and *N. caninum*) DNA using 11 different PCR protocols, detailed in Supplementary material Table S1. DNA extraction from frozen samples was achieved using a "four-step" method, namely, ReliaPrep™ gDNA Tissue Miniprep System (Promega Italia S.r.l. Milan, Italy), whereas DNA extraction from paraffin-embedded samples, previously purified using the QIAamp DNA FFPE Tissue Kit, was achieved using QIAamp® DNA Mini Kit (QIAGEN, Hilden, Germany).

For each positive PCR result, amplicons of the expected size were sequenced using the BigDye® Terminator kit v 3.1 (Applied Biosystems, Foster City, CA, USA) and analyzed on an ABI 3130 Genetic Analyzer (Applied Biosystems). The obtained sequences were curated manually if necessary and analyzed using BioEdit software, version 7.0.5.3 [25]. Generated DNA consensus sequences were aligned to appropriate reference sequences using MEGA X software (http://www.megasoftware.net/ accessed on 20 April 2021) [26], and compared with available sequences retrieved from the NCBI database through the BLAST tool (http://blast.ncbi.nlm.nih.gov/Blast.cgi, accessed on 20 February 2021).

Toxoplasma gondii strain genotyping analyses were carried out at the Complutense University of Madrid. DNA extracts from SD2 heart, muscle, and brain tissues were subjected to the widely used Mn-PCR restriction fragment length polymorphism (RFLP) method, with the markers *SAG1, SAG2 (5′–3′ SAG2*, and alt. *SAG2), SAG3, BTUB, GRA6, c22-8, c29-2, L358, PK1, Apico*, and *CS3* [27,28]. ToxoDB RFLP genotype was identified according to http://toxodb.org/toxo/ accessed on 20 April 2021.

3. Results

3.1. Naturally Infected Dolphins

Results of *postmortem*, routine investigations, along with anamnestic data, are summarized in Table 1.

3.2. Histology and Immunohistochemistry

Significant histopathological lesions detected for SD1 and SD2 are detailed in Table 1. Both SD1 and SD2 were diagnosed with severe and diffuse NS meningoencephalitis in association with the detection of protozoan tissue cysts (a single cyst for SD1 and two cysts for SD2). The cysts were morphologically distinct from *T. gondii*. In addition, SD2 was also diagnosed with *T. gondii*, whereas SD1 samples demonstrated no detectable antibodies for *T. gondii* nor detectable *T. gondii* DNA by PCR. No cysts were observed in the other target tissues investigated: muscle of SD2 and heart of both SD1 and SD2. *Morbillivirus*-specific antigens were detected in brain, urinary bladder, and muscle of SD2 by IHC, while SD1 did not show any specific staining.

All brain cysts observed in both animals stained positive with the polyclonal Ab raised for *T. gondii*. Since these data were not supported by molecular and serological investigations performed in SD1, an antigenic cross-reactivity among genetically related protozoa was hypothesized.

Table 1. Stranding data, body condition, most significant findings (gross and microscopic), pathogens detected, and the hypothesis of causa mortis in the two animals under study.

ID	YS	DC	NuS	Age/ Sex	Main Lesions (Gross and Microscopic)	Detected Pathogens	Cause of Death	Reference
SD1	2011	3	Poor	Adult M	Severe granulomatous pneumonia; fibrinous peritonitis; splenomegaly associated to chronic granulomatous splenitis; cholangiohepatitis; generalized lymphadenitis associated to lymphoid depletion; severe NS meningoencephalitis	*Photobacterium damselae* subsp. *damselae* (isolated from blowhole and lungs); *Campula palliata* (pancreas and liver); *Monorygma grimaldi* (musculature and peritoneum); *Skrjabinalius guevarai* (lung); anti-DMV antibodies (1:16) in blood serum (VN)	Infectious disease (parasitic and unknown agent)	[11]
SD2	2017	2	Moderate	Juvenile M	Skin ulcers; ulcerative glossitis; subcutaneous parasitic cysts; bronchointerstitial pneumonia; multifocal necrotizing hepatitis; cholangiohepatitis; splenomegaly and generalized lymphadenomegaly associated to multicentric lymphoid necrosis; interstitial nephritis; lymphadenitis; severe NS meningoencephalitis	*Phyllobotrium* spp (blubber); *Monorygma grimaldii* (musculature); *Brucella ceti* (isolation from CNS, lung, and spleen; PCR from brain, liver, lung spleen, and lymph nodes); DMV (PCR, IHC from CNS, spleen, kidney, tonsils, and lymph nodes, bladder and muscle); *Toxoplasma gondii* (PCR, IHC from CNS, lymph nodes, spleen, liver, and muscle); anti-DMV antibodies (1:8) in blood serum (VN); anti-*T. gondii* antibodies (>1:640) in blood serum, (1:160) in CSF, (1:80) in aqueous humor (IFAT). Severe immunosuppression (CAN = 0.688) [18]	Infectious disease (viral, bacterial, and parasitic)	[10]

YS = year of stranding; DC = decomposition code (2, fresh; 3, moderate autolysis); NuS = nutritional status; M = male; NS = nonsuppurative; DMV = *Dolphin morbillivirus*; VN = virus neutralization; CNS = central nervous system; PCR = polymerase chain reaction; IHC = immunohistochemistry; CFS = cerebrospinal fluid; IFAT = indirect fluorescent antibody technique; CAN = canonical variable.

3.3. PCR and Sequence Analysis

No biomolecular evidence of DMV, HV, *T. gondii* or *Brucella* spp. was found in SD1.

A systemic DMV infection was demonstrated in SD2, through PCR, in brain, lung, laryngeal tonsils, tracheobronchial lymph node, spleen, kidney, and bladder, and subsequently confirmed through amplicon sequencing and BLAST analysis.

However, in SD2, molecular data supported a coinfection with both *T. gondii*, in brain, liver, muscle, spleen, and tracheobronchial and pulmonary lymph nodes, and *Brucella* sp., by PCR detection, in brain, liver, lung, spleen, and tracheobronchial, pulmonary, and prescapular lymph nodes. The HV analysis was negative for SD2.

3.4. Serological, Toxicological, and Microbiological Analyses

Anti-*Morbillivirus* antibodies (1:16) were detected in serum of SD1, while anti-*Morbillivirus* (1:8 serum) and anti-*T. gondii* antibodies (>1:640 serum; 1:160 CSF; 1:80 aqueous humor) were detected in SD2. No evidence of anti-*Brucella* spp. antibodies was demonstrated in sera, cerebrospinal fluid, or aqueous humor samples from either of the dolphins.

For SD2, the levels of PCBs, HCB, and DDTs, expressed in $ng \cdot g^{-1}$ on a lipid weight basis (PCBs: 136810.3; DDTs: 69866.92; HCB: 153.6; canonical variable value (CAN) = 0.688), confirmed the presence of immunotoxic levels of OC pollutants (CAN > 0.47).

Photobacterium damselae subsp. *damselae* was isolated by microaerobic bacterial culture from blowhole and lungs of SD1; no other significant bacteria, including *Salmonella* spp., *Listeria* spp., and *Brucella* spp., were isolated. *Brucella ceti* was isolated from CNS, spleen, and lung of SD2; no other significant bacteria, including *Listeria* spp. and *Salmonella* spp., were isolated.

3.5. Light Microscopy Re-Examination: Histology and Immunohistochemistry

For both SD1 and SD2, several non-*Toxoplasma* protozoal tissue cysts were observed in all the CNS areas studied, with major involvement of the brain cortical areas. Overall, neural tissue cysts were round to oval in shape, from 27 to 119.3 μm in diameter, presented with a distinguishable, apparently smooth outer wall, and they were filled with mild basophilic bradyzoites (Figure 2). Within the range of the available magnifications, neither internal septations nor villous protrusions were visible by light microscopy. Moreover, it was not possible to detect the presence of free schizonts or merozoites in any of the brain sections examined. The histomorphological appearance was consistent with that of other apicomplexan coccidia, with most features resembling the genus *Sarcocystis*. No other cysts were observed in the heart of either case, while, in the skeletal muscle from SD2, only one tissue cyst morphologically resembling *T. gondii* was observed.

Sarcocystis-like tissue cysts in the brains of both SD1 and SD2 stained immunopositive with the anti-*S. neurona* polyclonal antiserum. Similar results were obtained with the anti-*S. falcatula* polyclonal antiserum whereas MAb 2G5 failed in labeling protozoal antigens. Overall, the application of the polyclonal antisera resulted in a negative staining of the cyst wall and in a sparse labeling of the enclosed bradyzoites (Figure 2).

3.6. Ultrastructural Description of Tissue Cysts

After processing of samples, no tissue cysts could be observed from the SD1 specimen-derived blocks, but a mature thin-walled (400 nm) cyst resembling *T. gondii* (Figure 3) was examined in dolphin SD2. Typical morphology with simple wall structure presenting vesicles, absence of septae, and small bradyzoites of 5.3 × 1.4 μm in size (n = 10) were observed.

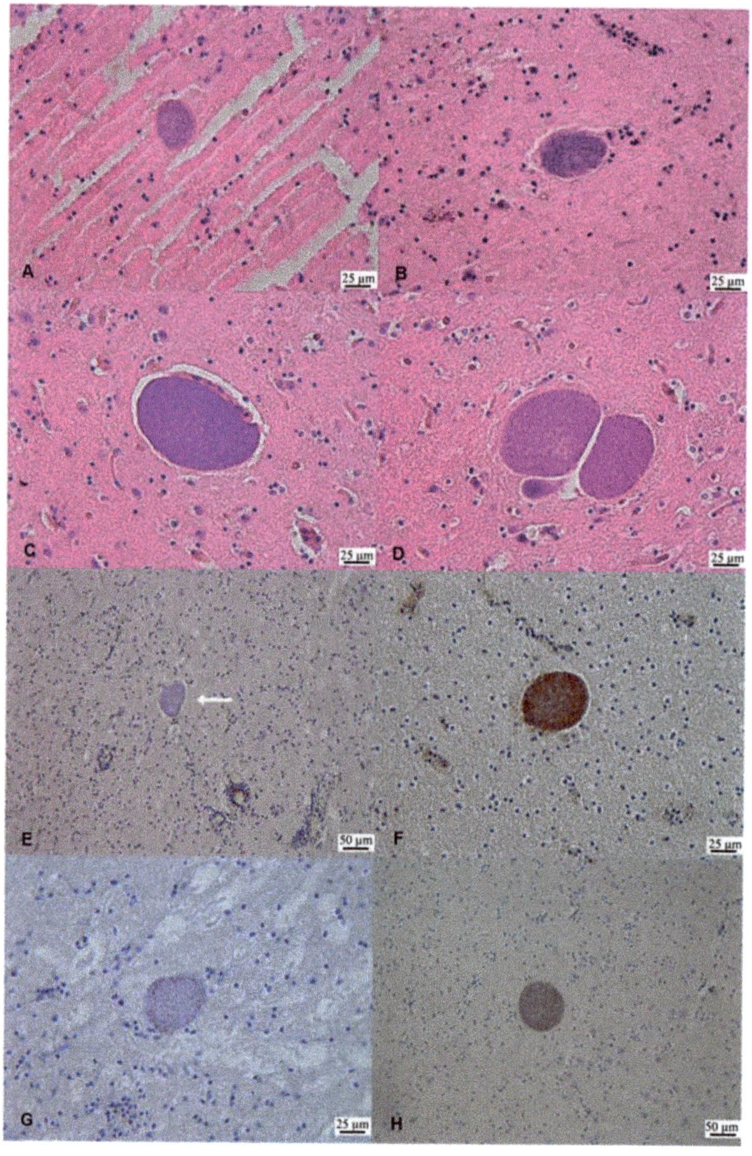

Figure 2. *Sarcocystis*-like tissue cysts in the brain of striped dolphins (*Stenella coeruleoalba*) SD1 and SD2 from Liguria, Italy. (**A**) Parietal cortex (SD1). Protozoan tissue cyst measuring 70 × 50 µm. H&E. (**B**) Occipital cortex (SD1). Protozoan tissue cyst measuring 44.6 × 58.1 µm. H&E. (**C**) Frontal cortex (SD2). Protozoan tissue cyst measuring 72.83 × 116.34 µm. H&E. (**D**) Basal ganglia (SD2). Protozoan tissue cysts measuring (left-right reading) 110 × 119.3 µm, 40 × 19.8 µm and 50 × 99.1 µm. H&E. (**E**) Mesencephalon (SD1). Negative immunostaining of a protozoan tissue cyst (arrow). Monoclonal Ab anti-*S. neurona*. (**F**) Cerebellum (SD1). Positive labeling of *Sarcocystis*-like tissue cyst. Polyclonal Ab anti-*S. falcatula*. (**G**) Mesencephalon (SD1). Positive labeling of a protozoan tissue cyst bradyzoites. Polyclonal Ab anti-*S. neurona*. (**H**) Parietal cortex (SD2). Positive immunostaining of a protozoan tissue cyst bradyzoites. Polyclonal Ab anti-*S. neurona*.

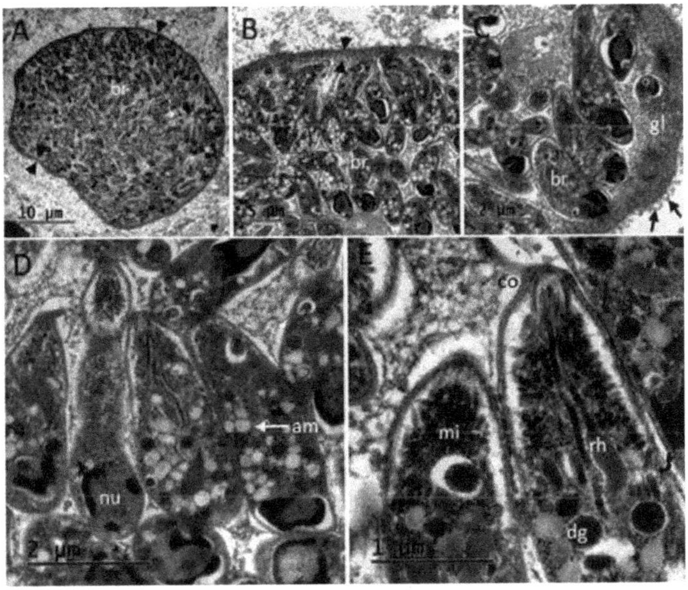

Figure 3. Transmission electron microscopy micrographs of the *Toxoplasma gondii* tissue cyst studied from central nervous system of striped dolphin (*Stenella coeruleoalba*) SD2 in Liguria, Italy. (**A**) Section of the thin-walled tissue cyst; note the cyst wall (arrowheads) and densely packaged bradyzoites (br). (**B**,**C**) Details of the simple and thin cyst wall (arrowheads) and granular layer (gl) presenting vesicles (arrows). (**D**,**E**) Ultrastructural details of bradyzoites, note: nucleus (nu), micronemes (mi), dense granules (dg), amylopectin granules (am), conoid (co), and rhoptries (rh).

3.7. Molecular Detection and Parasite Identification

The findings on PCR screening for tissue-cyst forming coccidia DNA are summarized in Supplementary material Table S2. PCR n.1 (18S region) and n.9 protocols showed the presence of cyst-forming coccidia DNA in at least one tissue of each animal, in agreement with histological findings. Sequencing of PCR products confirmed the absence of *T. gondii* DNA in SD1 and the presence of *T. gondii* in SD2. PCR n.5 and n.6 protocols showed the presence of *Sarcocystis* sp. DNA in the tissues from both SD1 and SD2. The *Sarcocystis* sp. DNA detection was confirmed by sequencing of the amplicons produced from DNA extracted from SD1 myocardial tissue and SD2 skeletal muscle tissue. BLAST® analysis retrieved different homology to *Sarcocystis* species infecting members of the family Bovidae, *S. hirsuta* (99.4%), and *S. buffalonis* (97.8%), respectively. These sequences were deposited in GenBank® with the following accession numbers: MW151248 (*Sarcocystis* sp.) and MW151249 (*Sarcocystis* sp.).

PCR n.7 and n.8 protocols resulted in the detection of Sarcocystidae-unspecific products in heart and brain tissues from SD2 followed by PCR n.8, and DNA sequencing which was used to confirm the presence of *T. gondii* DNA in SD2 heart and brain tissues. This finding was further documented by PCR n.10 (specific for *T. gondii* amplification). Molecular methods confirmed the presence of *T.gondii* in addition to the detection of *Sarcocystis* DNA. The molecular results suggest a potentially greater distribution of *T. gondii* parasites or higher concentration of DNA from *T. gondii* as compared to the *Sarcocystis* sp.

Phylogenetic analysis of the 18S rRNA *Sarcocystis* sp. sequence from SD2 indicated a *Sarcocystis buffalonis*-like organism (MW151249) when the sequence was compared to genetically similar species and the *Sarcocystis* spp. infecting major livestock species in Mediterranean Europe (Figure 4). The *S. hirsuta*-like sequence (MW151248) was not included in the tree because of its short length (122 bp). PCR n.11 screening for *N. caninum*

DNA resulted in none detected in both cases. All SD2 target organs tested for the presence of *T. gondii* DNA were strongly positive (100% homology with other *T. gondii* sequences deposited such as MH793505). Furthermore, all three DNA samples from SD2 were genotyped by PCR-RFLP method as ToxoDB genotype #3 showing type II alleles for all the markers except *Apico* (type I allele) (Table 2).

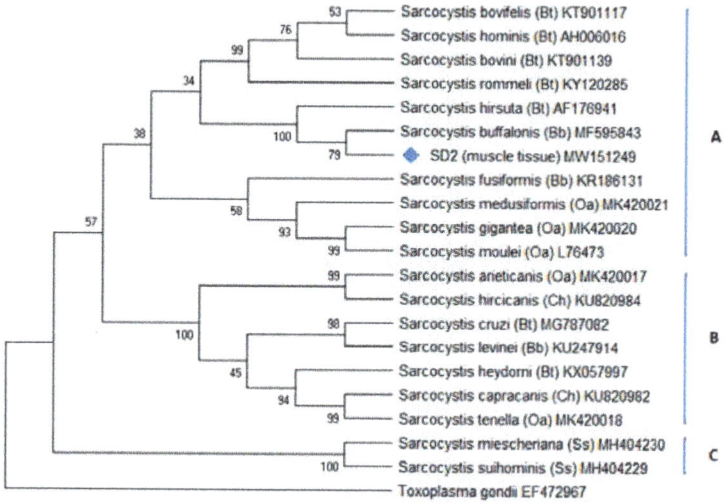

Figure 4. Phylogenetic positioning of the *Sarcocystis*-like organism found in muscle of striped dolphin (*Stenella coeruleoalba*) SD2 in Liguria, Italy. The evolutionary history was inferred using the maximum parsimony (MP) method. Tree n.1 out of three most parsimonious trees (length = 1532) is shown. The consistency index is (0.750000), the retention index is (0.820830), and the composite index is 0.689026 (0.615623) for all sites and parsimony-informative sites (in parentheses). The percentage of replicate trees in which the associated taxa clustered together in the bootstrap test (1000 replicates) is shown next to the branches [29].The MP tree was obtained using the subtree–pruning–regrafting (SPR) algorithm [30] with search level 1 in which the initial trees were obtained by the random addition of sequences (10 replicates). This analysis involved 21 nucleotide sequences from *Sarcocystis* species infecting domestic hosts that are raised in Europe (Bt, *Bos taurus*; Bb, *Bubalus bubalis*, Ch, *Capra hircus*; Oa, *Ovis aries*; Ss, *Sus scrofa*). In cluster B and C, a high or moderate bootstrap (BP) value at each node supported each group containing closely related *Sarcocystis* species with canids as the definitive host, respectively, whereas, in cluster A, some low BP values indicated that the phylogenetic position of *Sarcocystis* species with felids as definitive hosts is not conclusive (BP = 34–38%). These results are probably due to the fact that the 18S rRNA locus is not the most appropriate to infer phylogenetic relationships; moreover, the short length of the SD2 sequence obtained is a limitation identified here. Nonetheless, it should be noted that neighbor-joining and maximum-likelihood methods also resulted in phylogenetic trees in which *S. hirsuta*, *S. buffalonis*, and SD2 *Sarcocystis* sp. grouped together (data not shown). There were a total of 1974 positions in the final dataset. Evolutionary analyses were conducted in MEGA X [26].

Table 2. Results of genotyping analysis carried out on *Toxoplasma gondii* strain identified in dolphin SD2.

Isolate/ Sample	SAG1	3′-SAG2	5′-SAG2	Alt. SAG2	SAG3	BTUB	GRA6	c22-8	C29-2	L358	PK1	Apico	ToxoDB PCR-RFLP Genotype #
RH (ref. type I)	I	I/III	I/II	I	I	I	I	I	I	I	I	I	#1
Me-49 (ref. type II)	II/III	II	I/II	II	II	II	II	II	II	II	II	II	#1
NED (ref. type III)	II/III	I/III	III	III	III	III	III	III	III	III	III	III	#2
SD2 (muscle)	II/III	II	I/II	II	II	II	II	II	II	II	II	I	#3
SD2 (CNS)	II/III	II	I/II	II	II	II	II	II	II	II	II	I	#3
SD2 (Heart)	II/III	II	I/II	II	II	II	II	II	II	II	II	I	#3

4. Discussion

NS meningoencephalitis in dolphins is usually related to *B. ceti* infections, to viruses such as DMV and HV, and to protozoa, especially *T. gondii* [7,31]. In this study, involving two striped dolphins (*Stenella coeruleoalba*) stranded along the Ligurian coast of Italy, samples from SD2 demonstrated the presence of *T. gondii*, DMV, and *B. ceti*. This coinfection, alongside with the toxicological stress detected (CAN > 0.47, [18]), has been considered responsible for the cerebral impairment and the consequential animal's stranding [10]. In contrast, samples from SD1 lacked direct detection of commonly recognized neurotropic agents to explain the observed neuroinflammatory pattern present. SD1 demonstrated serological evidence of DMV infection, evident by a very low titer of antibodies, and suggestive of contact with the virus, rather than the disease, i.e., subclinical infection [32]. Therefore, a closer examination of the tissues for other potential neurotropic agents was undertaken for SD1 and SD2 for comparison.

From the histopathological examination, the morphological appearance of the unusual cysts observed in both animals was highly suggestive of a *Sarcocystis*-like coccidium. Although neither villous protrusions nor internal septations were observed, the protozoan tissue cysts were large (up to 116 µm in diameter), presented with a discernible thin outer wall, and the enclosed bradyzoites stained more basophilic than *T. gondii* ones.

As it is routinely done in suspected cases of toxoplasmosis, an immunohistochemical characterization of the unknown organism was attempted. The obtained results were inconclusive and consistent with the reviewed literature [20–22]. The reactivity of the two polyclonal *Sarcocystis* spp. antisera was mild and failed to label the tissue cyst wall. The absent reactivity of the cyst wall may be related to its maturity, as previously proposed [22]. This similar reactivity of both the anti-*S. neurona* and anti-*S. falcatula* antisera to the cysts is in agreement with what has been previously reported [20,21]. A possible broader spectrum of antigenic cross-reactivity between closely related *Sarcocystis* species and other Apicomplexa has been suggested [20,33].

It should be also noted that, during previous standard investigations, large *Sarcocystis*-like tissue cysts in the CNS of both animals stained positively to the anti-*T. gondii* PoAb, even in the absence of a molecular and serological confirmation of *T. gondii* in SD1. This finding suggests that common epitopes may be shared between these two protozoa and that the commercial anti-*T. gondii* PoAb is not specific in discriminating between these two protozoa, as previously documented by other authors [34], in cases of closely related cyst-forming apicomplexan parasites. The anti-*S. neurona* 2G5 monoclonal antibody did not stain the bradyzoites or the cyst wall of the protozoan tissue cyst in SD1 brain. Although this monoclonal antiserum is directed against a more conserved epitope among *S. neurona* strains and it is suitable to stain FFPE tissues [21], low reactivity was also observed by other authors [35] in the IHC investigations performed on the brains from California Sea otters (*Enhydra lutris nereis*) with PCR-confirmed *S. neurona* infection.

Therefore, the diagnostic value of the IHC staining is questionable and, in this study, could be limited in that both the monoclonal and polyclonal antibodies were raised against merozoite epitopes and may not be suitable to label the cyst wall or the enclosed bradyzoites [35].

As the IHC evaluations were insufficient for parasite identification, in order to have a final confirmation of the parasites detected, 11 different PCR protocols were employed in an attempt to molecularly characterize the tissue-cyst forming protozoa. *Sarcocystis* sp. infections, previously unreported in muscle tissue of Mediterranean cetaceans, were confirmed by PCR means in the myocardium and in the skeletal muscle.

PCR n.1 excluded parasite identification as *S. neurona* or *S. falcatula* in either SD1 or SD2. Moreover, the sequencing of the 18S rDNA amplicon to discern the presence of mixed protozoal infections [35] was nonspecific in SD1, resulting in exclusion of other known tissue cyst-forming coccidia, such as *T. gondii*, *N. caninum*, and *Hammondia hammondi*. *Neospora caninum* infection was also excluded by specific PCR n.11, and the thickness of the observed tissue cyst walls did not correspond to what is expected in such an organism.

The results obtained from SD2 tissues should be carefully interpreted. A majority of the DNA sequencing showed homology to *T. gondii* sequences deposited in GenBank®. SD2 was already proven coinfected with *T. gondii* by means of PCR [14] and IHC [6] and the massive infection by *T. gondii* could have masked the detection of other pathogens, including *Sarcocystis* sp. Perhaps, in SD2, DNA extraction from purified tissue cyst would be more effective as compared to genomic DNA extracted without isolating tissue cyst from the surrounding tissue [36].

PCR protocol n.6 succeeded in amplifying *Sarcocystis* sp. DNA sequences from target organs. A short sequence (122 bp), showing high BLAST homology with *S. hirsuta*, was obtained from SD1 myocardium, whereas a longer (188 bp), high-quality sequence was obtained from SD2 skeletal muscle showing 97.8% homology with *S. buffalonis* isolates.

To date, *S. hirsuta* and *S. buffalonis* have been reported only in cattle and water buffaloes [33] and seem to be strictly intermediate host-specific, like other livestock *Sarcocystis* species, with no reported infection in non-ruminant intermediate hosts. Furthermore, in the present study, parasite identification was achieved targeting the 18S rRNA gene but other authors [37,38] recommend the analysis of *cox1* gene to molecularly discriminate between *Sarcocystis* species-infecting hosts from the Bovidae family.

Although lacking additional natural cases or experimental infections, the possibility for marine wildlife to share *Sarcocystis* species with domestic animals cannot be ruled out completely, especially when intermediate hosts are phylogenetically related. Specifically, the order Cetacea is a sister-group to the family Hippopotamidae and to the Ruminantia taxa [39] which includes also the Bovidae family.

However, considering the results of the BLAST analysis in relationship to sequence quality and the host specificity of this protozoal family, it is most likely that a previously-undescribed *Sarcocystis*-like protozoa, within the family Sarcocystidae and phylogenetically related to species cycling in livestock, is infecting Mediterranean marine mammals. The lack of reports and of prevalence data on muscular sarcocystosis in the Mediterranean makes it difficult to determine the origin of this pathogen.

Since the highest degree of similarity was observed with bovine and bubaline *Sarcocystis* species, a land-to-sea transfer can be hypothesized for this protozoon even though, in the absence of an exhaustive characterization, a marine cycle cannot be discarded. Being that most *Sarcocystis* species are highly host-specific, the hypothesis of a two-host heteroxenous marine cycle, like the one proposed for *Sarcocystis balaenopteralis* [40], seems unlikely as dolphins are apex predators rather than prey in the Mediterranean food web.

To date, *S. neurona* is commonly reported in cases of muscular and neural sarcocystosis diagnosed in marine mammals in North America [1], but its presence has not yet been reported in Europe. However, a *S. neurona* infection in Mediterranean dolphins is highly improbable due to the lack of the definitive hosts (*D. virginiana* and *D. albiventris*) which are geographically confined to the New World.

In the Mediterranean basin, protozoal meningoencephalitis has been limited to reports associated with *T. gondii* subacute to chronic infections [6,7]. A *T. gondii* infection was also confirmed and fully characterized by means of TEM, IHC, and PCR in the present study. Multilocus RFLP PCR genotyping of the *T. gondii* strain infecting SD2 retrieved a type II PRU variant genotype (ToxoDB#3), which is common in felids, livestock, and wildlife around Europe [41,42]. To date, type II genotypes account for the totality of toxoplasmosis reports in Mediterranean cetaceans [43–45] (reviewed in Table 3). Although a marine cycle for this parasite cannot be precluded, the results from SD1 and SD2 support the role of protozoa as a land-base "pollutant" that have expanded the range of their intermediate hosts to the marine environment.

Table 3. Summary of the available literature reporting genotyping data on *Toxoplasma gondii* strains infecting dolphins.

Host	Location	N°. of Individuals	Condition	Genotype (n)	Method (Markers)	Isolate ID	Reference
Bottlenose dolphin (*Tursiops truncatus*)	South Carolina (USA)	3	Stranded	#1 (2/3); Unique (1/3)	PCR-RFLP (SAG1, SAG2, SAG3, BTUB, GRA6, c22-8, c29-2, L358, PK1, and Apico)	TgDoUs1-3	[46]
	Canada (born in Russia)	1	Captivity	#3 (1/1)	PCR-RFLP (B1, SAG1, SAG2, SAG3, BTUB, GRA6, c22-8, c29-2, L358, PK1, and Apico) +PCR-Seq (B1, SAG1)	TgDoCA1	[47]
Hector's dolphins (*Cephalorhynchus hectori*)	New Zealand	8	By caught/ stranded	#3 (7/8); Type II variant (Type I + II at L358 and Type I at Apico)	PCR-RFLP (SAG1, SAG2 (5′ + 3′), SAG3, GRA6, L358, PK1, and Apico)	No isolation	[48]
Striped dolphin (*Stenella coeruleoalba*)	Costa Rica	1	Stranded	#1 (1/1)	PCR-RFLP (SAG1, SAG2, SAG3, BTUB, GRA6, c22-8, c29-2, L358, PK1, and Apico)	TgSdCo1	[49]
	Italy	3	Stranded	Type II (2/3); Unique (1/3)	PCR-seq (B1, gra6 a nd uprt1)	TSL2, TSL3, and TSL6	[43]
	Italy	1	Stranded	#1 (1/1)	RFLP-PCR (SAG1, SAG2 (5′ + 3′), alt SAG2, SAG3, BTUB, GRA6)	No isolation	[45]

Previously, *Sarcocystis* spp. have not been observed in histopathological brain sections of stranded cetaceans. The only documented *Sarcocystis* sp. infection in a wild Mediterranean marine mammal is a hepatic sarcocystosis due to a *S. canis*-like protozoan infecting a striped dolphin stranded along the Spanish coast [8]. *Sarcocystis* spp. have already been reported in other marine mammals [50,51]. In aquatic species, *S. canis*-like infection causes a fatal and acute hepatitis with microscopic lesions confined to the liver [8,24,50]. Mature and immature schizonts are the protozoal stages observed during histopathological and TEM investigations, whereas tissue cysts have not been observed [8,50].

However, sarcocysts in muscle tissue are chronic lesions that have been incidentally observed in mysticetes and odontocetes from other geographical areas without associated pathology [52–56]. In previous reports, protozoal tissue cysts were attributed to *Sarcocystis* spp. only on the basis of light microscopy and TEM findings. In our study, muscular sarcocystosis was evidenced by means of PCR in each stranded dolphin. To the authors' best knowledge, no other *Sarcocystis* species have been previously observed or isolated in Mediterranean cetaceans. Moreover, the prevalence of muscular sarcocystosis in Mediterranean marine mammals is unknown as such infections are likely overlooked during routine investigations. Therefore, it is not possible to state whether or not tissue cysts are a common finding in muscle tissues and which species of *Sarcocystis* are prevalent in Mediterranean marine mammals.

The most likely hypothesis would be to consider the same protozoa in the muscle tissues as the etiologic agents of the cysts observed in the CNS. Nevertheless, we cannot discard the possibility of two different species infecting the same host. The morphology of the observed neural cysts is highly suggestive of a *Sarcocystis*-like coccidium. It was not feasible to perform a morphological comparison with the muscle tissue cysts because SD1 FFPE skeletal muscle was not available and only *T. gondii* muscular cysts were observed in SD2 skeletal muscle sections (*data not shown*), while no cysts were identified in the heart sections of either SD1 or SD2.

As the biomolecular investigations failed in amplifying specific *Sarcocystis* sequences in the CNS, further investigations are needed to confirm our putative diagnosis.

5. Conclusions

The present study represents the first description of a *Sarcocystis*-like infection in muscle tissue of dolphins from the Mediterranean basin along with the first report of *Sarcocystis*-like tissue cysts in the brain of stranded cetaceans.

The *T. gondii* strain detected belongs to a common genotype circulating in Europe, while the unknown organisms were genetically similar to *Sarcocystis* species infecting the Bovidae family. Such results might suggest a land-to-sea cycling of these Apicomplexan parasites and the need for further investigations.

Because of the novelty of these findings, special attention should be reserved for the differential diagnosis of protozoal infections when performing sanitary surveillance on stranded Mediterranean cetaceans, including collection and preservation of tissues to enable a panel of characterization studies.

Supplementary Materials: The following are available online at https://www.mdpi.com/article/10.3390/ani11051201/s1, Table S1: Selected PCR protocols used in the present study to detect DNA from tissue-cyst forming coccidian, Table S2: Results of the PCR protocols screening for tissue cyst-forming coccidian DNA in target organs.

Author Contributions: Conceptualization, F.G., E.S., A.F. and C.C.; formal analysis, F.G., U.R.-C., A.E.M., C.G. and R.C.-B.; funding acquisition, R.C.-B. and C.C.; investigation, B.I., A.P., K.V., E.B., P.G., L.M., C.E.D.F., M.G., F.V., T.A., S.P. and M.F.-E.; methodology, F.G., U.R.-C., A.E.M., C.G. and R.C.-B.; supervision, M.C., R.C.-B. and C.C.; writing—original draft, F.G. and U.R.-C.; writing—review and editing, A.E.M., M.C., R.C.-B. and C.C. All authors have read and agreed to the published version of the manuscript.

Funding: This research was funded by the Italian Ministry of Health [Ricerca Corrente 2016 IZS PLV 05/16 RC].

Institutional Review Board Statement: Not applicable.

Data Availability Statement: The data presented in this study are available within the article and the Supplementary Materials. The sequences generated in the present study were submitted to the GenBank database with the following accession numbers: MW151248-MW151249.

Acknowledgments: The authors are grateful to Milena Monnier and Laura Serracca for their support in biomolecular analyses, to Walter Mignone for necropsy investigations and to Agustín Rebollada-Merino (VISAVET Health Surveillance Centre, UCM) for his assistance with microscope images. Rafael Calero-Bernal is part of the TOXOSOURCES consortium supported by the funding from the European Union's Horizon 2020 Research and Innovation Program under the grant agreement No 773830: One Health European Joint Program.

Conflicts of Interest: The authors declare no conflict of interest. The funders had no role in the design of the study; in the collection, analyses, or interpretation of data; in the writing of the manuscript, or in the decision to publish the results.

References

1. Miller, M.A.; Shapiro, K.; Murray, M.J.; Haulena, M.; Raverty, S. Protozoan Parasites of Marine Mammals. In *CRC Handbook of Marine Mammals*; Gulland, F.M.D., Dierauf, L.A., Whitman, K.L., Eds.; CRC Press: New York, NY, USA, 2018; pp. 425–469; ISBN 9781498796873.
2. Gibson, A.K.; Raverty, S.; Lambourn, D.M.; Huggins, J.; Magargal, S.L.; Grigg, M.E. Polyparasitism is associated with increased disease severity in Toxoplasma gondii-infected marine sentinel species. *PLoS Negl. Trop. Dis.* **2011**, *5*, 14–16. [CrossRef]
3. Shapiro, K.; Miller, M.A.; Mazet, J. Temporal Association Between Land-Based Runoff Events and California Sea Otter (Enhydra Lutris Nereis) Protozoal Mortalities. *J. Wildl. Dis.* **2012**, *48*, 394–404. [CrossRef]
4. Van Bressem, M.; Raga, A.; Di Guardo, G.; Jepson, P.; Duignan, P.; Siebert, U.; Barrett, T.; Santos, M.; Moreno, I.; Siciliano, S.; et al. Emerging infectious diseases in cetaceans worldwide and the possible role of environmental stressors. *Dis. Aquat. Organ.* **2009**, *86*, 143–157. [CrossRef]
5. Barbosa, L.; Johnson, C.K.; Lambourn, D.M.; Gibson, A.K.; Haman, K.H.; Huggins, J.L.; Sweeny, A.R.; Sundar, N.; Raverty, S.A.; Grigg, M.E. A novel Sarcocystis neurona genotype XIII is associated with severe encephalitis in an unexpectedly broad range of marine mammals from the northeastern Pacific Ocean. *Int. J. Parasitol.* **2015**, *45*, 595–603. [CrossRef]
6. Di Guardo, G.; Proietto, U.; Di Francesco, C.E.; Marsilio, F.; Zaccaroni, A.; Scaravelli, D.; Mignone, W.; Garibaldi, F.; Kennedy, S.; Forster, F.; et al. Cerebral Toxoplasmosis in Striped Dolphins (Stenella coeruleoalba) Stranded Along the Ligurian Sea Coast of Italy. *Vet. Pathol.* **2010**, *47*, 245–253. [CrossRef]
7. Pintore, M.D.; Mignone, W.; Di Guardo, G.; Mazzariol, S.; Ballardini, M.; Florio, C.L.; Goria, M.; Romano, A.; Caracappa, S.; Giorda, F.; et al. Neuropathologic findings in cetaceans stranded in Italy (2002–2014). *J. Wildl. Dis.* **2018**, *54*. [CrossRef] [PubMed]
8. Resendes, A.R.; Juan-Sallés, C.; Almeria, S.; Majó, N.; Domingo, M.; Dubey, J.P. Hepatic sarcocystosis in a striped dolphin (Stenella coeruleoalba) from the Spanish Mediterranean coast. *J. Parasitol.* **2002**, *88*, 206–209. [CrossRef]
9. Geraci, J.R.; Lounsbury, V.J. *Marine Mammals Ashore: A Field Guide for Strandings*, 2nd ed.; National Aquarium: Baltimore, MD, USA, 2005; Volume 486, ISBN 0977460908.
10. Garofolo, G.; Petrella, A.; Lucifora, G.; Di Francesco, G.; Di Guardo, G.; Pautasso, A.; Iulini, B.; Varello, K.; Giorda, F.; Goria, M.; et al. Occurrence of Brucella ceti in striped dolphins from Italian Seas. *PLoS ONE* **2020**, *15*, e0240178. [CrossRef]
11. Giorda, F.; Ballardini, M.; Di Guardo, G.; Pintore, M.D.; Grattarola, C.; Iulini, B.; Mignone, W.; Goria, M.; Serracca, L.; Varello, K.; et al. Postmortem findings in cetaceans found stranded in the Pelagos sanctuary, Italy, 2007–2014. *J. Wildl. Dis.* **2017**, *53*, 795–803. [CrossRef] [PubMed]
12. Verna, F.; Giorda, F.; Miceli, I.; Rizzo, G.; Pautasso, A.; Romano, A.; Iulini, B.; Pintore, M.D.; Mignone, W.; Grattarola, C.; et al. Detection of morbillivirus infection by RT-PCR RFLP analysis in cetaceans and carnivores. *J. Virol. Methods* **2017**, *247*, 22–27. [CrossRef]
13. VanDevanter, D.R.; Warrener, P.; Bennett, L.; Schultz, E.R.; Coulter, S.; Garber, R.L.; Rose, T.M. Detection and analysis of diverse herpesviral species by consensus primer PCR. *J. Clin. Microbiol.* **1996**, *34*, 1666–1671. [CrossRef]
14. Vitale, M.; Galluzzo, P.; Currò, V.; Gozdzik, K.; Schillaci, D.; Di Marco Lo Presti, V. A high sensitive nested PCR for Toxoplasma gondii detection in animal and food samples. *J. Microb. Biochem. Technol.* **2013**, *05*, 39–41. [CrossRef]
15. Baily, G.G.; Krahn, J.B.; Drasar, B.S.; Stoker, N.G. Detection of Brucella melitensis and Brucella abortus by DNA amplification. *J. Trop. Med. Hyg.* **1992**, *95*, 271–275.
16. Hernández-Mora, G.; González-Barrientos, R.; Morales, J.A.; Chaves-Olarte, E.; Guzmán-Verri, C.; Baquero-Calvo, E.; De-Miguel, M.J.; Marín, C.M.; Blasco, J.M.; Moreno, E. Neurobrucellosis in stranded dolphins, Costa Rica. *Emerg. Infect. Dis.* **2008**, *14*, 1430–1433. [CrossRef]
17. Marsili, L.; Focardi, S. Chlorinated hydrocarbon (HCB, DDTs and PCBs) levels in cetaceans stranded along the Italian coasts: An overview. *Environ. Monit. Assess.* **1997**, *45*, 129–180. [CrossRef]
18. Marsili, L.; D'Agostino, A.; Bucalossi, D.; Malatesta, T.; Fossi, M.C. Theoretical models to evaluate hazard due to organochlorine compounds (OCs) in Mediterranean striped dolphin (Stenella coeruleoalba). *Chemosphere* **2004**, *56*, 791–801. [CrossRef]
19. World Organisation for Animal Health (OIE). *Manual of Diagnostic Tests and Vaccines for Terrestrial Animals (Terrestrial Manual)*, 8th ed.; World Organisation for Animal Health (OIE): Paris, France, 2018; ISBN 978-92-95108-18-9.
20. Dubey, J.P.; Garner, M.M.; Stetter, M.D.; Marsh, A.E.; Barr, B.C. Acute Sarcocystis Falcatula-Like Infection in a Carmine Bee-Eater (Merops Nubicus) and Immunohistochemical Cross Reactivity Between Sarcocystis Falcatula and Sarcocystis Neurona. *J. Parasitol.* **2001**, *87*, 824–832. [CrossRef]
21. Marsh, A.; Hyun, C.; Barr, B.C.; Tindall, R.; Lakritz, J. Characterization of monoclonal antibodies developed against Sarcocystis neurona. *Parasitol. Res.* **2002**, *88*, 501–506. [CrossRef] [PubMed]
22. Marsh, A.E.; Chaney, S.B.; Howe, D.K.; Saville, W.J.; Reed, S.M. Small sarcocysts can be a feature of experimental infections with Sarcocystis neurona merozoites. *Vet. Parasitol.* **2017**, *245*, 116–118. [CrossRef] [PubMed]
23. Chaney, S.B.; Marsh, A.E.; Lewis, S.; Carman, M.; Howe, D.K.; Saville, W.J.; Reed, S.M. Sarcocystis neurona manipulation using culture-derived merozoites for bradyzoite and sporocyst production. *Vet. Parasitol.* **2017**, *238*, 35–42. [CrossRef] [PubMed]

24. Dubey, J.P.; Calero-Bernal, R.; Rosenthal, B.M.; Speer, C.A.; Fayer, R. Unique Multiple-Host Sarcocystis Species. In *Sarcocystosis of Animals and Humans*, 2nd ed.; CRC Press: Boca Raton, FL, USA, 2015; pp. 125–168.
25. Hall, T.A. BIOEDIT: A user-friendly biological sequence alignment editor and analysis program for Windows 95/98/ NT. *Nucleic Acids Symp. Ser.* **1999**, *41*, 95–98. [CrossRef]
26. Kumar, S.; Stecher, G.; Li, M.; Knyaz, C.; Tamura, K. MEGA X: Molecular evolutionary genetics analysis across computing platforms. *Mol. Biol. Evol.* **2018**, *35*, 1547–1549. [CrossRef]
27. Su, C.; Shwab, E.K.; Zhou, P.; Zhu, X.Q.; Dubey, J.P. Moving towards an integrated approach to molecular detection and identification of Toxoplasma gondii. *Parasitology* **2010**, *137*, 1–11. [CrossRef]
28. Pena, H.F.J.; Gennari, S.M.; Dubey, J.P.; Su, C. Population structure and mouse-virulence of Toxoplasma gondii in Brazil. *Int. J. Parasitol.* **2008**, *38*, 561–569. [CrossRef] [PubMed]
29. Felsenstein, J. Confidence Limits on Phylogenies: An Approach Using the Bootstrap. *Evolution (N. Y.)* **1985**, *39*, 783. [CrossRef]
30. Masatoshi, N.; Sudhir, K. *Molecular Evolution and Phylogenetics*; Oxford University Press: Oxford, UK, 2000; ISBN 0 19 513584 9.
31. Sierra, E.; Sánchez, S.; Saliki, J.T.; Blas-Machado, U.; Arbelo, M.; Zucca, D.; Fernández, A. Retrospective study of etiologic agents associated with nonsuppurative meningoencephalitis in stranded cetaceans in the canary Islands. *J. Clin. Microbiol.* **2014**, *52*, 2390–2397. [CrossRef]
32. Van Bressem, M.-F.; Duignan, P.; Banyard, A.; Barbieri, M.; Colegrove, K.; De Guise, S.; Di Guardo, G.; Dobson, A.; Domingo, M.; Fauquier, D.; et al. Cetacean Morbillivirus: Current Knowledge and Future Directions. *Viruses* **2014**, *6*, 5145–5181. [CrossRef] [PubMed]
33. Dubey, J.P.; Calero-Bernal, R.; Rosenthal, B.M.; Speer, C.A.; Fayer, R. *Sarcocystosis of Animals and Humans*, 2nd ed.; CRC Press: Boca Raton, FL, USA, 2015; ISBN 9780429183188.
34. Sundermann, C.A.; Estridge, B.H.; Branton, M.S.; Bridgman, C.R.; Lindsay, D.S. Immunohistochemical diagnosis of Toxoplasma gondii: Potential for cross- reactivity with Neospora caninum. *J. Parasitol.* **1997**, *83*, 440–443. [CrossRef]
35. Miller, M.A.; Barr, B.C.; Nordhausen, R.; James, E.R.; Magargal, S.L.; Murray, M.; Conrad, P.A.; Toy-Choutka, S.; Jessup, D.A.; Grigg, M.E. Ultrastructural and molecular confirmation of the development of Sarcocystis neurona tissue cysts in the central nervous system of southern sea otters (Enhydra lutris nereis). *Int. J. Parasitol.* **2009**, *39*, 1363–1372. [CrossRef]
36. Dubey, J.P.; Calero-Bernal, R.; Rosenthal, B.M.; Speer, C.A.; Fayer, R. Techniques. In *Sarcocystosis of Animals and Humans*; CRC Press: Boca Raton, FL, USA, 2015; pp. 121–122.
37. Gjerde, B.; Hilali, M.; Abbas, I.E. Molecular differentiation of Sarcocystis buffalonis and Sarcocystis levinei in water buffaloes (Bubalus bubalis) from Sarcocystis hirsuta and Sarcocystis cruzi in cattle (Bos taurus). *Parasitol. Res.* **2016**, *115*, 2459–2471. [CrossRef]
38. Gjerde, B. Phylogenetic relationships among Sarcocystis species in cervids, cattle and sheep inferred from the mitochondrial cytochrome c oxidase subunit I gene. *Int. J. Parasitol.* **2013**, *43*, 579–591. [CrossRef]
39. Hassanin, A.; Delsuc, F.; Ropiquet, A.; Hammer, C.; Jansen Van Vuuren, B.; Matthee, C.; Ruiz-Garcia, M.; Catzeflis, F.; Areskoug, V.; Nguyen, T.T.; et al. Pattern and timing of diversification of Cetartiodactyla (Mammalia, Laurasiatheria), as revealed by a comprehensive analysis of mitochondrial genomes. *Comptes Rendus Biol.* **2012**, *335*, 32–50. [CrossRef]
40. Seilacher, A.; Reif, W.-E.; Wenk, P. The parasite connection in ecosystems and macroevolution. *Naturwissenschaften* **2007**, *94*, 155–169. [CrossRef]
41. Amouei, A.; Sarvi, S.; Sharif, M.; Aghayan, S.A.; Javidnia, J.; Mizani, A.; Moosazadeh, M.; Shams, N.; Hosseini, S.A.; Hosseininejad, Z.; et al. A systematic review of Toxoplasma gondii genotypes and feline: Geographical distribution trends. *Transbound. Emerg. Dis.* **2020**, *67*, 46–64. [CrossRef]
42. Fernández-Escobar, M.; Calero-Bernal, R.; Benavides, J.; Regidor-Cerrillo, J.; Guerrero-Molina, M.C.; Gutiérrez-Expósito, D.; Collantes-Fernández, E.; Ortega-Mora, L.M. Isolation and genetic characterization of Toxoplasma gondii in Spanish sheep flocks. *Parasites Vectors* **2020**, *13*, 1–14. [CrossRef]
43. Di Guardo, G.; Di Cesare, A.; Otranto, D.; Casalone, C.; Iulini, B.; Mignone, W.; Tittarelli, C.; Meloni, S.; Castagna, G.; Forster, F.; et al. Genotyping of Toxoplasma gondii isolates in meningo-encephalitis affected striped dolphins (Stenella coeruleoalba) from Italy. *Vet. Parasitol.* **2011**, *183*, 31–36. [CrossRef]
44. Marcer, F.; Marchiori, E.; Centelleghe, C.; Ajzenberg, D.; Gustinelli, A.; Meroni, V.; Mazzariol, S. Parasitological and pathological findings in fin whales Balaenoptera physalus stranded along Italian coastlines. *Dis. Aquat. Organ.* **2019**, *133*, 25–37. [CrossRef] [PubMed]
45. Terracciano, G.; Fichi, G.; Comentale, A.; Ricci, E.; Mancusi, C.; Perrucci, S. Dolphins stranded along the tuscan coastline (Central Italy) of the "pelagos sanctuary": A parasitological investigation. *Pathogens* **2020**, *9*, 612. [CrossRef] [PubMed]
46. Dubey, J.P.; Fair, P.A.; Sundar, N.; Velmurugan, G.; Kwok, O.C.H.H.; McFee, W.E.; Majumdar, D.; Su, C. Isolation of Toxoplasma gondii From Bottlenose Dolphins (Tursiops truncatus). *J. Parasitol.* **2008**, *94*, 821–823. [CrossRef] [PubMed]
47. Dubey, J.P.; Mergl, J.; Gehring, E.; Sundar, N.; Velmurugan, G.V.; Kwok, O.C.H.H.; Grigg, M.E.; Su, C.; Martineau, D. Toxoplasmosis in captive dolphins (Tursiops truncatus) and walrus (Odobenus rosmarus). *J. Parasitol.* **2009**, *95*, 82–85. [CrossRef] [PubMed]
48. Roe, W.D.; Howe, L.; Baker, E.J.; Burrows, L.; Hunter, S.A. An atypical genotype of Toxoplasma gondii as a cause of mortality in Hector's dolphins (Cephalorhynchus hectori). *Vet. Parasitol.* **2013**, *192*, 67–74. [CrossRef]

49. Dubey, J.P.; Morales, J.A.; Sundar, N.; Velmurugan, G.V.; González-Barrientos, C.R.; Hernández-Mora, G.; Su, C. Isolation and genetic characterization of Toxoplasma gondii from striped dolphin (Stenella coeruleoalba) from Costa Rica. *J. Parasitol.* **2007**, *93*, 710–711. [CrossRef]
50. Calero-Bernal, R.; Mauroo, N.F.; Hui, S.W.; Kuiken, T.; van de Bildt, M.W.G.; de Jong, A.W.; Osterhaus, A.D.M.E.; Sims, L.; Gendron-Fitzpatrick, A.; Carmena, D.; et al. Acute fatal sarcocystosis hepatitis in an Indo-Pacific bottlenose dolphin (Tursiops aduncus) in Hong Kong. *Vet. Parasitol.* **2017**, *235*, 64–68. [CrossRef]
51. Dubey, J.P.; Zarnke, R.; Thomas, N.J.; Wong, S.K.; Van Bonn, W.; Briggs, M.; Davis, J.W.; Ewing, R.; Mense, M.; Kwok, O.C.H.; et al. Toxoplasma gondii, Neospora caninum, Sarcocystis neurona, and Sarcocystis canis-like infections in marine mammals. *Vet. Parasitol.* **2003**, *116*, 275–296. [CrossRef]
52. De Guise, S.; Lagacé, A.; Girard, C.; Béland, P. Intramuscular Sarcocystis in two beluga whales and an Atlantic white-sided dolphin from the St. Lawrence estuary, Quebec, Canada. *J. VET Diagn. Investig.* **1993**, *5*, 296–300. [CrossRef] [PubMed]
53. Akao, S. A New Species of Sarcocystis Parasitic in the Whale Balaenoptera borealis. *J. Protozool.* **1970**, *17*, 290–294. [CrossRef]
54. Ewing, R.; Zaias, J.; Stamper, M.A.; Bossart, G.D.; Dubey, J.P. Prevalence of Sarcocystis sp. in stranded Atlantic white-sided dolphins (Lagenorhynchus acutus). *J. Wildl. Dis.* **2002**, *38*, 291–296. [CrossRef]
55. Sierra, E.; Espinosa de los Monteros, A.; Fernández, A.; Díaz-Delgado, J.; Suárez-Santana, C.; Arbelo, M.; Sierra, M.A.; Herráez, P. Muscle Pathology in Free-Ranging Stranded Cetaceans. *Vet. Pathol.* **2017**, *54*, 298–311. [CrossRef]
56. Munday, B.L.; Mason, R.W.; Hartley, W.J.; Presidente, P.J.; Obendorf, D. Sarcocystis and related organisms in Australian wildlife: I. Survey findings in mammals. *J. Wildl. Dis.* **1978**, *14*, 417–433. [CrossRef] [PubMed]

animals

MDPI

Article

The Role of Mustelids in the Transmission of *Sarcocystis* spp. Using Cattle as Intermediate Hosts

Petras Prakas *, Linas Balčiauskas, Evelina Juozaitytė-Ngugu and Dalius Butkauskas

Nature Research Centre, Akademijos Str. 2, LT-08412 Vilnius, Lithuania; linas.balciauskas@gamtc.lt (L.B.); evelina.ngugu@gamtc.lt (E.J.-N.); dalius.butkauskas@gamtc.lt (D.B.)

* Correspondence: petras.prakas@gamtc.lt

Simple Summary: Members of the genus *Sarcocystis* are worldwide distributed protozoan parasites. *Sarcocystis* infections cause great losses in economically important animals. There is a lack of studies on *Sarcocystis* in naturally infected wild predators, especially of the family Mustelidae which represent a presumably important group of definitive hosts of these parasites. The objective of the present study was to examine the small intestine samples of various mustelid species from Lithuania serving as a possible source of *Sarcocystis* spp. using cattle as intermediate hosts. Overall, 84 samples collected from five mustelid species were analyzed. Oocysts/sporocysts of *Sarcocystis* spp. were detected in 75 animals (89.3%). Using molecular methods four *Sarcocystis* spp., *S. bovifelis*, *S. cruzi*, *S. hirsuta* and *S. hominis* were identified, with the first two being the most prevalent. These results indicate that mustelids are involved in the transmission of *Sarcocystis* spp. using cattle as intermediate hosts. The determined high prevalence of *Sarcocystis* spp. rates cause concerns about food safety issues. To control the spread of infection, further studies on the way carcasses of cattle or beef waste become accessible to mustelids are needed.

Abstract: There is a lack of research on the role of mustelids in the transmission of various *Sarcocystis* spp. In the present study we tested the hypothesis that widespread mustelids in Lithuania could be involved in the transmission of *Sarcocystis* spp. using cattle as intermediate hosts. In 2016–2020, intestinal samples of 84 mustelids were examined. *Sarcocystis* spp. were identified by species-specific PCR targeting the *cox1* gene and subsequent sequencing. Under a light microscope, oocysts/sporocysts of *Sarcocystis* spp. were observed in 40 samples (47.6%), while using molecular methods, they were detected in 75 animals (89.3%). Four *Sarcocystis* spp. were identified in the intestinal samples of American mink (*Neovison vison*), Beech marten (*Martes foina*), European pine marten (*Martes martes*), European badger (*Meles meles*) and European polecat (*Mustela putorius*). The prevalence of predominant *Sarcocystis* spp., *S. bovifelis* (89.3%) and *S. cruzi* (73.8%) was significantly higher than that of *S. hirsuta* (3.6%) and *S. hominis* (1.2%). In an individual sample, most frequently two *Sarcocystis* spp. were identified (69.0%), then a single species (15.5%) and three species (4.8%). The present study provides strong evidence that mustelids serve as definitive hosts for *Sarcocystis* spp. using cattle as intermediate hosts.

Keywords: *Sarcocystis*; cattle; mustelidae; life cycle; *cox1*; molecular identification

Citation: Prakas, P.; Balčiauskas, L.; Juozaitytė-Ngugu, E.; Butkauskas, D. The Role of Mustelids in the Transmission of *Sarcocystis* spp. Using Cattle as Intermediate Hosts. *Animals* **2021**, *11*, 822. https://doi.org/10.3390/ani11030822

Academic Editors: Rafael Calero-Bernal and Ignacio Garcia-Bocanegra

Received: 15 February 2021
Accepted: 12 March 2021
Published: 15 March 2021

1. Introduction

Representatives of the genus *Sarcocystis* (Apicomplexa: Sarcocystidae) are cyst forming coccidians with an obligatory prey-predator two-host life cycle. Asexual multiplication with the formation of sarcocysts takes place in the extra-intestinal tissues of the intermediate host (IH), while sexual stages (oocysts-sporocysts) develop in the small intestine of the definitive host (DH) [1]. Predators and scavengers serve as DH for *Sarcocystis* spp., whereas prey animals become IH [2].

Members of the family Mustelidae may act as IH or DH for several *Sarcocystis* spp. The agent of equine protozoal myeloencephalitis, *S. neurona* was also detected in the muscles of a fisher (*Martes pennanti*), ferret (*Mustela putorius furo*) and American mink (*Neovison vison*) [3]. Additionally, eight species of *Sarcocystis* have been observed in the muscles of various mustelids [4]. Recently described *S. lutrae* [5] was identified in the muscles of several Carnivora families, Canidae, Mustelidae and Procyonidae [5–7]. The role of mustelids as DH of *Sarcocystis* spp. has not been investigated [8].

Mustelidae is the largest and most diverse family in the order of Carnivora in Lithuania, with nine species present [9]. Representatives of mustelids occur in all habitats, including the urban ones [10,11]. The broad habitat niches of the American mink, the Beech marten (*Martes foina*), European badger (*Meles meles*), European pine marten (*Martes martes*) and European polecat (*Mustela putorius*) are reflected in their diverse diets [10,11]. In general, members of the family Mustelidae are opportunistic predators and their diet consists of birds, various mammals, fish, amphibians, invertebrates, fruits, ungulate carcasses, plants and mushrooms [12–16]. In Lithuania, the food chains of mustelids, including cattle carrion, were not investigated in detail, with exception of the European pine marten [17]. Diet of this species in the cold period included 5.3% of carcasses of domestic animals according to the biomass consumed. Thus, far no studies on the role of mustelids in the transmission of *Sarcocystis* in Lithuania have been undertaken.

Recently, a high prevalence of *Sarcocystis* spp. in cattle from Lithuania has been recorded [18]. By performing trypsinization of the diaphragm muscles and species-specific PCR targeting the *cox1* (mitochondrial gene encoding subunit 1 of cytochrome c oxidase), *S. cruzi* was identified in 96.1% of the samples, *S. bovifelis* was detected in 71.6% of the samples, *S. hirsuta* was confirmed in 30.4% of the samples and *S. hominis* was observed in 13.7% of the samples [19]. Canids are DH for *S. cruzi*, humans are DH for *S. hominis*, whereas *S. hirsuta* and *S. bovifelis* are transmitted via felids [19]. The Eurasian lynx (*Lynx lynx*) is the only wild member of the felids in Lithuania [9]. However, this species is not abundant and there were approximately 160 lynx individuals in Lithuania in 2018 [20]. Thus, the high prevalence of *S. bovifelis* implies that it is not solely felids that contribute to the spread of this species. Therefore, we put forward the hypothesis that mustelids can act as DH of *S. bovifelis*. In order to test the assumption, the aim of the present study was to examine the small intestines of various mustelids from Lithuania for the presence of *Sarcocystis* spp. employing cattle as IH.

2. Materials and Methods

2.1. Sample Collection

Between 2016 and 2020, intestine samples of 84 mustelids (40 American mink, 4 Beech marten, 5 European badger, 20 European pine marten and 15 European polecat) were studied for the presence of *Sarcocystis* spp. The animals were collected from hunters, taxidermists, or biologists who found dead animals on the roadways in different regions of Lithuania (Figure 1).

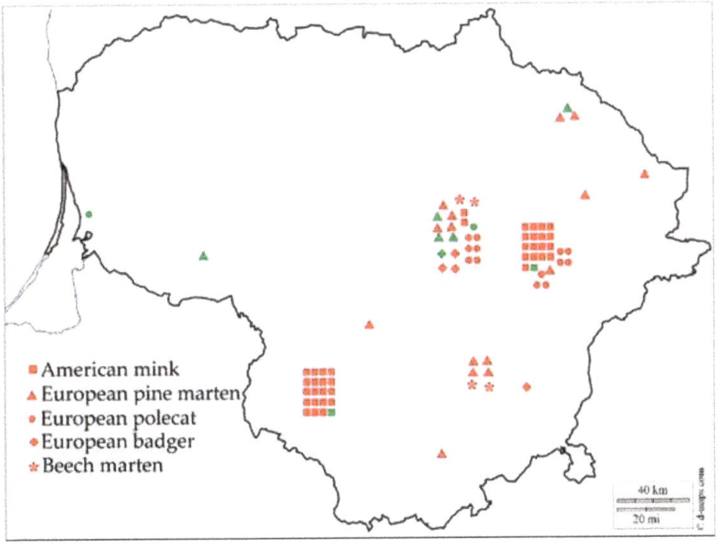

Figure 1. *Sarcocystis* spp. in the species of Mustelidae in Lithuania. Red color means positive individuals, green color represents negative individuals.

2.2. Examination of Intestines

Oocysts/sporocysts of *Sarcocystis* spp. were excreted from the entire intestine of each mustelids using a slightly modified Verma et al. [21] technique. At first, faeces of each intestine were squeezed and the entire intestine was cut lengthwise. The intestinal epithelium was lightly scraped with the help of a scalpel blade and suspended in 50 mL of water. The homogenate was centrifuged for 10 min at 1000 rpm, 25 °C in 50 mL centrifuge tubes. The supernatant was discarded and sediments were re-suspended in 50 mL water. Subsequently, the homogenate was centrifuged for 10 min at 1000 rpm, 25 °C and the supernatant was discarded. The examination of the sediments for oocysts/sporocysts under a light microscope was repeated. The 200 µL of re-suspended sediments were taken from each sample and used for DNA extraction. DNA was isolated from all mustelid samples.

2.3. Molecular Analysis

DNA extraction from mucosal suspension was performed using the GeneJET Genomic DNA Purification Kit (Thermo Fisher Scientific Baltics, Vilnius, Lithuania). *Sarcocystis* spp. were identified by nested PCR of partial *cox1* sequences. Primers used in the present study are listed in Table 1. PCRs were conducted in the final volume of 25 µL made of 12.5 µL of DreamTaq PCR Master Mix (Thermo Fisher Scientific, Vilnius, Lithuania), 0.5 µM of each primer, 0.04 µg template DNA and nuclease-free water. The first run of nested PCR began with one cycle at 95 °C for 5 min followed by 35 cycles of 94 °C for 45 s, 58–60 °C, depending on primer pair for 60 s and 72 °C for 80 s and ending with one cycle at 72 °C for 7 min. For the second PCR assay, 1 µL from the first PCR assay was used. Visualization, purification and sequencing of PCR products were carried out using a previously described protocol [22]. The obtained *cox1* sequences were compared with the Nucleotide BLAST program (megablast option) [23]. The *cox1* sequences generated in the present study are available in GenBank with Acc. No. MW595468–MW595608.

Table 1. The primers used for the nested PCR.

Species	Primer Name	Primer Sequence	Orientation	The Run of Nested PCR
S. bovifelis	SF1 [1]	ATGGCGTACAACAATCATAAAGAA	Forward	First
	SkatR	CAGGCTGAACAGHABTACGA	Reverse	First
	V2bo3	ATATTTACCGGTGCCGTACTTATGTT	Forward	Second
	V2bo4	GCCACATCATTGGTGCTTAGTCT	Reverse	Second
S. cruzi	SF1 [1]	ATGGCGTACAACAATCATAAAGAA	Forward	First
	SsunR2	GTGCCTCCCAGGCTGAAYAG	Reverse	First
	GsScruF	TGTATCTACTTACGGCAGGTATCTTT	Forward	Second
	GsScruR	CGTAGTTAGATCCATATCACTCGGTA	Reverse	Second
S. hirsuta	SF1 [1]	ATGGCGTACAACAATCATAAAGAA	Forward	First
	SkatR	CAGGCTGAACAGHABTACGA	Reverse	First
	GaHiEF [2]	GTTGTGCGGTATGAATTATCAACCT	Forward	Second
	GaHiER [2]	GGTAAGAACTGGAATGGTTAATATCAG	Reverse	Second
S. hominis	VohoF	GTGCGGTATGAACTGTCTACTGCT	Forward	First
	VohoR	AATACCTGCCCGGCCTTAAC	Reverse	First
	GaHoEF [2]	TCTCTGGTTTTGGTAACTACTTCGT	Forward	Second
	GaHoER [2]	CAGACACTGGGATATAATACCGAAC	Reverse	Second

[1] [24], [2] [19].

2.4. Statistical Tests

The prevalence and 95% CI for prevalence were calculated using OpenEpi epidemiological software [25], following the Wilson method for calculating score interval [26]. Differences in the prevalence of the identified *Sarcocystis* spp. were evaluated using the Chi-squared test, calculated in WinPepi, ver. 11.39 and using Upton's approximation for small and medium sample sizes [27]. Comparing the prevalence of *Sarcocystis* spp., the effect size was expressed according to adjusted Cohen's w [28].

3. Results

3.1. Differences in Prevalence of Sarcocystis spp. Using Microscopic and Molecular Methods

Based on microscopic examination, the prevalence of *Sarcocystis* spp. in mucosal scrapings was 47.6% (Table 2). Under a light microscope usually free sporocysts measuring 11.8 × 8.3 µm (7.1–14.5 × 6.5–10.9 µm; $n = 219$) were seen. Sporulated oocysts of *Sarcocystis* 17.7 × 13.1 µm (12.5–23.7 × 10.5–18.3 µm; $n = 100$) were also noticed. With the help of nested PCR and subsequent sequencing *Sarcocystis* spp. were confirmed in 75 animals (89.3%). In general, as compared with morphological examination, the detection rate of *Sarcocystis* spp. was significantly higher ($\chi^2 = 33.56$, $p < 0.0001$; adjusted Cohen's w = 0.709, large effect size) when a molecular method was employed. The molecular method yielded significantly more detections in the American mink, European polecat and European badger (Cohen's w = 1.083, 0.606 and 1.061, respectively, large effect size). Differences between the two methods in the Beech marten and European pine marten were not significant (Table 2). In one American mink and three Beech marten samples, oocysts/sporocysts were detected microscopically, however, these samples were negative for the examined *Sarcocystis* spp. using a molecular analysis.

Based on molecular analysis, the highest prevalence of *Sarcocystis* spp. was observed in the Beech marten, followed by the American mink and European polecat; however, even the lowest prevalence of *Sarcocystis* spp. detected in the European badger and European pine marten were 75% and higher (Table 2). The prevalence of *Sarcocystis* spp. observed in the Beech marten, American mink and European polecat did not differ statistically (species cluster with the highest prevalence). The prevalence of *Sarcocystis* spp. observed in the American mink was significantly higher ($\chi^2 = 5.09$, $p < 0.025$; Cohen's w = 0.435, medium effect size) than that detected in the European pine marten. Other differences were not significant and the effect size was either small or absent.

Table 2. Identification of *Sarcocystis* spp. oocysts/sporocysts in mustelids using microscopic and molecular examination.

Host Species	*N*	*Sarcocystis* spp. Positive Animals					
		Microscopic Analysis			Molecular Analysis		
		n	%	95% CI	N	%	95% CI
American mink	40	15	37.5	24.2–53.0	38	95.0 ***	83.5–98.6
Beech marten	4	3	75.0	30.1–95.4	4	100 NS	51.0–100.0
European pine marten	20	12	60.0	38.7–78.1	15	75.0 NS	53.1–88.8
European badger	5	1	20.0	36.2–62.5	4	80.0 *	37.6–96.4
European polecat	15	9	60.0	35.8–80.2	14	93.3 **	70.2–98.8
Total	84	40	47.6	37.3–58.2	75	89.3 ***	80.9–94.34

Significance of differences between methods is shown in superscript: * $p < 0.05$, ** $p < 0.01$, *** $p < 0.0001$, NS not significant.

3.2. Molecular Identification of *Sarcocystis* spp.

The comparison of sequences generated in the present study showed the presence of four *Sarcocystis* spp. (*S. bovifelis*, *S. cruzi*, *S. hirsuta* and *S. hominis*) in the analyzed samples of Mustelidae (Table 3).

Table 3. Intra- and inter-specific genetic variability of identified *Sarcocystis* spp.

Sarcocystis spp.	GenBank Accession No. (Length in bp)	Sequence Similarity (%)		
		Comparing Obtained Sequences	Comparing Isolates of the Same Species	Comparing Isolates with Other Closely Related Species
S. bovifelis	MW595468–MW595542 (361)	98.4–100	97.2–100% *S. bovifelis* (KT900961–KT900998, KC209690–KC209696, MK962347–MK962348, MT796903–MT796925)	92.5–94.5% *S. bovini* (KT900999–KT901022, LC171858)
S. cruzi	MW595543–MW595604 (556)	98.2–100	96.0–100% *S. cruzi* (KC209597–KC209600, KT901078–KT901095, LC171859–LC171862, MG787071–MG787076, MT796926–MT796945)	90.8–93.4% *S. pilosa* (KU753903–KU753910, LC349942, LC349966–LC349967, LC466196–LC466201, LC481077–LC481081, LC496070, MT070670– MT070677)
S. hirsuta	MW595605–MW595607 (461)	98.9–99.8	98.9–99.8% *S. hirsuta* (KC209634, KT901023–KT901077, LC171863, MT796946–MT796951, MT796958–MT7969)	95.6–96.3% *S. buffalonis* (KU247868–KU247873, MG792800–MG792802)
S. hominis	MW595608 (501)	-	97.6–99.0% *S. hominis* (MH021119, MK497840–MK497843, MT796961–MT796964)	87.1–87.8% *S. bovifelis*

3.3. Distribution of *Sarcocystis* spp. in the Intestine Samples of Mustelids

Irrespective of the host species, *S. bovifelis* in the examined samples was identified most often (Figure 2A). The prevalence of *S. bovifelis* (89.3%) was significantly higher than that of *S. cruzi* (73.8%, a small effect size), *S. hirsuta* (3.6%, a large effect size) and *S. hominis* (1.2%, a large effect size). The prevalence of *S. cruzi* was significantly higher than that of *S. hirsuta* (3.6%) and *S. hominis* (a large effect size both).

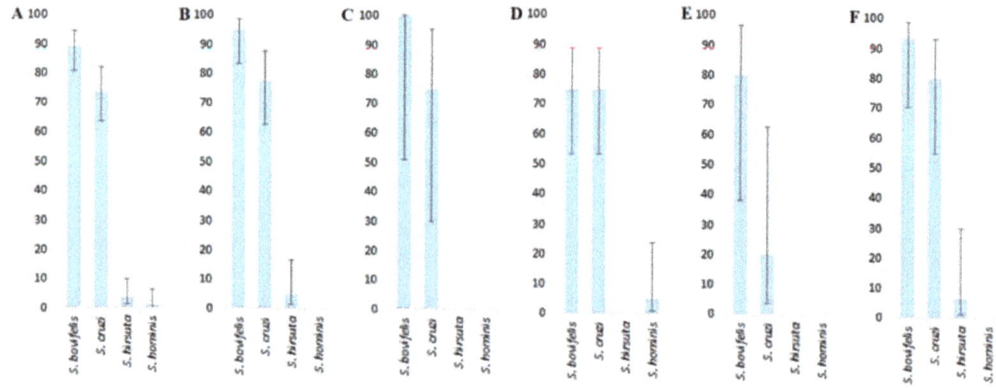

Figure 2. Prevalence of *Sarcocystis* spp. in the examined samples of mustelids. (**A**)—in the pooled sample of all host species, (**B**)—in American mink, (**C**)—in Beech marten, (**D**)—in European pine marten, (**E**)—in European badger, (**F**)—in European polecat. Differences of prevalence in A: *S. bovifelis* > *S. cruzi* (χ^2 = 6.65, p < 0.01; Cohen's w = 0.288), >*S. hirsuta* (χ^2 = 123.32, p < 0.001; w = 2.376) and >*S. hominis* (χ^2 = 130.79, p < 0.001; w = 2.688); *S. cruzi* > *S. hirsuta* (3.6%, χ^2 = 86.83, p < 0.001; w = 1.472) and >*S. hominis* (χ^2 = 93.94, p < 0.001; w = 1.604); in B: *S. bovifelis* >*S. cruzi* (χ^2 = 5.10, p < 0.025; w = 0.372); in E: *S. bovifelis* > *S. cruzi* (χ^2 = 3.24, p < 0.075; w = 1.064).

The prevalence of *S. bovifelis* was the highest, exceeding that of *S. cruzi* in the examined samples of the American mink (a medium effect size, Figure 2B) and European badger (a large effect size, Figure 2E). The prevalence of *S. bovifelis* and *S. cruzi* did not differ significantly in European polecat (Figure 2F) and Beech marten (Figure 2C); in European pine marten they were equal (Figure 2D). The prevalence of predominant *Sarcocystis* spp., *S. bovifelis* and *S. cruzi*, was significantly higher than that of *S. hirsuta* and *S. hominis*, in all host species (Figure 2B–F). Both predominant species were observed in all five examined host species. *Sarcocystis hirsuta* was identified in two American mink individuals and one European polecat individual; whereas *S. hominis* was confirmed in one European pine marten individual.

Up to three *Sarcocystis* spp. were identified in one host individual (Figure 3). No examined *Sarcocystis* spp. were found in approximately one tenth of the investigated animals (10.7%). The prevalence of single species infections was 15.5%; in all cases when a single species was detected in individual samples, it was *S. bovifelis*. Two *Sarcocystis* spp. (69.0%) were most frequently identified in one host individual and in all such cases it was *S. cruzi*/*S. bovifelis* co-infection. Three *Sarcocystis* spp. were confirmed in four animals (4.8%), one European polecat individual, one European pine marten individual and two American minks. In three of these cases, it was *S. bovifelis*/*S. cruzi*/*S. hirsuta* co-infection, in one case—*S. bovifelis*/*S. cruzi*/*S. hominis* co-infection.

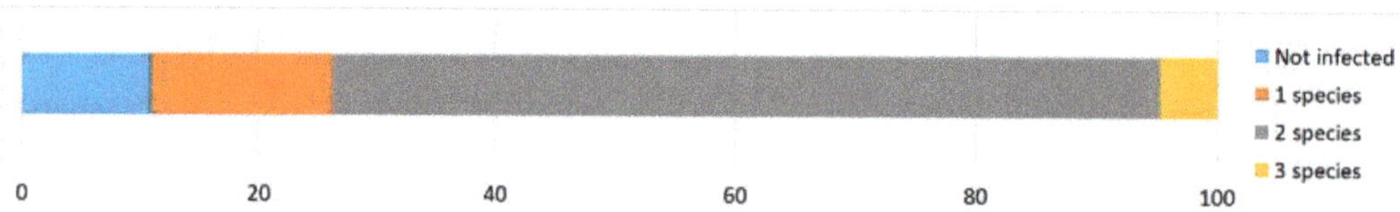

Figure 3. Distribution of the number of *Sarcocystis* spp. identified in the examined samples of mustelids.

4. Discussion

In the present study, high rates (89.3%) of *Sarcocystis* spp. employing cattle as IH were observed in mustelids from Lithuania. Under a light microscope oocysts/sporocysts were detected in 40 out of 84 samples (47.6%). In comparison, the presence of *Sarcocystis* spp. in 75 (89.3%) mucosal scrapings of mustelids were confirmed by molecular methods. Usually, molecular analysis is performed when oocysts/sporocysts of *Sarcocystis* spp. are microscopically detected in intestine mucosal or faecal samples [2,29–31]. However, the results of the present study reveal that molecular methods should be applied in testing all examined samples rather than only microscopically positive ones. No *Sarcocystis* spp. were identified in the mucosal scrapings of a single American mink and three European pine martens using species-specific PCR; however, oocysts/sporocysts were detected in these samples under a light microscope. Thus, these animals were most likely infected with oocysts/sporocysts of *Sarcocystis* spp., which employ other than cattle IH. There are a few reports on mustelids as DH for *Sarcocystis* spp. Transmission experiments have shown that mustelids are DH of several *Sarcocystis* spp., *S. campestris*, *S. muris*, *S. putorii*, *S. undulati* and *S. citellivulpes* (invalid species by Dubey [1]) using members of the order Rodentia as IH [8]. Further studies are needed to reveal the role of mustelids in the transmission of *Sarcocystis* spp. using various mammals and birds as IH.

Sarcocystis spp. identified in the present study, namely, *S. bovifelis*, *S. cruzi*, *S. hirsuta* and *S. hominis*, are specific to their IH [32]. Molecular data suggest that *S. cruzi* might occasionally infect water buffaloes (*Bubalus bubalis*) [33]. However, sheep, goats, pigs, horses and other domestic animals raised in Lithuania cannot serve as IH of the above-mentioned *Sarcocystis* spp. [1]. Of the Lithuanian wild fauna, only the European bison (*Bison bonasus*) can possibly act as an IH of some *Sarcocystis* spp. detected in this study [34–36]. However, the *B. bonasus* population in Lithuania is not large, it stands at less than 300 individuals and their distribution range does not intersect with the sites of our material on mustelids [9–11]. Therefore, it is impossible for *B. bonasus* to be responsible for the high rates of *S. bovifelis* and *S. cruzi* in the intestinal samples of mustelids.

The forest is considered a primary habitat of two mustelid species, European pine marten and European badger, though they are frequent visitors to the surrounding woodlots, meadows and riversides [9]. The habitat of the American mink is related to water—they inhabit banks of rivers, lakes and ponds. These mustelid species are not closely related to human settlements. Two other investigated mustelids, American mink and European polecat, are more often related to settlements than to other habitats, such as forests and shrubby areas [9]. Habitats preferred by mustelids in Lithuania are similar to those in other countries [37]. Diet peculiarities of the investigated mustelids are not directly related to the involvement of these species in the transmission of *Sarcocystis* spp. using cattle as IH. All the investigated mustelid species are opportunistic feeders. Among such diet sources as fruits, berries and other plant materials, invertebrates, fish, amphibians, birds and various mammals [12–17], only one source, namely, cattle carrion, or other sources of cattle meat may be related to *Sarcocystis* spp. we have identified. Mustelid species that we have investigated [12–17], with the exception of the American mink [38], use carrion of wild ungulates.

Cattle are too large prey for mustelids to hunt; therefore, mustelids become infected with *S. bovifelis*, *S. cruzi*, *S. hirsuta* and *S. hominis* species by scavenging carcasses of cattle. However, habitat distribution of the five investigated mustelid species in Lithuania (see above) should exclude contact with carrion of at least two species, American mink and European pine marten. Therefore, the first assumption about high rates of *Sarcocystis* spp. employing cattle as IH is related to food safety issues. In further studies we are going to examine in what way cattle carcasses or beef waste become accessible to mustelids in Lithuania. It is important to understand whether there are gaps in the management of anthropogenic carrion [39] and if this has already become a source of predictable resources accessible to mustelids. Improper carrion management may be related to (i) dumping sites, (ii) treatment of the waste from meat processing factories, especially small ones and located in the countryside and (iii) raw meat waste from homesteads and farms. The two last sources may be neighboring forests and water bodies, therefore becoming sources of possible infection and available even to the American mink and European pine marten, otherwise having no contact with cattle carrion.

Historically, the disclosure of DH of *Sarcocystis* spp. was performed by transmission experiments [40]. Among carnivorous mammals, transmission experiments of *Sarcocystis* spp. have mainly been carried out with dogs, foxes and cats [41,42]. Recently, molecular methods have been applied for the identification of *Sarcocystis* spp. from fecal or mucosal scraping samples of various wild predators or scavengers infected under natural conditions [2,29–31]. The present work is the first study of the molecular identification of *Sarcocystis* spp. in mustelids. Further molecular examination of oocysts/sporocysts detected in the intestine or fecal samples of mustelids can help to clarify the role of these carnivorous mammals in the transmission of *Sarcocystis* parasites.

It is well known that *Sarcocystis* spp. transmitted via canids cannot be spread via felids and vice versa [1]. However, there is a lack of data on whether *Sarcocystis* spp. transmitted via canids and/or felids can be spread via mustelids. It was demonstrated that mustelids and canids could serve as DH of *S. undulati* and *S. citellivulpes* [8,43], whereas mustelids and felids could act as DH for *S. muris* [8]. Two species, *S. bovifelis* (89.3%) and *S. cruzi* (73.8%), were most common in the analyzed intestinal samples of mustelids (Figure 2), whereas *S. hirusta* and *S. hominis* were confirmed in three and single samples, respectively. Canids serve as DH for *S. cruzi*, felids act as DH for *S. hirsuta* and *S. bovifelis* and humans are DH for *S. hominis* [19]. Thus, our results indicate that mustelids might be involved in the transmission of *Sarcocystis* spp. which were confirmed to be transmitted via canids and felids. Nevertheless, further detailed studies on this subject are required. Considering a low abundance of wild felids in Lithuania, we speculate that *S. hirsuta* is mainly transmitted via felids and *S. bovifelis* is mainly transmitted via mustelids. To test the hypothesis, the prevalence of *S. hirsuta* and *S. bovifelis* in muscles of cattle can be examined in European countries where wild felids are more prevalent. Estonia and Finland are the nearest countries with similar environments and with similar abundances of mustelids but with the high abundances of Eurasian lynx, while Germany or Belgium may be the reference countries with the European wildcat (*Felis silvestris*) populations [44].

5. Conclusions

Using a molecular analysis four *Sarcocystis* spp. employing cattle as IH (*S. bovifelis*, *S. cruzi*, *S. hirsuta* and *S. hominis*) were identified in the intestine mucosal scrapings of five Mustelidae species for the first time. Thus, the results of the present study indicate that a wide range of mustelids serve as DH of these *Sarcocystis* spp. Therefore, it is necessary to identify gaps in the management of cattle carrion and beef waste.

Author Contributions: Conceptualization, P.P. and D.B.; methodology, P.P.; software, L.B.; validation, D.B. and P.P.; formal analysis, L.B. and P.P.; investigation, E.J.-N.; resources D.B.; data curation, P.P.; writing—original draft preparation, P.P., L.B., E.J.-N. and D.B.; writing—review and editing, P.P., L.B., E.J.-N. and D.B.; visualization L.B. and E.J.-N.; supervision, P.P.; project administration, P.P. and D.B.; funding acquisition, D.B. All authors have read and agreed to the published version of the manuscript.

Funding: This work was funded by the Research Council of Lithuania (grant number S-MIP-20-24).

Institutional Review Board Statement: Not applicable.

Data Availability Statement: Data supporting the conclusions of this article are included in the article. The sequences generated in the present study were submitted to the GenBank database under accession numbers MW595468–MW595608.

Acknowledgments: This study was supported by the Open Access research infrastructure of the Nature Research Centre under the Lithuanian open access network initiative. The authors are grateful to Valentinas Pabrinkis (Nature Research Centre, Vilnius, Lithuania) who provided samples for the study.

Conflicts of Interest: The authors declare that they have no conflict of interest.

References

1. Dubey, J.P.; Calero-Bernal, R.; Rosenthal, B.M.; Speer, C.A.; Fayer, R. *Sarcocystosis of Animals and Humans*, 2nd ed.; CRC Press: Boca Raton, FL, USA, 2016.
2. Moré, G.; Maksimov, A.; Conraths, F.J.; Schares, G. Molecular Identification of *Sarcocystis* spp. in Foxes (*Vulpes vulpes*) and Raccoon Dogs (*Nyctereutes procyonoides*) from Germany. *Vet. Parasitol.* **2016**, *220*, 9–14. [CrossRef]
3. Britton, A.P.; Dubey, J.P.; Rosenthal, B.M. Rhinitis and Disseminated Disease in a Ferret (*Mustela putorius furo*) Naturally Infected with *Sarcocystis neurona*. *Vet. Parasitol.* **2010**, *169*, 226–231. [CrossRef]
4. Prakas, P.; Strazdaitė-Žielienė, Ž.; Rudaitytė-Lukošienė, E.; Servienė, E.; Butkauskas, D. Molecular Identification of *Sarcocystis lutrae* (Apicomplexa: Sarcocystidae) in Muscles of Five Species of the Family Mustelidae. *Parasitol. Res.* **2018**, *117*, 1989–1993. [CrossRef]
5. Gjerde, B.; Josefsen, T.D. Molecular Characterisation of *Sarcocystis lutrae* n. sp. and *Toxoplasma gondii* from the Musculature of Two Eurasian Otters (*Lutra lutra*) in Norway. *Parasitol. Res.* **2015**, *114*, 873–886. [CrossRef]
6. Kirillova, V.; Prakas, P.; Calero-Bernal, R.; Gavarāne, I.; Fernández-García, J.L.; Martínez-González, M.; Rudaitytė-Lukošienė, E.; Martinez-Estellez, M.A.H.; Butkauskas, D.; Kirjušina, M. Identification and Genetic Characterization of *Sarcocystis arctica* and *Sarcocystis lutrae* in Red Foxes (*Vulpes vulpes*) from Baltic States and Spain. *Parasites Vectors* **2018**, *11*, 173. [CrossRef] [PubMed]
7. Máca, O. Molecular Identification of *Sarcocystis lutrae* (Apicomplexa: Sarcocystidae) from the Raccoon Dog, *Nyctereutes procyonoides*, and the Common Raccoon, *Procyon lotor*, in the Czech Republic. *Parasites Vectors* **2020**, *13*, 231. [CrossRef] [PubMed]
8. Odening, K. The Present State of Species-Systematics in *Sarcocystis* Lankester, 1882 (Protista, Sporozoa, Coccidia). *Syst. Parasitol.* **1998**, *41*, 209–233. [CrossRef]
9. Atlas of Lithuanian Mammals. Available online: https://gamtostyrimai.lt/lt/users/viewGroup/id.24/pageId.26 (accessed on 29 December 2020).
10. Kontrimavičius, V.; Januškis, V.; Virbickas, J.; Augustauskas, J.; Eitminavičiūtė, I.; Kazlauskas, R.; Logminas, V.; Pileckis, S.; Prūsaitė, J.; Valenta, V.; et al. *Lietuvos Fauna. Žinduoliai*; Mokslas: Vilnius, Lithuania, 1988.
11. Balčiauskas, L.; Trakimas, G.; Juškaitis, R.; Ulevičius, A.; Balčiauskienė, L. *Atlas of Lithuanian Mammals, Amphibians and Reptiles*, 2nd ed.; Akstis: Vilnius, Lithuania, 1999.
12. Baghli, A.; Engel, E.; Verhagen, R. Feeding Habits and Trophic Niche Overlap of Two Sympatric Mustelidae, the Polecat *Mustela putorius* and the Beech Marten *Martes foina*. *Z. Jagdwiss.* **2002**, *48*, 217–225. [CrossRef]
13. Lanszki, J.; Heltai, M. Feeding Habits of Sympatric Mustelids in an Agricultural Area of Hungary. *Acta Zool. Acad. Sci. Hung.* **2011**, *57*, 291–304.
14. Newman, C.; Zhou, Y.B.; Buesching, C.D.; Kaneko, Y.; Macdonald, D.W. Contrasting Sociality in Two Widespread, Generalist, Mustelid Genera, *Meles* and *Martes*. *Mammal Study* **2011**, *36*, 169–188. [CrossRef]
15. Malecha, A.W.; Antczak, M. Diet of the European Polecat *Mustela putorius* in an Agricultural Area in Poland. *J. Vertebr. Biol.* **2013**, *62*, 48–53. [CrossRef]
16. Nováková, L.; Vohralík, V. Diet of *Martes foina* in Bohemia, Czech Republic (Carnivora: Mustelidae). *Lynx New Ser.* **2017**, *48*, 155–164. [CrossRef]
17. Baltrūnaitė, L. Diet Composition of the Red Fox (*Vulpes vulpes* L.), Pine Marten (*Martes martes* L.) and Raccoon Dog (*Nyctereutes procyonoides Gray*) in Clay Plain Landscape, Lithuania. *Acta Zool. Litu.* **2002**, *12*, 362–368. [CrossRef]
18. Januškevičius, V.; Januškevičienė, G.; Prakas, P.; Butkauskas, D.; Petkevičius, S. Prevalence and Intensity of *Sarcocystis* spp. Infection in Animals Slaughtered for Food in Lithuania. *Vet. Med. Czech* **2019**, *64*, 149–157. [CrossRef]
19. Prakas, P.; Strazdaitė-Žielienė, Ž.; Januškevičius, V.; Chiesa, F.; Baranauskaitė, A.; Rudaitytė-Lukošienė, E.; Servienė, E.; Petkevičius, S.; Butkauskas, D. Molecular Identification of Four *Sarcocystis* Species in Cattle from Lithuania, Including *S. hominis*, and Development of a rapid Molecular Detection Method. *Parasites Vectors* **2020**, *13*, 610. [CrossRef] [PubMed]
20. Balčiauskas, L.; Balčiauskienė, L.; Litvaitis, J.A.; Tijušas, E. Citizen Scientists Showed a Four-fold Increase of *Lynx* Numbers in Lithuania. *Sustainability* **2020**, *12*, 9777. [CrossRef]
21. Verma, S.K.; Lindsay, D.S.; Grigg, M.E.; Dubey, J.P. Isolation, Culture and Cryopreservation of *Sarcocystis* species. *Curr. Protoc. Microbiol.* **2017**, *45*, 11–127. [CrossRef] [PubMed]
22. Prakas, P.; Butkauskas, D.; Rudaitytė, E.; Kutkienė, L.; Sruoga, A.; Pūraitė, I. Morphological and Molecular Characterization of *Sarcocystis taeniata* and *Sarcocystis pilosa* n. sp. from the Sika Deer (*Cervus nippon*) in Lithuania. *Parasitol. Res.* **2016**, *115*, 3021–3032. [CrossRef] [PubMed]
23. Altschul, S.F.; Gish, W.; Miller, W.; Myers, E.W.; Lipman, D.J. Basic Local Alignment Search Tool. *J. Mol. Biol.* **1990**, *215*, 403–410. [CrossRef]
24. Gjerde, B. Phylogenetic Relationships among *Sarcocystis* Species in Cervids, Cattle and Sheep Inferred from the Mitochondrial Cytochrome c Oxidase Subunit I Gene. *Int. J. Parasitol.* **2013**, *43*, 579–591. [CrossRef]
25. Dean, A.G.; Sullivan, K.M.; Soe, M.M. OpenEpi: Open Source Epidemiologic Statistics for Public Health. Available online: www.OpenEpi.com (accessed on 19 January 2021).
26. Brown, L.D.; Cat, T.T.; DasGupta, A. Interval Estimation for a Proportion. *Stat. Sci.* **2001**, *16*, 101–133.
27. Abramson, J.H. WINPEPI Updated: Computer Programs for Epidemiologists, and their Teaching Potential. *Epidemiol. Perspect. Innov.* **2011**, *8*, 1. [CrossRef] [PubMed]

28. Thomas, J.R.; Salazar, W.; Landers, D.M. What is Missing in p < 05? Effect Size. *Res. Q. Exerc. Sport* **1991**, *62*, 344–348. [CrossRef] [PubMed]
29. Prakas, P.; Liaugaudaitė, S.; Kutkienė, L.; Sruoga, A.; Švažas, S. Molecular Identification of *Sarcocystis rileyi* Sporocysts in red Foxes (*Vulpes vulpes*) and Raccoon Dogs (*Nyctereutes procyonoides*) in Lithuania. *Parasitol. Res.* **2015**, *114*, 1671–1676. [CrossRef]
30. Basso, W.; Alvarez Rojas, C.A.; Buob, D.; Ruetten, M.; Deplazes, P. *Sarcocystis* Infection in Red Deer (*Cervus elaphus*) With Eosinophilic Myositis/Fasciitis in Switzerland and Involvement of Red Foxes (*Vulpes vulpes*) and Hunting Dogs in the Transmission. *Int. J. Parasitol. Parasites Wildl.* **2020**, *13*, 130–141. [CrossRef] [PubMed]
31. Irie, T.; Uraguchi, K.; Ito, T.; Yamazaki, A.; Takai, S.; Yagi, K. First Report of *Sarcocystis pilosa* Sporocysts in Feces from red fox, *Vulpes vulpes schrencki*, in Hokkaido, Japan. *Int. J. Parasitol. Parasites Wildl.* **2020**, *11*, 29–31. [CrossRef] [PubMed]
32. Gjerde, B. Molecular Characterisation of *Sarcocystis bovifelis*, *Sarcocystis bovini* n. sp., *Sarcocystis hirsuta* and *Sarcocystis cruzi* from Cattle (*Bos taurus*) and *Sarcocystis sinensis* from Water Buffaloes (*Bubalus bubalis*). *Parasitol. Res.* **2016**, *115*, 1473–1492. [CrossRef] [PubMed]
33. Gjerde, B.; Hilali, M.; Abbas, I.E. Molecular Differentiation of *Sarcocystis buffalonis* and *Sarcocystis levinei* in Water Buffaloes (*Bubalus bubalis*) from *Sarcocystis hirsuta* and *Sarcocystis cruzi* in Cattle (*Bos taurus*). *Parasitol. Res.* **2016**, *115*, 2459–2471. [CrossRef] [PubMed]
34. Odening, K.; Wesemeier, H.H.; Walter, G.; Bockhardt, I. The Wisent (*Bison bonasus*, Bovidae) as an Intermediate Host of Three *Sarcocystis* species (Apicomplexa: Sarcocystidae) of Cattle. *Folia Parasitol.* **1994**, *41*, 115–121.
35. Pyziel, A.M.; Demiaszkiewicz, A.W. *Sarcocystis cruzi* (Protozoa: Apicomplexa: Sarcocystidae) Infection in European Bison (*Bison bonasus*) from Białowieza Forest, Poland. *Wiad. Parazytol.* **2009**, *55*, 31–34.
36. Calero-Bernal, R.; Verma, S.K.; Seaton, C.T.; Sinnett, D.; Ball, E.; Dunams, D.; Rosenthal, B.M.; Dubey, J.P. *Sarcocystis cruzi* Infection in Wood Bison (*Bison bison athabascae*). *Vet. Parasitol.* **2015**, *210*, 102–105. [CrossRef] [PubMed]
37. Bright, P.W. Lessons from Lean Beasts: Conservation Biology of the Mustelids. *Mamm. Rev.* **2000**, *30*, 217–226. [CrossRef]
38. Zschille, J.; Stier, N.; Roth, M.; Mayer, R. Feeding Habits of Invasive American mink (*Neovison vison*) in Northern Germany—Potential Implications for Fishery and Waterfowl. *Acta Theriol.* **2013**, *59*, 25–34. [CrossRef]
39. Moreno-Opo, R.; Margalida, A. Human-Mediated Carrion: Effects on Ecological Processes. In *Carrion Ecology and Management. Wildlife Research Monographs*; Olea, P., Mateo-Tomás, P., Sánchez-Zapata, J., Eds.; Springer: Cham, Switzerland, 2019; Volume 2, pp. 183–211. [CrossRef]
40. Dahlgren, S.S.; Gjerde, B. The red fox (*Vulpes vulpes*) and the Arctic fox (*Vulpes lagopus*) are Definitive Hosts of *Sarcocystis alces* and *Sarcocystis hjorti* from Moose (*Alces alces*). *Parasitology* **2010**, *137*, 1547–1557. [CrossRef]
41. Khan, R.A.; Evans, L. Prevalence of *Sarcocystis* spp. in Two Subspecies of Caribou (*Rangifer tarandus*) in Newfoundland and Labrador, and Foxes (*Vulpes vulpes*), Wolves (*Canis lupus*), and Husky Dogs (*Canis familiaris*) as Potential Definitive hosts. *J. Parasitol.* **2006**, *92*, 662–663. [CrossRef] [PubMed]
42. Gjerde, B.; Hilali, M. Domestic cats (*Felis catus*) are Definitive Hosts for *Sarcocystis sinensis* from Water Buffaloes (*Bubalus bubalis*). *J. Vet. Med. Sci.* **2016**, *78*, 1217–1221. [CrossRef]
43. Pak, S.M.; Perminova, V.V.; Yeshtokina, N.V. *Sarcocystis citellivulpes* sp. n. from the Yellow Suslik *Citellus fulvus* Lichtenstain, 1923. In *Toksoplazmidy, Protozoologiya*; Beyer, T.V., Bezukladnikova, N.A., Galuzo, I.G., Konovalova, S.I., Pak, S.M., Eds.; Akademii Nauk Sovetskoi Sotsialisticheskoi Respubliki: Moscow, Russia, 1979; pp. 111–114.
44. Mitchell-Jones, A.J.; Amori, G.; Bogdanowicz, W.; Krystufek, B.; Reijnders, P.J.H.; Spitzenberger, F.; Stubbe, M.; Thissen, J.B.M.; Vohralik, V.; Zima, J. *The Atlas of European Mammals*, 1st ed.; Academic Press: London, UK, 1999.

animals

MDPI

Article

Trichinella spp. in Wild Boars (*Sus scrofa*), Brown Bears (*Ursus arctos*), Eurasian Lynxes (*Lynx lynx*) and Badgers (*Meles meles*) in Estonia, 2007–2014

Age Kärssin [1,2,*], Liidia Häkkinen [1], Annika Vilem [1,2], Pikka Jokelainen [2,3,4] and Brian Lassen [5]

1 Estonian Veterinary and Food Laboratory, 51006 Tartu, Estonia; liidia.hakkinen@vetlab.ee (L.H.); annika.vilem@vetlab.ee (A.V.)
2 Institute of Veterinary Medicine and Animal Sciences, Estonian University of Life Sciences, 51006 Tartu, Estonia; PIJO@ssi.dk
3 Laboratory of Parasitology, Department of Bacteria, Parasites and Fungi, Infectious Disease Preparedness, Statens Serum Institut, 2300 Copenhagen, Denmark
4 Faculty of Veterinary Medicine, University of Helsinki, 00014 Helsinki, Finland
5 Research Group for Global Capacity Building, Division for Global Surveillance, National Food Institute, Technical University of Denmark, 2800 Copenhagen, Denmark; brlas@food.dtu.dk
* Correspondence: age.karssin@vetlab.ee; Tel.: +372-7386-116

Simple Summary: Trichinellosis is an important foodborne zoonosis. In Estonia, *Trichinella* infections are endemic in wild animals. This paper summarizes findings of *Trichinella*-parasites during an 8-year period in Estonia in selected host species: wild boars, brown bears, Eurasian lynxes, and badgers. The results highlight that testing wildlife hunted for human consumption for *Trichinella* is important, and that there is room for improvement in the proportion of hunted animals tested.

Abstract: In this study, we summarize *Trichinella* findings from four wild, free-ranging host species from Estonia during 2007–2014. *Trichinella* spp. larvae were detected in 281 (0.9%, 95% confidence interval (CI) 0.8–1.0) of 30,566 wild boars (*Sus scrofa*), 63 (14.7%, 95% CI 11.6–18.3) of 429 brown bears (*Ursus arctos*), 59 (65.56%, 95% CI 55.3–74.8) of 90 Eurasian lynxes (*Lynx lynx*), and three (60.0%, 95% CI 18.2–92.7) of five badgers (*Meles meles*). All four European *Trichinella* species were detected: *T. britovi* in 0.7% of the wild boars, 7.2% of the brown bears, 45.6% of the lynxes, and 40.0% of the badgers; *T. nativa* in 0.1% of the wild boars, 5.8% of the brown bears, and 20.0% of the lynxes; *T. pseudospiralis* in 0.02% the wild boars; and *T. spiralis* in 0.03% of the wild boars and 4.4% of the lynxes. The results include the first description from Estonia of *T. britovi* in brown bear and badgers, *T. pseudospiralis* in wild boars, and *T. spiralis* in wild boars and lynxes. The results indicate high infection pressure in the sylvatic cycles across the years—illustrating continuous risk of spillover to domestic cycles and of transmission to humans.

Keywords: foodborne; game meat; *Trichinella*; wildlife; zoonosis

Citation: Kärssin, A.; Häkkinen, L.; Vilem, A.; Jokelainen, P.; Lassen, B. *Trichinella* spp. in Wild Boars (*Sus scrofa*), Brown Bears (*Ursus arctos*), Eurasian Lynxes (*Lynx lynx*) and Badgers (*Meles meles*) in Estonia, 2007–2014. *Animals* **2021**, *11*, 183. https://doi.org/10.3390/ani11010183

Received: 7 December 2020
Accepted: 11 January 2021
Published: 14 January 2021

1. Introduction

Trichinella spp. are zoonotic parasitic nematodes that can be transmitted to humans by consumption of undercooked or raw meat of infected animals. A multicriteria-based approach placed *Trichinella spiralis* as the third and *Trichinella* spp. other than *T. spiralis* as the fifth on a prioritization ranking list of foodborne parasites in Europe, and the fourth and the third, respectively, in Eastern Europe [1].

Meat of game animals, especially meat of wild boars (*Sus scrofa*), is considered one of the main sources of *Trichinella* infections for humans in Europe [2], and it is acknowledged as the main source in Estonia [3]. Cases of human trichinellosis have been reported from Estonia [4], and the proportion testing positive for antibodies against *Trichinella* spp. was 3.1% in the general adult human population and 4.9% among hunters [5].

Trichinella spp. are endemic in wildlife in Estonia [6]. A high proportion, 42.1%, of investigated wild boars that were hunted in 2012–2013 tested positive for antibodies against *Trichinella* [7], and the biomass of *Trichinella* has increased in the main reservoir hosts raccoon dogs (*Nyctereutes procyonoides*) and red foxes (*Vulpes vulpes*) [6,8]. To add to the knowledge on epidemiology of these zoonotic parasites, the aim of this study was to summarize *Trichinella* findings during 2007–2014 in selected sylvatic hosts that are hunted in Estonia: wild boars, brown bears (*Ursus arctos*), Eurasian lynxes (*Lynx lynx*), and badgers (*Meles meles*).

2. Materials and Methods

2.1. Ethics

No animals were killed for the purpose of this study. No data of the hunters were handled in this study.

2.2. Setting

Estonia is a Baltic country located in northeastern Europe. Altogether 158,670 wild boars, 348 brown bears, 797 lynxes and 1527 badgers were hunted in Estonia in 2007–2014 [9]. Meat of all these host species included in this study is eaten in the country.

2.3. Samples and Data

The muscle samples included in this study were collected from wild boars, brown bears, lynxes, and badgers, primarily from the predilection muscle groups (diaphragm, muscles of foreleg, or tongue), from across Estonia in 2007–2014 by hunters and by meat inspectors in game meat processing plants. The samples were sent to the Estonian Veterinary and Food Laboratory for *Trichinella* testing as part of meat inspection, either for primary or confirmatory testing.

Data on sex and age category of the animal, the county where the animal was hunted, and the year when the animal was hunted were extracted from the submission forms that accompanied the samples. The age category of wild boars, brown bears and lynxes was 'juvenile' for animals the hunters estimated to be up to 2 years of age and 'adult' for animals the hunters estimated to be over two years of age. The counties were categorized into eastern vs. western counties and southern vs. northern counties (Table 1).

Table 1. Prevalence of *Trichinella* infection in wild boars (*Sus scrofa*), brown bears (*Ursus arctos*), Eurasian lynxes (*Lynx lynx*) and badgers (*Meles meles*) hunted in Estonia, 2007–2014, by sex, age category, region, and year. Univariable odds to test positive in comparison to the reference sex (male), age category (juvenile), east-west category (eastern counties), south-north category (southern counties), and year (2007) are shown for each variable, and larval burden data and the *Trichinella* species identified are summarized.

Variable	N Tested [a]	n pos [a] (n pos [a, b])	Prevalence (95% CI) [a]	Odds Ratio (95% CI) [a]	*p*-Value [a]	Median lpg [a,b]	Mean lpg [a,b]	Range lpg [a,b]	*Trichinella* Species Identified (n Animals) [a,b]
					Wild boars				
Sex									
Male	2631	65 (66)	2.5 (1.9–3.1)	reference	–	1.43	11.89	0.02–101.08	Tb (52), Tn (6), Tp (1), Tb+Tn (2), Tb+Ts (1), Tspp (4)
Female	1386	45 (47)	3.2 (2.4–4.3)	1.3 (0.9–1.9)	0.157	0.58	10.59	0.02–100.00	Tb (40), Tn (2), Tb+Ts (1), Tspp (4)
Unknown	26,549	171 (183)	0.6 (0.6–0.7)			2.12 [c]	16.30 [c]	0.01–654.50 [c]	Tb (123), Tn (7), Ts (5), Tp (5), Tb+Tn (3), Tb+Ts (1), Tspp (39)
Age category									
Juvenile	3992	67 (68)	1.7 (1.3–2.1)	reference	–	0.82	12.56	0.02–100.00	Tb (54), Tn (5), Tb+Ts (1), Tspp (8)
Adult	2057	60 (62)	2.9 (2.3–3.7)	1.7 (1.2–2.4)	0.003 **	1.57	9.87	0.02–101.08	Tb (53), Tn (2), Tb+Tn (2), Tb+Ts (1), Tspp (4)
Unknown	24,517	154 (166)	0.6 (0.5–0.7)			2.18 [c]	16.86 [c]	0.01–654.50 [c]	Tb (108), Tn (8), Ts (5), Tp (6), Tb+Tn (3), Tb+Ts (1), Tspp (35)
Region									
Eastern counties	8449	76 (79)	0.9 (0.7–1.1)	reference	–	0.83	11.17	0.02–230.88	Tb (47), Tn (10), Ts (1), Tb+Tn (2), Tspp (19)
Western counties	10,054	196 (208)	1.95 (1.7–2.2)	2.2 (1.7–2.9)	<0.001 ***	2.44	15.99	0.01–654.50	Tb (162), Tn (5), Tp (6), Ts (4) Tb+Tn (3), Tb+Ts (3), Tspp (25)
Southern counties	8204	103 (115)	1.3 (1.0–1.5)	reference	–	2.58	12.95	0.01–190.00	Tb (75), Tn (7), Tp (6), Ts (3), Tb+Tn (4), Tspp (20)
Northern counties	10,299	169 (172)	1.6 (1.4–1.9)	1.3 (1.0–1.7)	0.026 *	1.43	15.89	0.02–654.50	Tb (134), Tn (8), Ts (2), Tb+Tn (1), Tb+Ts (3), Tspp (24)
Unknown	12,063	9 (9)	0.1 (0.04–0.1)			1.73 [c]	3.25 [c]	0.06–12.54 [c]	Tb (6), Tspp (3)
Year									
2007	2422	10 (12)	0.4 (0.2–0.7)	reference	–	2.55	26.15	0.10–100.00	Tb (4), Tspp (8)
2008	2758	10 (13)	0.4 (0.2–0.6)	0.9 (0.4–2.2)	0.774	6.00 [c]	13.01 [c]	0.42–58.00 [c]	Tb (3), Tspp (10)
2009	4380	30 (37)	0.7 (0.5–1.0)	1.7 (0.8–3.6)	0.160	2.40	8.92	0.01–52.15	Tb (20), Tp (2), Ts (5), Tb+Ts (1), Tspp (9)
2010	3598	26 (27)	0.7 (0.5–1.0)	1.8 (0.9–3.8)	0.127	2.68	13.72	0.02–100.00	Tb (22), Tn (1), Tb+Tn (1), Tspp (3)
2011	2713	33 (35)	1.2 (0.9–1.7)	3.0 (1.5–6.3)	0.001 ***	3.44	17.88	0.02–230.88	Tb (25), Tn (3), Tb+Tn (2), Tspp (5)
2012	3986	26 (26)	0.7 (0.4–0.9)	1.6 (0.8–3.4)	0.217	0.89	31.40	0.02–654.50	Tb (20), Tn (1), Tp (2), Tspp (3)
2013	4715	77 (77)	1.6 (1.3–2.0)	4.0 (2.1–8.2)	<0.001 ***	1.60	9.07	0.02–60.00	Tb (66), Tn (4), Tp (1), Tb+Tn (2), Tspp (4)
2014	5994	69 (69)	1.2 (0.9–1.4)	2.8 (1.5–5.8)	<0.001 ***	0.82	13.59	0.02–191.36	Tb (55), Tn (6), Tp (1), Tb+Ts (2), Tspp (5)
Wild boars total	30,566	281 (296)	0.9 (0.8–1.0)			1.64 [c]	14.40 [c]	0.01–654.50 [c]	Tb (215), Tn (15), Tp (6), Ts (5), Tb+Tn (5), Tb+Ts (3), Tspp (47)

Table 1. *Cont.*

Variable	N Tested [a]	n pos [a] (n pos [a, b])	Prevalence (95% CI) [a]	Odds Ratio (95% CI) [a]	*p*-Value [a]	Median lpg [a,b]	Mean lpg [a,b]	Range lpg [a,b]	*Trichinella* Species Identified (n Animals) [a,b]
Brown bears									
Sex									
Male	26	13 (13)	50.0 (31.3–68.7)	reference	-	0.20	1.17	0.02–10.96	Tb (7), Tn (1), Tspp (5)
Female	14	9 (9)	64.3 (37.6–85.6)	1.8 (0.5–7.3)	0.413	3.48	4.81	0.02–11.02	Tb (4), Tn (4), Tb+Tn (1)
Unknown	389	41 (42)	10.5 (7.8–13.9)			0.54	4.78	0.02–81.96	Tb (18), Tn (17), Tb+Tn (2), Tspp (5)
Age category									
Juvenile	22	2 (2)	9.1 (1.55–26.9)	reference	-	4.53	4.53	0.70–8.36	Tb (1), Tn (1)
Adult	87	19 (19)	21.8 (14.1–31.4)	2.8 (0.7–19.0)	0.185	0.86	3.00	0.02–11.02	Tb (10), Tn (5), Tb+Tn (2), Tspp (2)
Unknown	320	42 (43)	13.1 (9.75–17.2)			0.24	4.50	0.02–81.96	Tb (18), Tn (16), Tb+Tn (1), Tspp (8)
Region									
Eastern counties	304	48 (48)	15.8 (12.0–20.2)	reference	–	0.42	4.78	0.03–81.96	Tb (20), Tn (18), Tspp (10)
Western counties	72	12 (13)	16.7 (9.4–26.6)	1.1 (0.5–2.1)	0.839	0.13	2.06	0.02–10.76	Tb (7), Tn (3), Tb+Tn (3)
Southern counties	141	16 (17)	11.35 (6.9–17.4)	reference	–	0.14	6.27	0.02–81.96	Tb (10), Tn (4), Tb+Tn (1), Tspp (2)
Northern counties	235	44 (44)	18.7 (14.1–24.1)	1.8 (1.0–3.4)	0.058	0.77	3.40	0.02–29.96	Tb (17), Tn (17), Tb+Tn (2), Tspp (2)
Unknown	53	3 (3)	5.7 (1.5–14.6)			0.76	1.11	0.20–2.36	Tb (2), Tn (1)
Year									
2007	46	8 (8)	17.4 (8.4–30.4)	reference	–	0.35	0.95	0.03–4.10	Tb (2), Tn (1), Tspp (5)
2008	50	5 (5)	10.0 (3.8–20.8)	0.5 (0.1–1.8)	0.311	0.60	1.01	0.06–2.03	Tb (3), Tspp (2)
2009	51	7 (8)	13.7 (6.2–25.3)	0.8 (0.2–2.3)	0.631	0.44	4.71	0.06–15.90	Tb (3), Tn (3), Tspp (2)
2010	64	9 (9)	14.1 (7.1–24.2)	0.8 (0.3–2.3)	0.641	0.20	1.53	0.02–9.48	Tb (7), Tn (2)
2011	64	8 (8)	12.5 (6.0–22.4)	0.7 (0.2–2.0)	0.487	3.44	5.53	0.04–16.34	Tb (5), Tn (3)
2012	74	8 (8)	12.2 (6.1–21.1)	0.6 (0.2–1.7)	0.320	0.85	4.61	0.02–28.96	Tb (1), Tn (5), Tb+Tn (2)
2013	42	15 (15)	35.7 (22.4–51.0)	2.6 (1.0–7.4)	0.057	0.19	2.43	0.06–11.02	Tb (7), Tn (6), Tb+Tn (1), Tspp (1)
2014	38	3 (3)	7.9 (2.1–20.0)	0.4 (0.1–1.6)	0.220	0.10	27.39	0.10–81.96	Tb (1), Tn (2)
Brown bears total	429	63 (64)	14.7 (11.6–18.3)			0.44	4.11	0.02–81.96	Tb (29), Tn (22), Tb+Tn (3), Tspp (10)

Table 1. *Cont.*

Variable	N Tested [a]	n pos [a] (n pos [a, b])	Prevalence (95% CI) [a]	Odds Ratio (95% CI) [a]	*p*-Value [a]	Median lpg [a,b]	Mean lpg [a,b]	Range lpg [a,b]	*Trichinella* Species Identified (n Animals) [a,b]
Lynxes									
Sex									
Male	14	14 (14)	100.0 (80.7–100.0)	reference	–	2.38	6.51	0.20–28.00	Tb (9), Tn (1), Tb+Tn (4)
Female	10	8 (8)	80.0 (48.1–96.5)	–	0.163	0.90	4.14	0.38–20.10	Tb (3), Tn (2), Ts (1), Tb+Tn (1), Tspp (1)
Unknown	66	37 (41)	56.1 (44.0–67.7)			1.20 [c]	3.53 [c]	0.02–21.40 [c]	Tb (15), Tn (1), Ts (3), Tb+Tn (10), Tb+Ts (1), Tspp (12)
Age category									
Juvenile	15	7 (7)	46.7 (23.2–71.3)	reference	–	14.88	10.97	0.20–28.00	Tb (3), Ts (1), Tb+Tn (3)
Adult	17	14 (14)	82.4 (59.1–95.3)	5.0 (1.0–30.2)	0.045 *	1.91	2.78	0.04–9.22	Tb (6), Tn (2), Ts (1), Tb+Tn (4), Tspp (1)
Unknown	58	38 (42)	65.5 (52.7–76.9)			0.99 [c]	3.65 [c]	0.02–21.40 [c]	Tb (18), Tn (2), Ts (2), Tb+Tn (8), Tb+Ts (1), Tspp (1)
Region									
Eastern counties	60	36 (38)	60.0 (47.3–71.8)	reference	–	1.19 [c]	2.76 [c]	0.02–21.40 [c]	Tb (16), Tn (1), Ts (1), Tb+Tn (8), Tb+Ts (1), Tspp (11)
Western counties	27	23 (25)	85.2 (68.0–95.1)	3.8 (1.2–14.2)	0.020 *	2.34	6.49	0.1–28.00	Tb (11), Tn (3), Ts (3), Tb+Tn (7), Tspp (1)
Southern counties	33	18 (21)	54.5 (37.5–70.8)	reference	–	2.42	3.66	0.02–14.60	Tb (9), Tn (1), Ts (2), Tb+Tn (9)
Northern counties	54	41 (42)	75.9 (63.2–85.9)	2.6 (1.0–6.7)	0.045 *	1.14 [c]	4.62 [c]	0.16–28.00 [c]	Tb (18), Tn (3), Ts (2), Tb+Tn (6), Tb+Ts (1), Tspp (12)
Unknown	3	0 (0)	0.0 (0.0–63.2)			–	–	–	–
Year									
2007	10	5 (6)	50.0 (21.2–78.8)	reference	–	1.45 [c]	6.25 [c]	0.70–21.40 [c]	Tb+Tn (1), Tspp (5)
2008	12	5 (5)	41.7 (17.2–69.8)	0.7 (0.1–4.2)	0.721	0.90	2.18	0.50–7.60	Tb (3), Tspp (2)
2009	20	14 (16)	70.0 (47.7–86.8)	2.3 (0.5–11.8)	0.321	0.50	2.08	0.10–14.88	Tb (5), Ts (4), Tb+Tn (3), Tb+Ts (1), Tspp (3)
2010	17	11 (12)	64.7 (40.5–84.3)	1.8 (0.35–9.5)	0.487	0.82	2.73	0.20–14.60	Tb (9), Tn (1), Tb+Tn (2)

Table 1. *Cont.*

Variable	N Tested [a]	n pos [a] (n pos [a, b])	Prevalence (95% CI) [a]	Odds Ratio (95% CI) [a]	*p*-Value [a]	Median lpg [a,b]	Mean lpg [a,b]	Range lpg [a,b]	*Trichinella* Species Identified (n Animals) [a,b]
2011	11	7 (7)	63.6 (33.6–87.2)	1.7 (0.3–10.9)	0.567	2.48	2.48	0.02–4.82	Tb (4), Tn (1), Tb+Tn (2)
2012	11	10 (10)	90.9 (62.7–99.6)	8.9 (0.9–259.1)	0.059	8.39	10.10	0.04–28.00	Tb (4), Tn (1), Tb+Tn (3), Tspp (2)
2013	9	7 (7)	77.8 (43.8–96.1)	3.3 (0.4–33.1)	0.260	6.24	6.23	0.38–12.10	Tb (2), Tn (1), Tb+Tn (4)
Lynxes total	90	59 (63)	65.6 (55.3–74.8)			1.42 [c]	4.29 [c]	0.02–28.00 [c]	Tb (27), Tn (4), Ts (4), Tb+Tn (15), Tb+Ts (1), Tspp (12)
					Badgers				
Sex									
Male	1	1	100.0 (5.0–100.0)	–	–	20.96	20.96	20.96	Tb (1)
Female	1	0	0.0 (0.0–95.0)	–	–	–	–	–	–
Unknown	3	2	66.7 (13.2–98.3)			7.09	7.09	2.90–11.28	Tb (1), Tspp (1)
Region									
Eastern counties	3	2	66.7 (13.2–98.3)	–	–	10.09	10.09	10.09	Tb (1), Tspp (1)
Western counties	2	1	50.0 (2.5–97.5)	–	–	20.96	20.96	20.96	Tb (1)
Southern counties	3	2	66.7 (13.2–98.3)	–	–	11.93	11.93	11.93	Tb (1), Tspp (1)
Northern counties	2	1	50.0 (2.5–97.5)	–	–	11.28	11.28	11.28	Tb (1)
Year									
2008	1	1	100.0 (5.0–100.0)	–	–	2.90	2.90	2.90	Tb (1)
2013	2	2	100.0 (22.4–100.0)	–	–	16.12	16.12	11.28–20.96	Tb (1), Tspp (1)
2014	2	0	0.0 (0.0–77.6)	–	–	–	–	–	–
Badgers total	5	3 (3)	60.0 (18.2–92.7)			11.28	11.71	2.90–20.96	Tb (2), Tspp (1)

N Tested: total number of animals tested; [a] Animals tested as primary testing; [b] Positive animals tested for confirmatory purposes.; [c] No data on larval burden for one wild boar and two lynxes; 95% CI: 95% confidence interval, Mid-P Exact; n pos: number of *Trichinella* positive animals; lpg: number of *Trichinella* larvae per gram of muscle tissue; n Animals: number of animals; *p*-value: two-tailed Mid-P Exact: *: $p \leq 0.05$; **: $p \leq 0.01$; ***: $p \leq 0.001$; Eastern counties: Ida-Virumaa, Jõgevamaa, Järvamaa, Lääne-Virumaa, Põlvamaa, Tartumaa, Valgamaa; Võrumaa; Western counties: Harjumaa, Hiiumaa, Läänemaa, Pärnumaa, Raplamaa, Saaremaa, Viljandimaa; Southern counties: Jõgevamaa, Põlvamaa, Pärnumaa, Saaremaa, Tartumaa, Valgamaa, Viljandimaa, Võrumaa; Northern counties: Harjumaa, Hiiumaa, Ida-Virumaa, Järvamaa, Läänemaa, Lääne-Virumaa, Raplamaa; Tb: *Trichinella britovi*; Tn: *Trichinella nativa*; Tp: *Trichinella pseudospiralis;* Ts: *Trichinella spiralis*; Tb+Tn: mixed infection with *Trichinella britovi* and *Trichinella nativa*; Tb+Ts: mixed infection with *Trichinella britovi* and *Trichinella spiralis;* Tspp: *Trichinella* species, no species-level result.

2.4. Artificial Digestion

The artificial digestion of the samples was carried out at the Estonian Veterinary and Food Laboratory, which is the national reference laboratory for parasites with its three regional laboratories. The laboratories are accredited by ISO 17025 and authorized as official laboratories for *Trichinella* digestion analyses according to EU 2075/2005 Annex I Chapter 1 [10]. One of the regional laboratories used Stomacher method according to EU 2075/2005 Annex I Chapter II [10] until 2009.

The testing included both primary and confirmatory testing. Other laboratories performing *Trichinella* testing send positive samples to the national reference laboratory for confirmation and species identification.

A minimum of 10 g muscle tissue was used from each animal in a pooled sample and 50 g for an individual sample, with the exception that in 2007, one regional laboratory used 5 g muscle tissue as the minimum in pooled samples for up to 20 animals. If a pooled sample was positive, the pool was divided into smaller pools and individual samples were tested to identify the infected animals.

If larvae were found, they were identified morphologically, counted, washed with tap water, and stored in ethanol, according to the procedures recommended by the European Union Reference Laboratory for Parasites [11]. The number of larvae per gram of muscle tissue (lpg) was calculated for each positive animal.

2.5. Molecular Analysis

Larvae collected in 2007–2010 were identified to the species level at the European Union Reference Laboratory [11], and larvae collected since 2011 at the Estonian Veterinary and Food Laboratory. The same multiplex PCR [12] was used for all the analyses. The method does not include sequencing.

2.6. Statistical Analyses

Only results from primary testing were used for prevalence estimates. Animals were excluded from statistical analyses if their individual infection status could not be determined, due to testing as part of a pooled sample followed by unsuccessful identification of the infected individuals.

We compared the prevalence estimates by sex, by age group, by eastern vs western counties and by southern vs northern counties, using two-by-two tables. In addition, we report univariable odds ratios for testing positive for *Trichinella*, using these same dichotomous variables, as well as counties as dummy variables and years as dummy variables.

For the statistical analyses, we used Microsoft Excel, OpenEpi and R [13,14]. We report 95% confidence intervals (95% CI, Mid-P Exact) for proportions. Associations were considered statistically significant if two-tailed $p < 0.05$.

3. Results

The proportion of animals included in this study from the officially reported hunting bag of the study period was 19.3% for wild boars, 123.6% for brown bears, 12.0% for lynxes and 0.3% for badgers [9]. A total of 44 wild boars and two lynxes were excluded from statistical analyses, because their individual infection status could not be determined. The final sample was 31,090 animals tested as primary testing (Table S1), and 20 positive animals (15 wild boars, one brown bear and four lynxes) that had been tested for confirmatory purposes. Data on larval burden were missing for one wild boar and two lynxes. For *Trichinella* spp. species identification, altogether 426 larval samples were tested; *Trichinella* species was not determined in 70 (16.4%) of the larval samples.

Of the altogether 31,090 animals tested as primary testing, 406 (1.3%, 95% CI 1.2–1.4) were positive for *Trichinella* spp. larvae. Altogether 281 (0.9%, 95% CI 0.8–1.0) of the 30,566 wild boars, 63 (14.7%, 95% CI 11.6–18.3) of the 429 brown bears, 59 (65.6%, 95% CI 55.3–74.8) of the 90 lynxes, and three (60.0%, 95% CI 18.2–92.7) of the five badgers were

Trichinella positive (Table 1). In wild boars and lynxes, a higher prevalence was observed in adults than in juveniles ($p = 0.003$ and $p = 0.045$, respectively) (Table 1). In wild boars, the prevalence was higher in the western counties than in the eastern counties ($p < 0.001$), and in the northern counties than in the southern counties ($p = 0.026$) (Table 1). The prevalence in lynxes was higher in the eastern counties than in the western counties ($p = 0.020$), and in the northern counties than in the southern counties ($p = 0.045$). The prevalence varied by year from 0.4% to 1.6% in wild boars, from 7.9% to 36.7% in brown bears, from 41.7% to 90.9% in lynxes, and from 0.0% to 100.0% in badgers (Figure 1, Table 1). In wild boars, the prevalence was higher in 2011 ($p = 0.001$), 2013 ($p < 0.001$), and 2014 ($p < 0.001$) than it was in 2007 (Table 1).

The larval burden appeared generally higher in wild boars than in brown bears and lynxes (Figure 1, Table 1). Nine wild boars had more than 100 lpg.

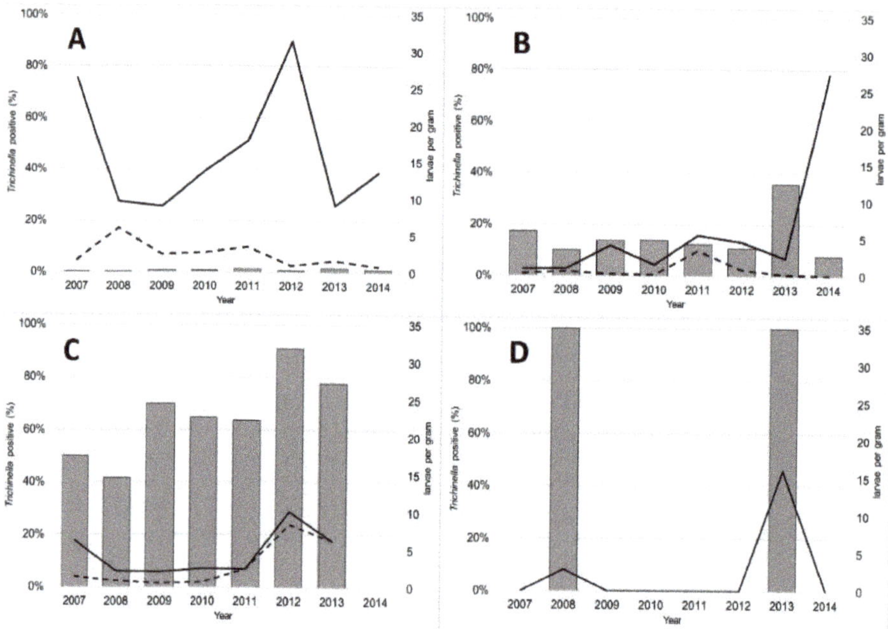

Figure 1. Percentage of *Trichinella* spp. positive animals; Mean (solid line) and median (dashed line) number of larvae per gram tissue in tested wild boars (*Sus scrofa*, **A**), brown bears (*Ursus arctos*, **B**), Eurasian lynxes (*Lynx lynx*, **C**) and badgers (*Meles meles*, **D**), 2007–2014, Estonia.

Mono-species *Trichinella* infection was found in 97.5% (95% CI 95.0–99.0) of the wild boars, 94.3% (95% CI 85.4–98.5) of the brown bears, 69.4% (95% CI 55.5–81.0) of the lynxes, and all badgers that were positive and had the *Trichinella* species identified. The *Trichinella* species that were detected are shown by county and by year in Figure 2, Table 1 and Table S2. The isolates of 2007–2010 were deposited in International *Trichinella* Reference Centre [11].

Trichinella britovi was the most common *Trichinella* species found in all the investigated host species. It was found in animals from all counties (Figure 2, Table S2), in 0.7% (95% CI 0.6–0.8) of wild boars, 7.2 % (95% CI 5.1–10.0) of brown bears, 45.6% (95% CI 35.5–55.9) of lynxes, and 40.0% (95% CI 7.4–81.8) of badgers (Table 2). *Trichinella britovi* infections were found in 31 brown bears: in five hunted in Ida-Virumaa in 2007, 2010 and 2011; in four hunted in Harjumaa in 2010, 2011, 2012, and 2013; in four hunted in Järvamaa in 2008 and 2013; in four hunted in Jõgevamaa in 2011 and 2013; in two hunted in Läänemaa in 2009 and 2010; in four hunted in Lääne-Virumaa in 2009, 2010 and 2013; in one hunted in Põlvamaa in 2012; in two hunted in Pärnumaa in 2008 and 2012; and in three hunted in Tartumaa in 2009, 2013 and 2014; and in two badgers, which had been hunted in Lääne-Virumaa and Viljandimaa in 2013—these are the first confirmed findings of this parasite species in these host species in Estonia (this study; [15]).

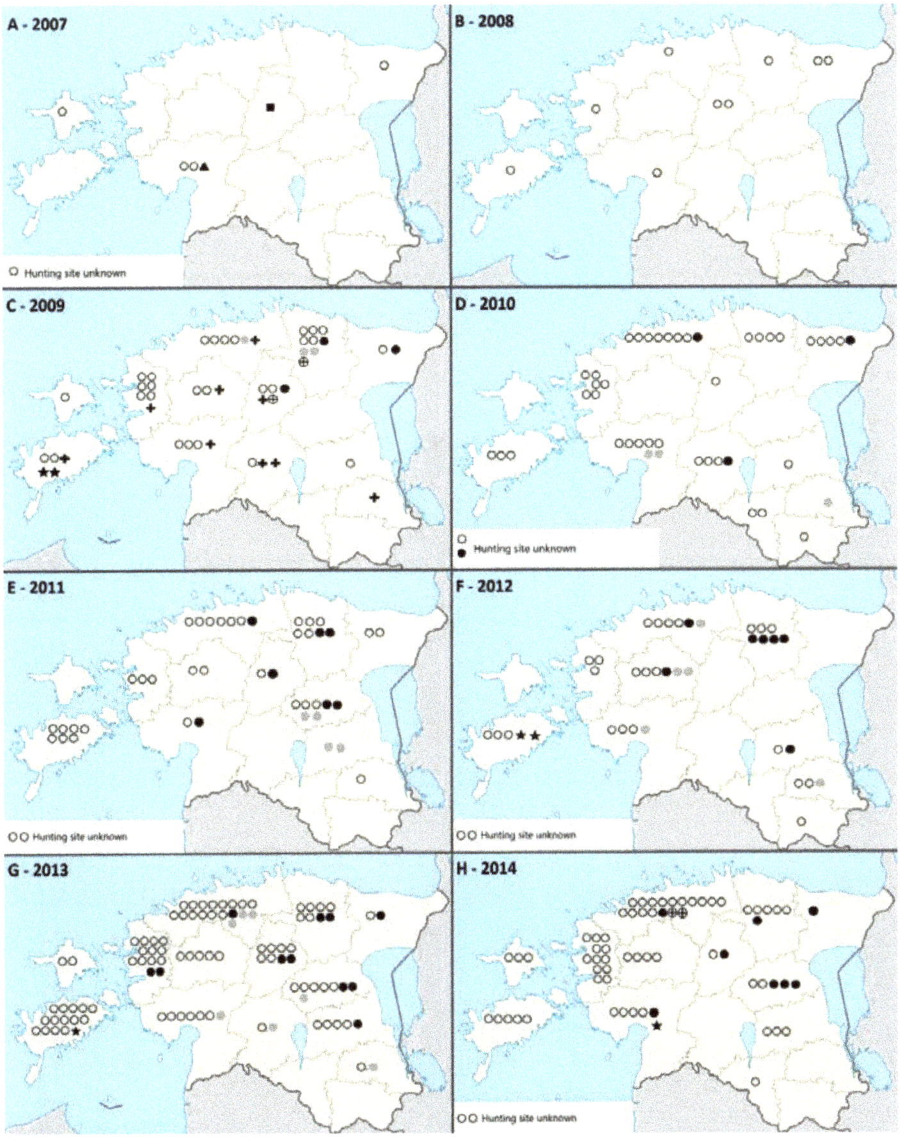

Figure 2. *Trichinella* species distribution in the tested animals that were positive and *Trichinella* species identification was successful, 2007–2014 (**A–H**, respectively), Estonia. *T. britovi*—ring, *T. nativa*—black dot, *T. pseudospiralis*—star; *T. spiralis*—cross, *T. britovi* and *T. nativa*—grey dot, *T. britovi* and *T. spiralis*—ring with cross.

Table 2. *Trichinella* species identified in wild boars (*Sus scrofa*), brown bears (*Ursus arctos*), Eurasian lynxes (*Lynx lynx*) and badgers (*Meles meles*) hunted in 2007–2014 in Estonia.

	Wild Boars (n = 30,566 [a] +15 [b])				Brown Bears (n = 429 [a] +1 [b])				Lynxes (n = 90 [a] +4 [b])				Badgers (n = 5 [a])			
Trichinella Species	n pos [a] (n pos [a,b])	Prevalence % (95% CI) [a]	% (95% CI) of *Trichinella* Positive [a,b]	Median; Mean (Range) of lpg [a,b]	n pos [a] (n pos [a,b])	Prevalence % (95% CI) [a]	% (95% CI) of *Trichinella* Positive [a,b]	Median; Mean (Range) of lpg [a,b]	n pos [a] (n pos [a,b])	Prevalence % (95% CI) [a]	% (95% CI) of *Trichinella* Positive [a,b]	Median; Mean (Range) of lpg [a,b]	n pos[a] (n pos [a,b])	Prevalence % (95% CI) [a]	% (95% CI) of *Trichinella* Positive [a,b]	Median; Mean (Range) of lpg [a,b]
T. britovi only	209 (215)	0.7 (0.6–0.8)	72.6 (67.3–77.5)	2.06; 16.30 (0.02–654.5)	28 (29)	6.5 (4.5–9.2)	45.3 (33.5–57.6)	0.48; 5.88 (0.10–81.96)	26 (27)	28.9 (20.2–38.9)	42.9 (31.1–55.3)	1.48; 4.71 (0.06–28.00)	2 (2)	40.0 (7.4–81.8)	66.7 (13.2–98.3)	16.12; 16.12 (11.28–20.96)
T. nativa only	15 (15)	0.05 (0.03–0.08)	5.1 (3.0–8.0)	0.82; 6.69 (0.04–54.32)	22 (22)	5.1 (3.3–7.5)	34.4 (23.5–46.6)	0.60; 3.40 (0.02–28.96)	4 (4)	4.4 (1.4–10.4)	6.3 (2.1–14.6)	3.53; 6.80 (0.02–20.10)	–	–	–	–
T. pseudospiralis only	6 (6)	0.02 (0.01–0.04)	2.0 (0.8–4.2)	1.63; 5.52 (0.18–23.34)	–	–	–	–	–	–	–	–	–	–	–	–
T. spiralis only	5 (5)	0.02 (0.01–0.04)	1.7 (0.6–3.7)	4.76; 4.51 (0.20–7.65)	–	–	–	–	3 (4)	3.3 (0.9–8.8)	6.3 (2.1–14.6)	3.61; 5.58 (0.23–14.88)	–	–	–	–
T. britovi and *T. nativa*	5 (5)	0.02 (0.01–0.04)	1.7 (0.6–3.7)	1.38; 10.58 (0.04–45.20)	3 (3)	0.7 (0.2–1.9)	4.7 (1.2–12.2)	3.48; 2.41 (0.02–3.74)	14 (15)	15.6 (9.1–24.2)	23.8 (14.5–35.5)	2.00; 4.62 (0.04–17.40)	–	–	–	–
T. britovi and *T. spiralis*	3 (3)	0.01 (0.00–0.03)	1.0 (0.3–2.7)	0.24; 2.67 (0.16–7.60)	–	–	–	–	1 (1)	1.1 (0.1–5.4)	1.6 (0.1–7.6)	0.20	–	–	–	–
T. britovi Total [c]	217 (223)	0.7 (0.6–0.8)	75.3 (70.2–80.0)	1.88; 15.99 (0.02–654.50)	31 (32)	7.2 (5.1–10.0)	50.0 (37.9–51.4)	0.54; 5.56 (0.02–81.96)	41 (43)	45.6 (35.5–55.9)	68.3 (56.0–78.8)	1.58; 4.23 (0.04–28.00)	2 (2)	40.0 (7.4–81.8)	66.7 (13.2–98.3)	16.12; 16.12 (11.28–20.96)
T. nativa Total [c]	20 (20)	0.07 (0.04–0.10)	6.8 (4.3–10.1)	1.14; 7.66 (0.04–54.32)	25 (25)	5.8 (3.9–8.4)	39.1 (27.7–51.4)	0.76; 3.28 (0.02–28.96)	18 (19)	20.0 (12.7–29.2)	30.2 (19.8–42.3)	2.00; 5.08 (0.02–20.10)		–	–	–
T. pseudospiralis total [c]	6 (6)	0.02 (0.01–0.04)	2.0 (0.8–4.2)	3.70; 6.21 (0.18–23.34)	–	–	–	–	–	–	–	–	–	–	–	–
T. spiralis Total [c]	8 (8)	0.03 (0.01–0.05)	2.7 (1.3–5.1)	4.26; 3.82 (0.16–7.65)	–	–	–	–	4 (5)	4.4 (1.4–10.4)	7.9 (3.0–16.7)	3.60; 4.51 (0.20–14.88)		–	–	–
Species-level result total	243 (249)	0.8 (0.7–0.9)	84.1 (79.6–88.0)	1.76; 15.07 (0.02–654.50)	53 (54)	12.4 (9.5–15.7)	84.4 (73.9–91.8)	0.60; 4.76 (0.02–81.96)	48 (51)	53.3 (43.0–63.5)	81.0 (69.9–89.3)	1.64; 4.54 (0.02–28.00)	2 (2)	40.0 (7.4–81.8)	66.7 (13.2–98.3)	16.12; 16.12 (11.28–20.96)
No species-level result	38 (47)	0.1 (0.1–0.2)	15.9 (12.0–20.4)	0.70; 10.55 (0.01–190.00) [d]	10 (10)	2.3 (1.2–4.1)	15.6 (8.2–26.1)	0.23; 0.68 (0.03–2.03)	11 (12)	12.2 (6.6–20.3)	19.0 (10.8–30.1)	0.65; 3.02 (0.16–21.40) [d]	1 (1)	20.0 (1.0–66.6)	33.3 (1.7–86.8)	2.9
Total	281 (296)	0.9 (0.8–1.0)	100.0	1.64; 14.40 (0.01–654.50) [d]	63 (64)	14.7 (11.6–18.3)	100.0	0.42; 4.05 (0.02–81.96)	59 (63)	65.6 (55.3–74.8)	100.0	1.45; 4.46 (0.02–28.00) [d]	3 (3)	60.0 (18.2–92.7)	100.0	11.28; 11.71 (290–20.96)

[a] Animals tested as primary testing; [b] Positive animals tested for confirmatory purposes; [c] The number of animals with the *Trichinella* species, either as the only species (mono-species infection) or in mixed infection; [d] No data on larval burden for one wild boar and two lynxes; 95% CI: 95% confidence interval, Mid-P Exact; n pos: number of *Trichinella* positive animals; lpg: number of *Trichinella* larvae per gram of muscle tissue.

The second most common *Trichinella* species was *T. nativa*, was found in all the investigated host species except badgers. Infected animals originated from 11 of the 15 counties; no findings were detected on the islands Hiiumaa and Saaremaa, and the southeastern counties Võrumaa and Valgamaa (Fig. 2, Table S2). *Trichinella nativa* was found in 0.1% (95% CI 0.0–0.1) of wild boars, 5.8 % (95% CI 3.9–8.4) of brown bears, and 20.0% (95% CI 12.7–29.2) of lynxes (Table 2).

Trichinella pseudospiralis was found in 2009 for the first time in wild boars in Estonia (Table 1; [11,16]). During the study period, this species was found in 0.02% of wild boars (Table 2); the prevalence was highest in Saaremaa, 0.2% (95% CI 0.1–0.5; Table S2).

The first *Trichinella spiralis* finding in a game animal in Estonia was identified in a lynx hunted in 2008 (shipped and tested in 2009), and further findings were detected in wild boars and lynxes hunted in 2009. The species was found in 0.03% (95% CI 0.0–0.05) of wild boars and 4.4% (95% CI 1.4–10.4) of lynxes (Table 2). It was detected in nine counties: Harjumaa, Järvamaa, Läänemaa, Lääne-Virumaa, Põlvamaa, Pärnumaa, Raplamaa, Saaremaa, and Viljandimaa (Table S2).

4. Discussion

The high number of observations in this study add substantially to the knowledge on epidemiology of *Trichinella* spp. in Estonia and highlight that wildlife, including game animals, have a key role in it. *Trichinella* spp. are important zoonotic parasites in the country and the region [17], and it is crucial that the One Health approaches addressing them cover not only domestic animals and humans, but also wildlife.

It should be emphasized that hunted animals are always a convenience sample: hunting periods affect the age of the animals included in the sample, and the representativeness of a hunter-harvested sample is challenging to evaluate. Moreover, it should be noted that e.g., animals injured in traffic accidents or hunted illegally are not included in the official hunting statistics. This could explain the higher number of brown bears in our sample than in the hunting bags.

The sampling was done by hunters and veterinary inspectors, who were advised to sample from the predilection muscle groups, if these were available [10]. The sampling was not supervised by the authors, and possible variation in sample material may have affected the results to the direction of underestimation of the prevalence and in particular of the larval burden. The predilection muscle groups vary by host species [10], and for detailed comparisons, sampling the exact same muscle within each host species would be optimal.

The background information on the animals was provided by the hunters, and the authors had no means to validate these data. Misclassification of some animals to wrong age category or sex remain possible, and no data were provided for many animals (Table 1; Table S1).

The methodology we used is harmonized at international level and thus yields comparable results. The prevalence estimates reported in this study are generally in line with results from previous studies focusing on these host species in Estonia, which estimated the prevalence to be 0.3–1.0% in wild boars, 29.4% in brown bears, and 47.4–50% in lynxes [8,18]. The proportion of badgers that tested positive in this study was significantly higher (60.0%, three of five, $p = 0.006$) than an earlier estimate for badgers hunted in 1965–2000 (6.7%, six of 89) [19].

While the results of this study are not directly comparable with those from other countries due to different sampling and study designs, it is clear that *Trichinella* parasites thrive in the region. The prevalence in wild boars in this study was lower than that observed in Latvia [20], but higher than that in Poland [21]. The prevalence in brown bears and lynxes was higher than that in Finland [22], while the prevalence in lynxes was lower than that in Latvia [8,23]. The prevalence in badgers is considered low in several countries [24], however the proportion of positives in this study was substantial, in line with what has been observed in Latvia, and higher than that in Finland [22,23].

The results of this study confirm that *T. britovi* has been winning host-terrain, while *T. nativa* is well-established in whole mainland Estonia. It is noteworthy that *T. pseudospiralis* was found in animals from the southwestern part of the country. Several studies have described an increase of *T. pseudospiralis* findings in wild boar samples in Europe [25]. One possible vector of *T. pseudospiralis* are predatory birds, including migratory birds [25]. In Estonia, the findings of *T. pseudospiralis* have been made in animals killed near the sea or wetlands areas, which are good nesting sites for birds. Further research focusing on the potential host species living in these specific environments could provide insight into the role of birds in the epidemiology of *T. pseudospiralis*. Another noteworthy finding was *T. spiralis* from a lynx killed in 2009 in Järvamaa county, approximately 30 km from where a human trichinellosis outbreak was documented ten years earlier, and where *T. spiralis* was found in a domestic pig during the investigation [18]. *Trichinella spiralis* could be infecting wildlife in Estonia similarly as described by Oksanen and others [22], as spillover from the domestic cycle. That we did not find freeze-sensitive *T. spiralis* in the main reservoir animal host species in our previous study [6] might be explained by two consecutive colder years before 2011/2012 [26].

In contrast to our previous epidemiological study in the reservoir hosts raccoon dogs and red foxes, where no obvious geographical differences in *Trichinella* prevalence were seen [6], geographical differences in the prevalence were observed in wild boars and lynxes in this study. Interestingly, we previously demonstrated a higher seroprevalence in wild boars in the southwestern part of the country [7], and the results of this study confirm a higher infection prevalence in western and southern counties. The geographical variation may be due to several factors, including climate, temperature, and snow cover [16,20].

The results of this study exemplify that wild boars can serve as an indicator for *Trichinella* spp. monitoring, being annually hunted in high numbers and routinely tested for *Trichinella*. Wild boars have been popular game in Estonia after their population rapidly increased since the second half of the 1990s, supported by relatively mild winters and supplementary feeding [27–30]. Importantly, the results of this study reflect the situation before the African swine fever (ASF) outbreak in Estonia, which started in September 2014, and will thus serve as baseline data for future studies that could evaluate how the ASF-related changes affected the wild boar population and the parasites these animals can host.

The results of this study highlight that testing wildlife hunted for human consumption for *Trichinella* remains important, and that there is room for improvement in the proportion of hunted animals tested. Wildlife are important for epidemiology of *Trichinella* spp. in Estonia, and hunting wild game for human consumption provides a potential transmission route to humans.

5. Conclusions

In Estonia, *Trichinella* infections are common in wildlife, including in game animals hunted for human consumption. High infection pressure was evident in sylvatic cycles, and the risk for spillover to and from domestic cycles and transmission to humans remain relevant.

Supplementary Materials: The following are available online at https://www.mdpi.com/2076-2615/11/1/183/s1: Table S1: Primary testing included in the study: number of wild boars (*Sus scrofa*), brown bears (*Ursus arctos*), Eurasian lynxes (*Lynx lynx*) and badgers (*Meles meles*) tested for *Trichinella* in Estonia, 2007–2014, by sex, age category, county, and by year. Table S2: Prevalence of *Trichinella* infection in wild boars (*Sus scrofa*), brown bears (*Ursus arctos*), Eurasian lynxes (*Lynx lynx*) and badgers (*Meles meles*) hunted in Estonia, 2007–2014, by county. Univariable odds to test positive in comparison to the reference county (Harjumaa), and larval burden data and the *Trichinella* species identified are summarized.

Author Contributions: Conceptualization, A.K., B.L. and P.J.; methodology, A.K., B.L. and P.J.; investigation, A.K., L.H. and A.V.; data curation, A.K., B.L. and P.J.; writing—original draft preparation, A.K., B.L. and P.J.; writing—review and editing, B.L and P.J.; supervision, B.L. and P.J. All authors contributed to the writing process. All authors have read and agreed to the published version of the manuscript.

Funding: This research received no external funding.

Institutional Review Board Statement: Not applicable.

Informed Consent Statement: Not applicable.

Data Availability Statement: Data is contained within the article or supplementary material.

Acknowledgments: We thank the hunters and veterinary inspectors for collecting the samples and providing the background information about the animals. We thank Ingrid Ott and Moonika Musting from Tartu VFL, Ülla Rajamets and Kristel Lilles from Tallinn VFL, Eda Laas and Merle Reiman from Rakvere VFL, Mai Truutsi from Saaremaa VFL for performing laboratory analyses and collecting data, and Katrin Lõhmus for helping with the laboratory database. We thank European Union Reference Laboratory for Parasites for the species identification of *Trichinella* larvae collected in 2007–2010, and Edoardo Pozio and Gianluca Marucci for the excellent training in the multiplex PCR method.

Conflicts of Interest: The authors declare no conflict of interest.

References

1. Bouwknegt, M.; Devleesschauwer, B.; Graham, H.; Robertson, L.J.; Van Der Giessen, J.W.; Participants, T.E.-F.W. Prioritisation of food-borne parasites in Europe, 2016. *Eurosurveillance* **2018**, *23*, 17–00161. [CrossRef]
2. The European Union summary report on trends and sources of zoonoses, zoonotic agents and food-borne outbreaks in 2017. *EFSA J.* **2018**, *16*, e05500. [CrossRef]
3. Health Protection Inspectorate. *Communicable Disease Statistics in Estonia. Part 9. Trichinellosis*; Jõgiste, A., Varjas, J., Märtin, J., Aro, T., Kutsar, K., Eds.; Health Protection Inspectorate: Tallinn, Estonia, 2000; pp. 88–93. ISBN 9985918673.
4. European Food Safety Authority and European Centre for Disease Prevention and Control. The European Union summary report on trends and sources of zoonoses, zoonotic agents and food-borne outbreaks in 2015. *EFSA J.* **2016**, *14*, 135–145. [CrossRef]
5. Lassen, B.; Janson, M.; Viltrop, A.; Neare, K.; Hütt, P.; Golovljova, I.; Tummeleht, L.; Jokelainen, P. Serological Evidence of Exposure to Globally Relevant Zoonotic Parasites in the Estonian Population. *PLoS ONE* **2016**, *11*, e0164142:1–e0164142:13. [CrossRef]
6. Kärssin, A.; Häkkinen, L.; Niin, E.; Peik, K.; Vilem, A.; Jokelainen, P.; Lassen, B. *Trichinella* spp. biomass has increased in raccoon dogs (*Nyctereutes procyonoides*) and red foxes (*Vulpes vulpes*) in Estonia. *Parasites Vectors* **2017**, *10*, 609:1–609:12. [CrossRef]
7. Kärssin, A.; Velström, K.; Gómez-Morales, M.A.; Saar, T.; Jokelainen, P.; Lassen, B. Cross-Sectional Study of Anti-*Trichinella* Antibody Prevalence in Domestic Pigs and Hunted Wild Boars in Estonia. *Vector Borne Zoonotic Dis.* **2016**, *16*, 604–610. [CrossRef] [PubMed]
8. Malakauskas, A.; Paulauskas, V.; Järvis, T.; Keidans, P.; Eddi, C.; Kapel, C.M. Molecular epidemiology of *Trichinella* spp. in three Baltic countries: Lithuania, Latvia, and Estonia. *Parasitol. Res.* **2007**, *100*, 687–693. [CrossRef] [PubMed]
9. Estonian Environment Agency. Hunting Statistics. 2019. Available online: https://www.keskkonnaagentuur.ee/et/ulukiseireulevaated (accessed on 29 April 2019).
10. European Commission. Commission Regulation (EC) No 2075/2005 of 5 December 2005 laying down specific rules on official controls for *Trichinella* in meat. *Off. J. Eur. Union* **2005**, *48*, 452–474.
11. International Trichinella Reference Centre. Database of Trichinella isolates. Available online: https://trichinella.iss.it/Trichinella/Database (accessed on 25 September 2019).
12. Pozio, E.; La Rosa, G. *Trichinella* . In *Molecular Detection of Foodborne Pathogens*; Liu, D., Ed.; CRC Press, Taylor & Francis Group: Boca Raton, FL, USA, 2010; pp. 851–863.
13. Dean, A.G.; Sullivan, K.M.; Soe, M.M. OpenEpi: Open Source Epidemiologic Statistics for Public Health, Version 3.01. 2019. Available online: www.OpenEpi.com (accessed on 8 May 2020).
14. R Core Team. R: A language and environment for statistical computing. R Foundation for Statistical Computing, Vienna, Austria. 2019. Available online: https://www.R-project.org (accessed on 13 November 2019).
15. Veterinary and Food Laboratory. Available online: http://www.vetlab.ee (accessed on 13 November 2019).
16. Pozio, E. Adaptation of *Trichinella* spp. for survival in cold climates. *Food Waterborne Parasitol.* **2016**, *4*, 4–12. [CrossRef]
17. Deksne, G.; Davidson, R.K.; Buchmann, K.; Kärssin, A.; Kirjušina, M.; Gavarāne, I.; Miller, A.L.; Pálsdóttir, G.R.; Robertson, L.J.; Mørk, T.; et al. Parasites in the changing world—Ten timely examples from the Nordic-Baltic region. *Parasite Epidemiol. Control* **2020**, *10*, e00150:1–e00150:11. [CrossRef]
18. Järvis, T.; Miller, I.; Pozio, E. Epidemiological studies on animal and human trichinellosis in Estonia. *Parasite* **2001**, *8*, 86–87. [CrossRef] [PubMed]
19. Miller, I. Trichinellosis in Estonia: Epidemiology, Diagnosis and Control. Ph.D. Thesis, Estonian University of Life Sciences, Tartu, Estonia, 2003.

20. Kirjušina, M.; Deksne, G.; Marucci, G.; Bakasejevs, E.; Jahundoviča, I.; Daukšte, A.; Zdankovska, A.; Běrziņa, Z.; Esite, Z.; Bella, A.; et al. A 38-year study on *Trichinella* spp. in wild boar (*Sus scrofa*) of Latvia shows a stable incidence with an increased parasite biomass in the last decade. *Parasites Vectors* **2015**, *8*, 137–138. [CrossRef]
21. Flis, M.; Grela, E.R.; Gugała, D. Epizootic and epidemiological situation of *Trichinella* sp. infection in Poland in 2006–2015 in view of wild boar population dynamics. *J. Vet. Res.* **2017**, *61*, 181–187. [CrossRef] [PubMed]
22. Oksanen, A.; Interisano, M.; Isomursu, M.; Heikkinen, P.; Tonanzi, D.; Oivanen, L.; Pozio, E. *Trichinella spiralis* prevalence among wildlife of a boreal region rapidly reduced in the absence of spillover from the domestic cycle. *Vet. Parasitol.* **2018**, *262*, 1–5. [CrossRef]
23. Deksne, G.; Segliņa, Z.; Jahundoviča, I.; Esīte, Z.; Bakasejevs, E.; Bagrade, G.; Keidāne, D.; Interisano, M.; Marucci, G.; Tonanzi, D.; et al. High prevalence of *Trichinella* spp. in sylvatic carnivore mammals of Latvia. *Vet. Parasitol.* **2016**, *231*, 118–123. [CrossRef]
24. Moskwa, B.; Goździk, K.; Bień, J.; Bogdaszewski, M.; Cabaj, W. Molecular identification of *Trichinella britovi* in martens (*Martes martes*) and badgers (*Meles meles*); new host records in Poland. *Acta Parasitol.* **2012**, *57*, 402–405. [CrossRef]
25. Pozio, E. *Trichinella pseudospiralis* an elusive nematode. *Vet. Parasitol.* **2016**, *231*, 97–101. [CrossRef]
26. Rjazin, J.; Pärn, O.; Alari, V. CLASSIFYING THE ICE SEASONS 1982-2016 USING THE WEIGHTED ICE DAYS NUMBER AS A NEW WINTER SEVERITY CHARACTERISTIC. *EUREKA: Phys. Eng.* **2017**, *5*, 49–56. [CrossRef]
27. Veeroja, R.; Männil, P. Population Development and Reproduction of Wild Boar (*Sus scrofa*) in Estonia. *Wildl. Biol. Pract.* **2013**, *10*, 17–21. [CrossRef]
28. Oja, R.; Kaasik, A.; Valdmann, H. Winter severity or supplementary feeding—which matters more for wild boar? *Acta Thériol.* **2014**, *59*, 553–559. [CrossRef]
29. Oja, R.; Velström, K.; Moks, E.; Jokelainen, P.; Lassen, B. How does supplementary feeding affect endoparasite infection in wild boar? *Parasitol. Res.* **2017**, *116*, 2131–2137. [CrossRef] [PubMed]
30. Statistics Estonia. Hunting Statistics. Available online: http://andmebaas.stat.ee/Index.aspx (accessed on 26 October 2020).

animals

MDPI

Article

Distribution and Host Selection of Tropical Rat Mite, *Ornithonyssus bacoti*, in Yunnan Province of Southwest China

Peng-Wu Yin [1,2], Xian-Guo Guo [1,2,*], Dao-Chao Jin [1], Rong Fan [2], Cheng-Fu Zhao [2], Zhi-Wei Zhang [2], Xiao-Bin Huang [2] and Ke-Yu Mao [2]

1 Institute of Entomology, Guizhou University, and the Provincial Key Laboratory for Agricultural Pest Management in Mountainous Region, Guiyang 550025, China; pengwuyin@vip.163.com (P.-W.Y.); jindaochao@163.com (D.-C.J.)
2 Vector Laboratory, Institute of Pathogens and Vectors, Yunnan Provincial Key Laboratory for Zoonosis Control and Prevention, Dali University, Dali 671000, China; fanfanlook1@outlook.com (R.F.); zchengf2020@163.com (C.-F.Z.); dalizzw@aliyun.com (Z.-W.Z.); huangxb633@nenu.edu.cn (X.-B.H.); 15887382535@163.com (K.-Y.M.)
* Correspondence: xianguoguo@yahoo.com; Tel.: +86-872-2257-104

Simple Summary: The tropical rat mite (*Ornithonyssus bacoti*) is a transmission vector of rickettsia pox and a potential vector of hemorrhagic fever with renal syndrome (HFRS). This article reports the distribution and host selection of *O. bacoti* in Yunnan Province of Southwest China. The original data came from the investigations in 39 counties of Yunnan. The prevalence (P_M), mean abundance (*MA*) and mean intensity (*MI*) were calculated to reflect the infestations of the dominant rat hosts with *O. bacoti* mites. The patchiness index and Taylor's power law were used to measure the spatial distribution of the mites. A total of 4121 *O. bacoti* mites were identified from 15 species of small mammal hosts in 27 of the 39 investigated counties, and 99.20% of them (4088/4121) were found on rodents. The majority of total *O. bacoti* mites was found in the flatland landscape (91.28%) and indoor habitat (73.48%). Moreover, 51.78% and 40.09% of *O. bacoti* mites were identified from *Rattus tanezumi* and *R. norvegicus*, the two synanthropic rat species. The mites had some host-specificity, with a preference to two dominant hosts (*R. tanezumi* and *R. norvegicus*), and they were of aggregated distribution on *R. tanezumi*.

Abstract: (1) Background: As a species of gamasid mite, the tropical rat mite (*Ornithonyssus bacoti*) is a common ectoparasite on rodents and some other small mammals. Besides stinging humans to cause dermatitis, *O. bacoti* can be a vector of rickettsia pox and a potential vector of hemorrhagic fever with renal syndrome (HFRS). (2) Objective: The present study was conducted to understand the host selection of *O. bacoti* on different animal hosts and the distribution in different environmental gradients in Yunnan Province of Southwest China. (3) Methods: The original data came from the investigations in 39 counties of Yunnan, between 1990 and 2015. The animal hosts, rodents and some other small mammals were mainly trapped with mouse traps. The *O. bacoti* mites on the body surface of animal hosts were collected and identified in a conventional way. The constituent ratio (*Cr*), prevalence (P_M), mean abundance (*MA*) and mean intensity (*MI*) were used to reflect infestations of animal hosts with *O. bacoti* mites. The patchiness index and Taylor's power law were used to measure the spatial distribution pattern of *O. bacoti* mites on their hosts. (4) Results: A total of 4121 tropical rat mites (*O. bacoti*) were identified from 15 species and 14,739 individuals of hosts, and 99.20% of them were found on rodents. More than half of *O. bacoti* mites (51.78%) were identified from the Asian house rat (*Rattus tanezumi*), and 40.09% of the mites from the Norway rat (*R. norvegicus*) ($p < 0.05$). The infestations of *R. tanezumi* ($P_M = 7.61\%$, $MA = 0.40$ and $MI = 5.31$) and *R. norvegicus* ($P_M = 10.98$, $MA = 1.14$ and $MI = 10.39$) with *O. bacoti* mites were significantly higher than those of other host species ($p < 0.05$). The infestations of two dominant rat hosts (*R. tanezumi* and *R. norvegicus*) with *O. bacoti* mites varied in different environmental gradients (latitudes, longitudes, altitudes, landscapes and habitats) and on different sexes and ages of the hosts. The prevalence of juvenile *R. norvegicus* rats with *O. bacoti* mites ($P_M = 12.90\%$) was significantly higher than that of adult rats ($P_M = 9.62\%$) ($p < 0.05$). The prevalence ($P_M = 38.46\%$) and mean abundance ($MA = 2.28$ mites/host) of *R. tanezumi*

Citation: Yin, P.-W.; Guo, X.-G.; Jin, D.-C.; Fan, R.; Zhao, C.-F.; Zhang, Z.-W.; Huang, X.-B.; Mao, K.-Y. Distribution and Host Selection of Tropical Rat Mite, *Ornithonyssus bacoti*, in Yunnan Province of Southwest China. *Animals* **2021**, *11*, 110. https://doi.org/10.3390/ani11010110

Received: 20 October 2020
Accepted: 30 December 2020
Published: 7 January 2021

rats with *O. bacoti* mites in the high latitude were higher than those in the low latitudes ($p < 0.05$). The majority of the total collected 4121 *O. bacoti* mites was found in the flatland landscape (91.28%) and indoor habitat (73.48%) ($p < 0.05$). The P_M (10.66%) and *MA* (0.49 mites/host) of *R. tanezumi* rats with *O. bacoti* mites were significantly higher in the indoor habitat than in the outdoor habitat ($p < 0.05$). The tropical rat mites showed an aggregated distribution pattern on their first dominant host, *R. tanezumi*. **Conclusion:** The tropical rat mite (*O. bacoti*) is a widely distributed species of gamasid mite in Yunnan Province, Southwest China, and its dominant hosts are two synanthropic species of rats, *R. tanezumi* and *R. norvegicus*. It is mainly distributed in the flatland landscape and indoor habitat. It has some host-specificity, with a preference to rodents, especially *R. tanezumi* and *R. norvegicus*. The *O. bacoti* mites are of aggregated distribution on *R. tanezumi* rats.

Keywords: Acari; gamasid mite; *Ornithonyssus bacoti*; distribution; host selection; Yunnan; China

1. Introduction

As a worldwide species of gamasid mite, the tropical rat mite (*Ornithonyssus bacoti*) is widely distributed in nearly all parts of the world, except the Arctic and Antarctic regions [1,2]. It is a common ectoparasite on rodents (rats, mice and voles) and some other small mammals (e.g., insectivores and tree shrews), frequently occurring on the body surface and in the nests of its hosts. *Ornithonyssus bacoti* is the most important species of gamasid mites associated with medicine, and nearly all its stages (larvae, protonymphs, deutonymphs and adults) in the life cycle can invade and sting animal hosts for blood meal [3–5]. The dermatitis caused by the stinging of *O. bacoti* mites is frequently reported throughout the world [5–16]. Besides directly stinging humans, *O. bacoti* is associated with the transmission of some zoonoses. Together with another species of gamasid mite (*Allodermanyssus sanguineus*), the mite *O. bacoti* is the confirmed vector of rickettsia pox, caused by the pathogen *Rickettsia akari* [1,17,18]. Rickettsia pox is a zoonosis associated with rodents, and it is mainly prevalent in the United States, Canada, Russia, Ukraine, India and Egypt, etc. [17,19–21]. Besides being the intermediate host of the animal parasite *Litomosoides carinii*, the mite *O. bacoti* has been proved to be the potential vector of hemorrhagic fever with renal syndrome (HFRS) caused by hantavirus [22–25].

To date, there have been a lot of studies on the tropical rat mite, *O. bacoti*. Early on in the 1940s, some scholars began to feed *O. bacoti* in the laboratory and designed some special devices suitable for *O. bacoti* and some other sucking mites [26,27]. Many previous publications on *O. bacoti* and some other species of gamasid mites were about the mites' ultrastructure [1,8,28–30], chromosome karyotype, gene sequencing, phylogeny and control [1,29,31–34], but only few studies were about the mite ecology. To date, there have been no systematically ecological studies on *O. bacoti* in Yunnan Province of Southwest China. Between 1990 and 2015, our research group made a long-term field investigation and accumulated abundant original data on gamasid mites in Yunnan. To take advantage of the previous investigation, the present paper analyzed the distribution and host selection of *O. bacoti* in Yunnan for the first time, which is an attempt to get more knowledge about *O. bacoti* and to enrich the ecological information of the mite.

2. Materials and Methods

2.1. Collection and Identification of *Ornithonyssus bacoti* and Its Animal Hosts

The original data came from a long-term field investigation in 39 counties of Yunnan Province in Southwest China, between 1990 and 2015, and the investigated 39 counties were shown in Figure 1, in Results. A stratified sampling investigation was made in different geographical localities, latitudes, longitudes, altitude, landscapes and habitats. To capture animal hosts, mouse traps were placed in the indoor and outdoor habitats of each investigation site, in the evening, and then checked the following morning. The indoor habitats covered houses, barns, stables, etc., and the outdoor habitats covered paddy fields, cornfields, bush habitats, woodlands, etc. [35]. Each trapped host was euthanized through anesthesia with ether (cotton balls soaked with ether), within a closed container. Under the anesthesia, the gamasid mites on the body surface of each host were all collected and then preserved in Eppendorf tubes with 70% of ethanol. After the mite collection, each animal host was identified into species according to its morphological features [4,36,37]. Through the dehydration and clarification, the collected gamasid mites were mounted onto glass slides with Hoyer's medium and they were then identified into species under microscopes [38,39]. Based on the identification of gamasid mites, the tropical rat mite (*O. bacoti*) was chosen as the target of the present study. In the animal euthanasia, most rodent pests in agriculture and forestry were anaesthetized to death because the local government encourages people to kill and eradicate them. Some non-pest small mammals (e.g., weasels, moles and some squirrels) were anaesthetized only for two to five minutes, according to their body size, and they were finally released to the wild when they woke up. The capturing of rodents and other small mammals was officially permitted by the local authority of wildlife service in Yunnan Province, China. The use of animals (including animal euthanasia) for research was officially approved by the Animals' Ethics Committee of Dali University, and the permitted number was DLDXLL2020-1104. The specimens of voucher mites and representative animal hosts are deposited in the specimen repository of the Institute of Pathogens and Vectors, Dali University, Yunnan, China.

2.2. Infestation Statistics

In a conventional way, the constituent ratio (*Cr*), prevalence (P_M), mean abundance (*MA*) and mean intensity (*MI*) were calculated to reflect the infestations of dominant hosts with tropical rat mites (*O. bacoti*) [4,40–42]. In the present study, *Cr* represents the percentage of *O. bacoti* mites, P_M the percentage of infested hosts by the mites, *MA* the number of the mites on each examined host and *MI* the number of the mites on each infested host.

2.3. Analysis on Host Selection and Distribution

The infestations of dominant animal hosts with *O. bacoti* mites were compared on different sexes and ages of the hosts to reflect the host selection of the mites. The infestations were compared in different latitudes and longitudes, to reflect the mite's horizontal distribution, and compared in different altitudes, to reflect the mite's vertical distribution. The latitudes and longitudes were divided into three gradients, and the altitudes were divided into four gradients. The three latitude gradients include <24° N, 24° N–26° N and >26° N, and the four longitude gradients are <100° E, 100° E–102° E, 102° E–104° E and >104° E. The four altitude gradients are <1000 m, 1000–2000 m, 2001–3000 m and >3000 m.

2.4. Analysis of Spatial Distribution Pattern

The patchiness index (m^*/m) and Taylor's power were used to measure the spatial distribution pattern of tropical rat mites (*O. bacoti*) among different individuals of its dominant hosts [43,44]. The formulae of patchiness index and Taylor's power are listed as follows.

$$m*/m = \frac{m + (\frac{\sigma^2}{m} - 1)}{m};\ \lg\sigma^2 = \lg a + b\lg m \quad (1)$$

In the above formulae, m^*/m = patchiness index, m = mean of *O. bacoti* mites on its dominant hosts and σ^2 = variance of the mites; lg a = intercept on the Y-axis, and b = the slope or regression coefficient. When $m^*/m > 1$, $a > 1$ and $b > 1$ or $a > 1$, $b = 1$, the spatial distribution pattern is determined as the aggregated distribution; when $m^*/m = 1$, $a = 1$ and $b = 1$, the random distribution; when $m^*/m < 1$, $a < 1$, $b < 1$, the uniform (or even) distribution [45,46].

2.5. Analysis on Interspecific Association

Based on a contingency table (see Table 10 in Results), the association coefficient (V) was used to measure the interspecific association between the tropical rat mite (*O. bacoti*) and some other related species of gamasid mites on the dominant animal hosts. In the contingency table, *O. bacoti* was defined as "species X", while the other related mite species was defined as "species Y". The association coefficient (V) is as follows.

$$V = \frac{ad - bc}{\sqrt{(a+b)(c+d)(a+c)(b+d)}} \quad (2)$$

In the above formula, V = the association coefficient between species X and Y; a = the host individuals on which species X and Y simultaneously occur; b = the host individuals on which species Y occurs, but species X does not occur; c = the host individuals on which species X occurs, but species Y does not occur; and d = the host individuals on which both species X and Y do not occur. When $V > 0$ and $p < 0.05$, the interspecific relationship between species X and Y is determined as the positive association; when $V < 0$ and $p < 0.05$, we have the negative association; P is the significance probability in Chi-square test (χ^2).

2.6. Significance Test

Chi-square test (χ^2) was used to test the significance of Cr, P_M and V. Nonparametric test was used to test the significance of MA and MI. All the statistical analyses were performed with R software version 3.5.3.

3. Results

3.1. Abundance of *Ornithonyssus bacoti*

A total of 139,111 gamasid mites were collected from 74 species and 17,638 individuals of animal hosts, rodents and some other small mammals, in 39 counties of Yunnan. Of 139,111 collected gamasid mites, 137,210 of them were identified as 156 species and 39 genera in 13 families, and the remaining 1901 mites remained unidentified because of blemished, obscure and damaged structures or suspected new species. A total of 4121 tropical rat mites (*O. bacoti*) were identified from 15 species and 14,739 individuals of hosts, and they accounted for only 2.96% (4121/139,111) of the total mites. The identified 4121 *O. bacoti* mites were distributed in 27 counties (Figure 1).

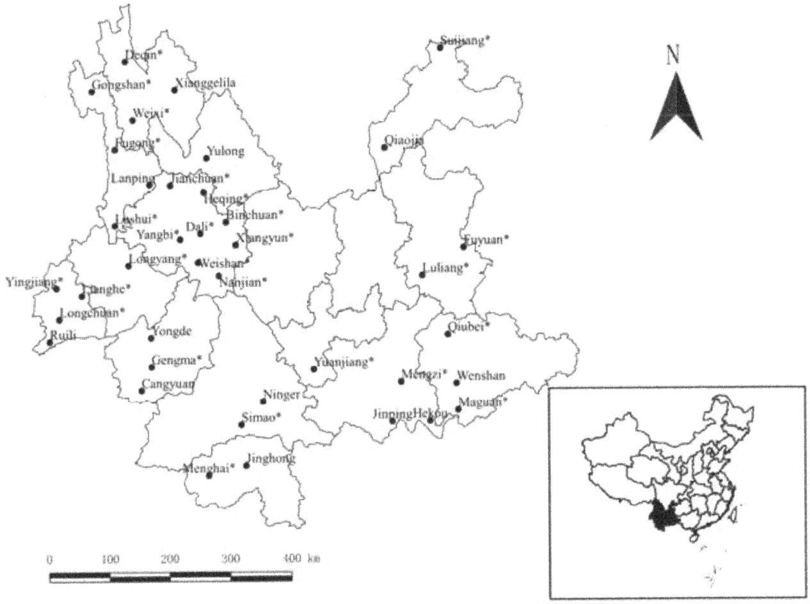

Figure 1. The 39 investigated counties and 27 counties with tropical rat mites (*Ornithonyssus bacoti*) collected (marked with "*") in Yunnan Province, Southwest China (1990–2015).

3.2. Host Selection of *Ornithonyssus bacoti*

The identified 4121 *O. bacoti* mites came from 15 species, 8 genera and 4 families of animal hosts in 3 orders, Rodentia, Soricomorpha and Scandetia. On the order level of animal hosts, 99.20% of *O. bacoti* mites (4088/4121) were found on the order Rodentia, which was significantly higher than that on Soricomorpha and Scandetia ($p < 0.05$). The percentages of *O. bacoti* mites found on the family Muridae (4088/4121 = 99.20%) and the genus *Rattus* (3953/4121 = 95.92%) were the highest of four host families and eight genera ($p < 0.05$). On the species level of hosts, the majority of *O. bacoti* mites was identified from two dominant rat hosts, with 51.78% of the mites on the Asian house rat (*Rattus tanezumi*) and 40.09% of the mites on the Norway rat (*R. norvegicus*) ($p < 0.05$). Of the 15 species of hosts, the infestations of the dominant rat hosts (*R. tanezumi* and *R. norvegicus*) with *O. bacoti* mites were significantly higher than those of other 13 species of hosts ($p < 0.05$) (Table 1). Positive linear correlations existed among P_M, *MA* and *MI*, with $r = 0.685$ between *MA* and *MI*, $r = 0.646$ between *MA* and P_M and $r = 0.332$ between *MI* and *PM* ($p < 0.05$) (Figure 2).

Table 1. Infestations of two dominant rat hosts (*Rattus tanezumi* and *R. norvegicus*) with tropical rat mites (*Ornithonyssus bacoti*) in Yunnan Province, Southwest China (1990–2015).

Host Species	Host Individuals		Infestations of Hosts with *O. bacoti* Mites				
	Examined Hosts	Infested Hosts	Mite Individuals	Constituent Ratios of Mites (*Cr*, %)	Prevalence (P_M, %)	Mean Abundance (*MA*)	Mean Intensity (*MI*)
Rattus tanezumi	5285	402	2134	51.78	7.61	0.40	5.31
Rattus norvegicus	1448	159	1652	40.09	10.98	1.14	10.39
Other 13 host species	8006	68	335	8.13	0.85	0.04	4.93
Total *	14,739	629	4121	100.00	4.27	0.28	6.55

*Annotation: The other 13 species of animal hosts are *Rattus nitidus, Mus musculus, R. brunneusculus (R. sladeni), Suncus murinus, Crocidura attenuata, Apodemus draco, Niviventer andersoni, Tupaia belangeri, A. chevrieri, N. confucianus, Eothenomys miletus, M. caroli* and *A. peninsulae.*

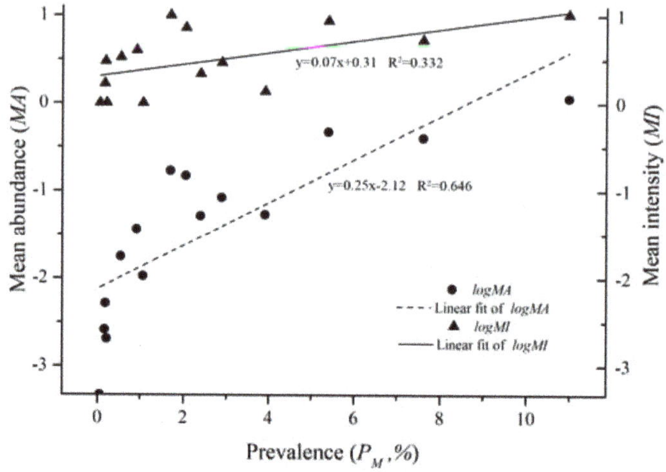

Figure 2. The linear regression between P_M, *MA* and *MI* of 15 host species with *O. bacoti* mites in Yunnan Province, Southwest China (1990–2015).

Different sexes and ages of two dominant rat hosts (*R. tanezumi* and *R. norvegicus*) showed some differences in the infestations with *O. bacoti* mites. The infestations of male rats (*R. tanezumi* and *R. norvegicus*) with *O. bacoti* mites were slightly higher than those of female rats, but the differences were of no statistical significance ($p > 0.05$) (Tables 2 and 3). The prevalence of juvenile *R. norvegicus* rats with the mites (P_M = 12.90%) was significantly higher than that of adult rats (P_M = 9.62%) ($p < 0.05$). The mean abundance (*MA* mites/host) and mean intensity (*MI* mites/host) of juvenile *R. norvegicus* rats with the mites were slightly higher than those of adult rats, but the differences were of no statistical significance ($p > 0.05$). The infestations of juvenile *R. tanezumi* rats with the mites (P_M, *MA* and *MI*) were also slightly higher than those of adult rats, but the differences were of no statistical significance ($p > 0.05$) (Tables 2 and 3).

Table 2. Infestations of different sexes and ages of *R. tanezumi* rats with *O. bacoti* mites in Yunnan Province, Southwest China (1990–2015).

Sexes and Ages of Hosts	Examined Hosts	Infested Hosts	Infestations of Different Sexes and Ages of Rat Hosts with *O. bacoti* Mites				
			Mite Individuals	Constituent Ratios of the Mites (*Cr*, %)	Prevalence (P_M, %)	Mean Abundance (*MA*)	Mean Intensity (*MI*)
Males	2540	207	1157	72.54	8.15	0.46	5.59
Females	2534	194	974	61.07	7.66	0.38	5.02
Total *	5074	401	2131	100.00	7.90	0.42	5.31
Juveniles	1227	99	536	25.15	8.07	0.44	5.41
Adults	3847	302	1595	74.85	7.85	0.41	5.28
Total *	5074	401	2131	100.00	7.90	0.42	5.31

* Annotation: The animal hosts without records of sexes and ages were not included in the above table.

Table 3. Infestations of different sexes and ages of *R. norvegicus* rats with *O. bacoti* mites in Yunnan Province, Southwest China (1990–2015).

Sexes and Ages of Hosts	Examined Hosts	Infested Hosts	Infestations of Different Sexes and Ages of Rat Hosts with *O. bacoti* Mites				
			Mite Individuals	Constituent Ratios of the Mites (Cr, %)	Prevalence (P_M, %)	Mean Abundance (MA)	Mean Intensity (MI)
Males	734	82	919	55.63	11.17	1.25	11.21
Females	712	77	733	44.37	10.81	1.03	9.52
Total *	1446	159	1652	100.00	11.00	1.14	10.39
Juveniles	659	85	1191	72.09	12.90	1.81	14.01
Adults	769	74	461	27.91	9.62	0.60	6.23
Total *	1428	159	1652	100.00	11.13	1.16	10.39

* Annotation: The animal hosts without records of sexes and ages were not included in the above table.

3.3. Horizontal Distribution of *Ornithonyssus bacoti*

The infestations of two dominant rat hosts (*R. tanezumi* and *R. norvegicus*) with *O. bacoti* mites showed some differences in different latitudes and longitudes (horizontal distribution). The P_M (38.46%) and MA (2.28 mites/host) of *R. tanezumi* with *O. bacoti* mites, together with MA (2.04 mites/host) of *R. norvegicus* with the mites, were higher in the high latitude (>26° N) than in other latitudes ($p < 0.05$) (Tables 4 and 5). The MA (0.63 mites/host) of *R. tanezumi* rats with the mites was higher in the longitude <100° E than in other three longitudes ($p < 0.05$), and the P_M (16.81%) and MA (2.10 mites/host) of *R. norvegicus* with the mites were higher in the longitude 100° E–102° E than in other three longitudes ($p < 0.05$) (Tables 4 and 5).

Table 4. Infestations of *R. tanezumi* rats with *O. bacoti* mites in different latitudes and longitudes of Yunnan Province, Southwest China (1990–2015).

Latitudes and Longitudes	Examined Hosts	Infested Hosts	Infestations of Rat Hosts with *O. bacoti* Mites in Different Latitudes and Longitudes				
			Mite Individuals	Constituent Ratios of the Mites (Cr, %)	Prevalence (P_M, %)	Mean Abundance (MA)	Mean Intensity (MI)
Low latitude <24° N	2967	73	345	16.17	2.46	0.12	4.73
Middle latitude 24–26° N	2279	314	1700	79.66	13.78	0.75	5.41
High latitude >26° N	39	15	89	4.17	38.46	2.28	5.93
Total *	5285	402	2134	100	7.61	0.40	5.31
Longitude <100° E	1789	185	1136	53.23	10.34	0.63	6.14
Longitude 100° E–102° E	3164	193	920	43.11	6.10	0.29	4.77
Longitude 102° E–104° E	205	8	18	0.84	3.90	0.09	2.25
Longitude >104° E	127	16	60	2.81	12.60	0.47	3.75
Total *	5285	402	2134	100	7.61	0.40	5.31

* Annotation: The animal hosts without records of latitudes and longitudes were not included in the above table.

Table 5. Infestations of *R. norvegicus* rats with *O. bacoti* mites in different latitudes and longitudes of Yunnan Province, Southwest China (1990–2015).

Latitudes and Longitudes	Examined Hosts	Infested Hosts	Infestations of Rat Hosts with *O. bacoti* Mites in Different Latitudes and Longitudes				
			Mite Individuals	Constituent Ratios of the Mites (*Cr*, %)	Prevalence (P_M, %)	Mean Abundance (*MA*)	Mean Intensity (*MI*)
Low latitude <24° N	117	8	25	1.51	6.84	0.21	3.13
Middle latitude 24–26° N	1264	141	1490	90.19	11.16	1.18	10.57
High latitude >26° N	67	10	137	8.29	14.93	2.04	13.70
Total *	1448	159	1652	100.00	10.98	1.14	10.39
Longitude <100° E	200	17	149	9.02	8.50	0.75	8.76
Longitude 100° E–102° E	690	116	1447	87.59	16.81	2.10	12.47
Longitude 102° E–104° E	259	25	53	3.21	9.65	0.20	2.12
Longitude >104° E	299	1	3	0.18	0.33	0.01	3.00
Total *	1448	159	1652	100	10.98	1.14	10.39

* Annotation: The animal hosts without records of latitudes and longitudes were not included in the above table.

3.4. Vertical Distribution of *Ornithonyssus bacoti*

The infestations of two dominant rat hosts (*R. tanezumi* and *R. norvegicus*) with *O. bacoti* mites showed some differences in different altitudes (vertical distribution). The P_M (27.27%) and *MA* (0.82 mites/host) of *R. tanezumi* rats with *O. bacoti* mites were highest above 3000 m, but *MI* (7.62 mites/host) was highest below 1000 m ($p < 0.05$). The P_M (13.40%), *MA* (0.77 mites/host) and *MI* (5.74 mites/host) of *R. norvegicus* rats with the mites were highest at 1000–2000 m ($p < 0.05$) (Table 6).

Table 6. Infestations of *R. tanezumi* and *R. norvegicus* rats with *O. bacoti* mites in different altitudes of Yunnan Province, Southwest China (1990–2015).

Dominant Rat Hosts and Altitudes (Meters)	Examined Hosts	Infested Hosts	Infestations of Rat Hosts with *O. bacoti* mites in Different Altitudes				
			Mite Individuals	Constituent Ratios of the Mites (*Cr*, %)	Prevalence (P_M, %)	Mean Abundance (*MA*)	Mean Intensity (*MI*)
R. tanezumi							
<1000	932	94	716	47.32	10.09	0.77	7.62
1000–2000	1548	210	763	50.43	13.57	0.49	3.63
2001–3000	195	13	25	1.65	6.67	0.13	1.92
>3000	11	3	9	0.59	27.27	0.82	3.00
Total *	2686	320	1513	100.00	11.91	0.56	4.73
R. norvegicus							
<1000	15	0	0	0	0	0	-
1000–2000	291	39	224	85.82	13.40	0.77	5.74
2001–3000	358	23	37	14.18	6.42	0.10	1.61
>3000	28	0	0	0	0	0	-
Total *	692	62	261	100.00	8.96	0.38	4.21

* Annotation: The animal hosts without records of altitudes were not included in the above table.

3.5. Landscape and Habitat Distribution of *Ornithonyssus bacoti*

The majority of total collected 4121 *O. bacoti* mites was found in the flatland landscape (1894/2075 = 91.28%) and indoor habitat (3028/4121 = 73.48%) ($p < 0.05$). The infestations of two dominant rat hosts (*R. tanezumi* and *R. norvegicus*) with *O. bacoti* mites showed some differences in two kinds of landscapes (mountainous and flatland landscapes), but the differences were of no statistical significance ($p > 0.05$) (Table 7). The infestations of *R. tanezumi* and *R. norvegicus* with the mites also showed some differences in two kinds of habitats. The P_M (10.66%) and *MA* (0.49 mites/host) of *R. tanezumi* rats with the mites were significantly higher in the indoor habitat than in the outdoor habitat ($p < 0.05$). The P_M (11.50%), *MA* (1.30 mites/host) and *MI* (11.31 mites/host) of *R. norvegicus* rats with the mites were higher in the indoor habitat than in the outdoor habitat, but the differences were of no statistical significance ($p > 0.05$) (Table 8).

Table 7. Infestations of *R. tanezumi* and *R. norvegicus* rats with *O. bacoti* mites in different landscapes of Yunnan Province, Southwest China (1990–2015).

Dominant Rat Hosts and Landscapes	Examined Hosts	Infested Hosts	Infestations of Rat Hosts with *O. bacoti* Mites in Different Landscapes				
			Mite Individuals	Constituent Ratios of the Mites (*Cr*, %)	Prevalence (P_M, %)	Mean Abundance (*MA*)	Mean Intensity (*MI*)
R. tanezumi							
Flatland landscape	2094	63	487	81.03	3.01	0.23	7.73
Mountainous landscape	879	31	114	18.97	3.53	0.13	3.68
Total *	2973	94	601	100.00	3.16	0.20	6.39
R. norvegicus							
Flatland landscape	940	104	1282	100.00	11.06	1.36	12.33
Mountainous landscape	110	0	0	0	0	0	-
Total *	1050	104	1282	100.00	9.90	1.22	12.33

* Annotation: The animal hosts without records of landscapes were not included in the above table.

Table 8. Infestations of *R. tanezumi* and *R. norvegicus* rats with *O. bacoti* mites in different habitats of Yunnan Province, Southwest China (1990–2015).

Dominant Rat Hosts and Habitats	Examined Hosts	Infested Hosts	Infestations of Rat Hosts with *O. bacoti* Mites in Different Habitats				
			Mite Individuals	Constituent Ratios of the Mites (*Cr*, %)	Prevalence (P_M, %)	Mean Abundance (*MA*)	Mean Intensity (*MI*)
R. tanezumi							
Indoor habitats	2607	278	1274	59.70	10.66	0.49	4.58
Outdoor habitats	2678	124	860	40.30	4.63	0.32	6.94
Total *	5285	402	2134	100.00	7.61	0.40	5.31
R. norvegicus							
Indoor habitats	1191	137	1549	93.77	11.50	1.30	11.31
Outdoor habitats	241	22	103	6.23	9.13	0.43	4.68
Total *	1432	159	1652	100.00	11.10	1.15	10.39

* Annotation: The animal hosts without records of habitats were not included in the above table.

3.6. Spatial Distribution Pattern and Interspecific Association

A total of 5285 Asian house rats (*R. tanezumi*), the first dominant host of the tropical rat mites (*O. bacoti*), were captured in 35 of 39 investigated counties, but there were only 24 counties where *R. tanezumi* harbored *O. bacoti* mites. To establish a linear regression equation based on Taylor's power law, the 24 counties were recombined as four "sample units", according to their adjacent locations, and then the mean (m) and variance (σ^2) of *O. bacoti* mites on *R. tanezumi* rats in each sample unit were calculated (Table 9). According to the calculated m and σ^2, a linear regression equation was established as $\lg \sigma^2 = \lg 39.30 + 1.42 \lg m$, where both a (39.30) and b (1.42) were beyond 1 ($a > 1$, $b > 1$), the border value for determining the aggregated distribution. The calculated patchiness index (m^*/m) in each sample unit was also higher than 1 ($m^*/m > 1$), the border value for determining the aggregated distribution (Table 9). On the body surface of *R. tanezumi* rats, there were a lot of *L. nuttalli* mites (the other species of gamasid mites) that co-occurred with *O. bacoti* mites, and therefore the interspecific association between *O. bacoti* and *L. nuttalli* was studied. The result showed that there was a slight negative association between *O. bacoti* and *L. nuttalli* ($V = -0.0794$, $V < 0$, $p < 0.05$) (Table 10).

Table 9. The mean (m), variance (σ^2) and patchiness index (m^*/m) of *O. bacoti* mites on *R. tanezumi* rats in each recombined sample unit of Yunnan, Southwest China (1990–2015).

Sample Units	Individuals of *R. tanezumi* Rats	Individuals of *O. bacoti* Mites	Mean (m)	Variance (σ^2)	Patchiness index (m^*/m)
1	1071	871	0.81	68.98	104.06
2	235	740	3.15	106.28	11.40
3	1329	375	0.28	7.66	93.65
4	2650	148	0.06	0.45	126.36
Total	5285	2134	-	-	-

Annotation: Each sample unit represents the following counties: 1 = Njian + Dali + Binchuan +Yangbi + Xiangyun + Weishan + Jianchuan + Heqing; 2 = Lushui + Fugong + Weixi + Gongshan + Lijiang; 3 = Gengma + Lianghe + Longchuan + Yingjiang + Longyang + Ruili + Yongde + Cangyuan; 4 = Luliang + Fuyuan + Maguan + Suijiang + Menghai + Yuanjiang + Simao + Mengzi + Jinghong + Ninger + Jinping + Hekou + Qiubei + Wenshan.

Table 10. The contingency table for measuring the interspecific association between *O. bacoti* mites and *L. nuttalli* mites on the body surface of *R. tanezumi rats* in Yunnan, Southwest China (1990–2015).

Items		*O. bacoti* Mites (Species X)		Total
		+	−	
L. nuttalli mites (Species Y)	+	79 (a)	1646 (b)	1725 ($a + b$)
	−	323 (c)	3237 (d)	3560 ($c + d$)
Total		402 ($a + c$)	4883 ($b + d$)	5285 (n)

4. Discussion

In laboratories, the tropical rat mite (*O. bacoti*) is often found on the experimental rats and mice, and it does a great harm to experimental animals [6,47,48]. Therefore, it is important to make a systematic study on *O. bacoti*. Some previous ecological studies of gamasid mites mainly focused on some local species surveys, faunal studies and community investigations [4,36,49]. Although some local investigations on the fauna and community of gamasid mites included the constituent ratio of *O. bacoti*, there were few independent and systematic studies of *O. bacoti* [16,38,50]. The present study systematically analyzed the distribution and host selection of *O. bacoti* in Yunnan Province of Southwest China for the first time. The original data came from a long-term investigation between 1990 and 2015, and the investigated 39 counties covered the different localities of Yunnan Province, Southwest China. The tropical rat mite (*O. bacoti*) was found in 27 of 39 investigated counties (Figure 1), and it suggests that *O. bacoti* is a widely distributed species of gamasid mites in Yunnan.

The present study showed that 99.20% of tropical rat mites (*O. bacoti*) were found on rodents (the order Rodentia), even though three orders of hosts (Rodentia, Soricomorpha and Scandetia) harbored the mites. Although *O. bacoti* mites occurred on different categories of hosts (15 species, 8 genera, 4 families and 3 orders), most of them were identified from two dominant rat species, the Asian house rat (*R. tanezumi*) and the Norway rat or brown rat (*R. norvegicus*). The infestations of *R. tanezumi* and *R. norvegicus* with *O. bacoti* mites were significantly higher than those of other 13 host species. The results suggest that *O. bacoti* has some host-specificity and it has a preference to *R. tanezumi* and *R. norvegicus* in Yunnan. The higher prevalence (P_M) of juvenile *R. norvegicus* rats with *O. bacoti* mites than that of adult rats (Table 2) may imply the preference of the mites to juvenile hosts. Rodents are closely related to human beings, and they are the infection source and reservoir hosts of many zoonotic diseases [18,51]. The rodent-preference of *O. bacoti* would increase the potential risk of the mite's attacking humans and spreading some zoonoses. The Asian house rat (*R. tanezumi*) and the Norway rat (*R. norvegicus*) are two major species of rodents associated with human settlements in Yunnan Province and some other places of China [52,53]. *Rattus tanezumi* (often called *R. flavipectus* in China) is widely distributed in the vast areas south of the Yangtze River, in Southern China. It is a very common rodent species in residential areas (especially the indoor habitats) in Central and Southern Yunnan [54,55]. *Rattus norvegicus* is widely distributed in the whole China, and it is also a very common rodent species in residential areas (especially the indoor habitats) in Yunnan, often co-occurring in the same areas with *R. tanezumi* [35,56–58]. The previous studies revealed that the main hosts of *O. bacoti* included some synanthropic rats and mice with humans (especially *R. norvegicus*) and experimental rats and mice [3,7,18]. The most commonly used laboratory rat is descended from *R. norvegicus*, which retains many biological characteristics of its ancestor *R. norvegicus* [59,60]. The frequent occurrence of *O. bacoti* on *R. tanezumi* and *R. norvegicus* would highly increase the risk of the mites' attacking humans and spreading some zoonoses.

Rattus tanezumi and *R. norvegicus* were two dominant rat hosts of *O. bacoti* mites, and therefore the present paper analyzed the infestations of the two host species with the mites in different horizontal gradients (latitudes and longitudes) and vertical gradients (altitudes). The results showed that the infestations of the rats (*R. tanezumi* and *R. norvegicus*) with *O. bacoti* mites showed some differences in different horizontal and vertical gradients. Some infestation indices (P_M and *MA*) were higher in the high latitude (>26° N) and low longitudes (<100° E and 100° E–102° E) than in other latitudes and longitudes (Tables 4 and 5). The P_M and *MA* of *R. tanezumi* rats with the mites were highest above 3000 m, but *MI* was highest below 1000 m. The P_M, *MA* and *MI* of *R. norvegicus* rats with the mites were highest at 1000–2000 m (Table 6). The results indicated an unstable fluctuation in different vertical gradients. The climates in Yunnan province greatly vary from region to region because of complex topography and altitude gradients. Even within the same latitude or longitude gradient zone, the climate at a mountainous site with higher altitude may be very different from that at a flatland site with lower altitude [61–63]. The different infestations of the rats (*R. tanezumi* and *R. norvegicus*) with *O. bacoti* mites in different horizontal and vertical gradients may be related to different climates (temperature, humidity and rainfall, etc.) in different geographical localities. However, it is difficult to explain the unstable fluctuation of the mite infestations in different horizontal and vertical gradients, and more research studies still remain to be conducted.

Located in the southwest of China, Yunnan is a mountainous province where mountainous landscapes with higher altitude and lower temperature account for 84% of the whole territory, and flatland landscapes with lower altitude and higher temperature are often embedded in mountainous landscapes [63,64]. Although the flatland landscape only takes a small portion of the whole territory, the majority of *O. bacoti* mites (91.28%) was found in the flatland landscape, and this suggests that *O. bacoti* is mainly distributed in the flatland landscape. The infestations of the rats (*R. tanezumi* and *R. norvegicus*) with the mites showed some differences in mountainous and flatland landscapes, but the differences were of no statistical significance (Table 7). In habitat distribution, 73.48% of *O. bacoti* mites were collected in the indoor habitat. The P_M and *MA* of *R. tanezumi* rats with the mites were significantly higher in the indoor habitat than in the outdoor habitat (Table 8), indicating the preference of the mites for the indoor habitat. The previous study showed that the optimum temperature for the development of *O. bacoti* was about 25 °C ± 5 °C, and higher than 30 °C or lower than 20 °C was not suitable for the mites' development [65]. The outdoor habitat in the present study involved a series of different sub-habitats, or microhabitats, such as cultivated farmlands (e.g., paddy fields and cornfields) and uncultivated bush areas and woodlands; the micro-climates in the outdoor habitat are often unstable. In comparison with the outdoor habitat, the indoor habitat is a relatively closed environment with a relatively stable and warm temperature and low humidity [65,66]. The stable and warm micro-climate with relatively low humidity in the indoor habitat may be more suitable to the growth and reproduction of *O. bacoti* mites. The frequent occurrence of *O. bacoti* in the indoor habitat would highly increase the risk of the mites' invading and stinging humans. When rats and mice are not available for *O. bacoti* mites to suck the blood of, the mites in the indoor habitat may quickly move onto humans for the blood meal and then expand their range of activity [1,67,68].

The measurement of spatial distribution pattern of a certain population is one of important issues in arthropod ecology [69,70]. There are usually three types of spatial distribution patterns: uniform (or even) distribution, random distribution and aggregated distribution [52,71–73]. There are a variety of statistical methods to measure the spatial distribution pattern of a certain population, and the patchiness index and Taylor's power law are two of them [43,44,74]. According to the statistics of the patchiness index and Taylor's power law, tropical rat mites (*O. bacoti*) were determined to be of aggregated distribution on *R. tanezumi*, the first dominant host. The aggregated distribution indicates that the mite infestation is not even among different hosts. Some hosts may harbor a large number of mites, forming a clump of mites on their body surface, while some other hosts may have no or very few mites on their body surface. The aggregated distribution pattern of *O. bacoti* in the present study is consistent with that of some other ectoparasites, such as chigger mites; this is a common phenomenon in many parasites [38,52,71]. The aggregated distribution may be beneficial to the survival, mating and defense of the parasites [4,38,72,73].

The analysis of the interspecific relationship between any two different species is also an important issue in animal ecology [75,76]. The association coefficient (V) used in the present study is a simple way to measure the interspecific relationship between any two species [77–80]. The negative value of the association coefficient ($V = -0.0794$) may imply that there is a slight negative association between *O. bacoti* and *L. nuttalli*, but the value of "$V = -0.0794$" was very close to "0", and more research is still needed.

5. Conclusions

The tropical rat mite (*O. bacoti*) is a widely distributed species of gamasid mite in Yunnan Province, Southwest China, and its dominant hosts are two synanthropic species of rats, *R. tanezumi* and *R. norvegicus*. It is mainly distributed in the flatland landscape and indoor habitat. It has some host-specificity, with a preference to rodents, especially *R. tanezumi* and *R. norvegicus*. The *O. bacoti* mites are of aggregated distribution on *R. tanezumi* rats.

Author Contributions: Conceptualization, funding acquisition, project administration, resources, supervision, validation, and writing—review & editing, X.-G.G.; supervision, D.-C.J.; data curation, formal analysis, software, visualization, and writing—original draft, P.-W.Y.; investigation, R.F., C.-F.Z., Z.-W.Z. and K.-Y.M.; methodology, X.-B.H. All authors have read and agreed to the published version of the manuscript.

Funding: The project was supported by the National Natural Science Foundation of China (Nos. 81960380 and 81672055), to Xian-Guo Guo, and the Innovation Team of Vector Biology, Dali University (No. ZKLX2019104).

Data Availability Statement: The experimental data used to support the findings of this study are available from the corresponding author request.

Acknowledgments: Up until now, more than 60 people have joined this study, including field investigation, collection of gamasid mites, specimen making and identification of the mites. Here we would like to express our sincere thanks to the following people, who have made special contributions to the field investigation and laboratory work: Qiao-Hua Wang, Yong Zhang, Cong-Hua Gao, Nan Zhao, Jian-Chang He, Guo-Li Li, Xue-Song He, Yun-Ji Zou, De-Cai Ouyang and Shuang-Lin Wang, some colleagues and college students.

Conflicts of Interest: The authors declare no conflict of interest.

References

1. Beck, W.; Folster-Holst, R. Tropical Rat Mites (*Ornithonyssus bacoti*)—Serious Ectoparasites. *J. Dtsch. Dermatol. Ges.* **2009**, *7*, 667–670. [CrossRef]
2. Beck, W.; Pantchev, N. *Parasitäre Zoonosen: Bild-Text-Atlas*; Schlütersche: Rostock, Germany, 2010.
3. Nath, A.; Venkataramana, K.; Fölster-Holst, R.; Beck, W. Epizootiology, Treatment and Control of Tropical Rat-Mite Infestation in a Breeding Colony of Swiss Mice under Temperate Climate of Nilgiris Hill-India. *Anim. Sci.* **2013**, *7*, 114–120.
4. Huang, L.Q.; Guo, X.G.; Speakman, J.R.; Dong, W.G. Analysis of Gamasid Mites (Acari: Mesostigmata) Associated with the Asian House Rat, *Rattus tanezumi* (Rodentia: Muridae) in Yunnan Province, Southwest China. *Parasitol. Res.* **2013**, *112*, 1967–1972. [CrossRef] [PubMed]
5. Beck, W. Occurrence of a House-Infesting Tropical Rat Mite (*Ornithonyssus bacoti*) on Murides and Human Beings. *Travel Med. Infect. Dis.* **2008**, *6*, 245–249. [CrossRef]
6. Chung, S.L.; Hwang, S.J.; Kwon, S.B.; Kim, D.W.; Jun, J.B.; Cho, B.K. Outbreak of Rat Mite Dermatitis in Medical Students. *Int. J. Dermatol.* **1998**, *37*, 591–594. [CrossRef] [PubMed]
7. Creel, N.B.; Crowe, M.A.; Mullen, G.R. Pet Hamsters as a Source of Rat Mite Dermatitis. *Cutis* **2003**, *71*, 457–461.
8. Engel, P.M.; Welzel, J.; Maass, M.; Schramm, U.; Wolff, H.H. Tropical Rat Mite Dermatitis: Case Report and Review. *Clin. Infect. Dis.* **1998**, *27*, 1465–1469. [CrossRef]
9. Mumcuoglu, Y.; Buchheim, E. Dermatitis Caused by the Tropical Rat Mite (*Ornithonyssus bacoti*) in Switzerland. Case Report. *Schweiz. Med. Wochenschr.* **1983**, *113*, 793–795.
10. Rosen, S.; Yeruham, I.; Braverman, Y. Dermatitis in Humans Associated with the Mites *Pyemotes tritici*, *Dermanyssus gallinae*, *Ornithonyssus bacoti* and *Androlaelaps casalis* in Israel. *Med. Vet. Entomol.* **2002**, *16*, 442–444. [CrossRef]
11. Skirnisson, K. The Tropical Rat Mite *Ornithonyssus bacoti* Attacks Humans in Iceland. *Laeknabladid* **2001**, *87*, 991–993.
12. Beck, W.; Pfister, K. Occurrence of a House-Infesting Tropical Rat Mite (*Ornithonyssus bacoti*) on Murides and Human Beings in Munich: 3 Case Reports. *Wien. Klin. Wochenschr.* **2004**, *116*, 65–68. [PubMed]
13. Baumstark, J.; Beck, W.; Hofmann, H. Outbreak of Tropical Rat Mite (*Ornithonyssus bacoti*) Dermatitis in a Home for Disabled Persons. *Dermatology* **2007**, *215*, 66–68. [CrossRef] [PubMed]
14. Fiechter, R.; Grimm, F.; Müller, G.; Schnyder, M. Cumulation of *Ornithonyssus bacoti* (Tropical Rat Mite) Infestations of Pet Rodents and Their Owners in the Canton of Zürich and Graubünden. *Schweiz. Arch. Tierheilkd.* **2011**, *153*, 79–85. [CrossRef] [PubMed]
15. Cafiero, M.; Raele, D.; Mancini, G.; Galante, D. Dermatitis by Tropical Rat Mite, *Ornithonyssus bacoti* (Mesostigmata, Macronyssidae) in Italian City-Dwellers: A Diagnostic Challenge. *J. Eur. Acad. Dermatol. Venereol.* **2016**, *30*, 1231–1233. [CrossRef] [PubMed]
16. D'Ovidio, D.; Noviello, E.; Santoro, D. Prevalence and Zoonotic Risk of Tropical Rat Mite (*Ornithonyssus bacoti*) in Exotic Companion Mammals in Southern Italy. *Vet. Dermatol.* **2018**, *29*, 522-e174. [CrossRef]
17. Reeves, W.K.; Loftis, A.D.; Szumlas, D.E.; Abbassy, M.M.; Helmy, I.M.; Hanafi, H.A.; Dasch, G.A. Rickettsial Pathogens in the Tropical Rat Mite *Ornithonyssus bacoti* (Acari: Macronyssidae) from Egyptian Rats (*Rattus* spp.). *Exp. Appl. Acarol.* **2007**, *41*, 101–107. [CrossRef]
18. Bhuyan, P.J.; Nath, A.J. Record of Tropical Rat Mite, *Ornithonyssus bacoti* (Acari: Mesostigmata: Macronyssidae) from Domestic and Peridomestic Rodents (*Rattus rattus*) in Nilgiris, Tamil Nadu, India. *J. Arthropod Borne Dis.* **2016**, *10*, 98–101.
19. Walker, D.H. Rickettsiae. In *Medical Microbiology*, 4th ed.; Baron, S., Ed.; University of Texas Medical Branch at Galveston: Galveston, TX, USA, 1996; Chapter 38.
20. Schmidt, S.; Essbauer, S.S.; Mayer-Scholl, A.; Poppert, S.; Schmidt-Chanasit, J.; Klempa, B.; Henning, K.; Schares, G.; Groschup, M.H.; Spitzenberger, F.; et al. Multiple Infections of Rodents with Zoonotic Pathogens in Austria. *Vector Borne Zoonotic Dis.* **2014**, *14*, 467–475. [CrossRef]
21. Leibler, J.H.; Zakhour, C.M.; Gadhoke, P.; Gaeta, J.M. Zoonotic and Vector-Borne Infections among Urban Homeless and Marginalized People in the United States and Europe, 1990–2014. *Vector Borne Zoonotic Dis.* **2016**, *16*, 435–444. [CrossRef]
22. Siddiqui, M.A.; Kershaw, W.E. Host-Parasite Relations in Cotton Rat Filariasis. II: The Quantitative Transmission of *Litomosoides carinii* to Delhi and Carworth Strains of White Rats, Including the Effect of Age. *Ann. Trop. Med. Parasitol.* **1976**, *70*, 313–322. [CrossRef]
23. Renz, A.; Wenk, P. Intracellular Development of the Cotton-Rat Filaria *Litomosoides carinii* in the Vector Mite *Ornithonyssus bacoti*. *Trans. R Soc. Trop. Med. Hyg.* **1981**, *75*, 166–168. [CrossRef]
24. Notarnicola, J.; Bain, O.; Navone, G.T. Two New Species of *Litomosoides* (Nematoda: Filarioidea) in Sigmodontines (Rodentia: Muridae) from Rio de La Plata Marshland, Argentina. *J. Parasitol.* **2000**, *86*, 1318–1325. [CrossRef]
25. Arumugam, S.; Pfarr, K.M.; Hoerauf, A. Infection of the Intermediate Mite Host with *Wolbachia*-depleted *Litomosoides sigmodontis* Microfilariae: Impaired L1 to L3 Development and Subsequent Sex-Ratio Distortion in Adult Worms. *Int. J. Parasitol.* **2008**, *38*, 981–987. [CrossRef] [PubMed]
26. Williams, R.W. The Laboratory Rearing of the Tropical Rat Mite, *Liponyssus bacoti* (Hirst). *J. Parasitol.* **1946**, *32*, 252–256. [CrossRef]
27. Bertram, D.S. An Apparatus for Collecting Blood-Sucking Mites. *Ann. Trop. Med. Parasitol.* **1946**, *40*, 209–214. [CrossRef] [PubMed]
28. Green, E.; Baker, C. Observations on the Micromorphology of the Tropical Rat Mite *Ornithonyssus bacoti* (Hirst) as Revealed by Scanning Electron Microscopy. *J. South Afr. Vet. Assoc.* **1996**, *67*, 128–132.

29. Bradley, J.M.; Mascarelli, P.E.; Trull, C.L.; Maggi, R.G.; Breitschwerdt, E.B. *Bartonella henselae* Infections in an Owner and Two Papillon Dogs Exposed to Tropical Rat Mites (*Ornithonyssus bacoti*). *Vector Borne Zoonotic Dis.* **2014**, *14*, 703–709. [CrossRef]
30. Nath, A.J.; Islam, S.; Sahu, S. Use of Scanning Electron Microscopy to Confirm the Identity of Tropical Rat Mite (*Ornithonyssus bacoti*): The Cause of Rat Mite Dermatitis. *J. Parasit. Dis.* **2016**, *40*, 161–165. [CrossRef]
31. Nieri-Bastos, F.A.; Labruna, M.B.; Marcili, A.; Durden, L.A.; Mendoza-Uribe, L.; Barros-Battesti, D.M. Morphological and Molecular Analysis of *Ornithonyssus* spp. (Acari: Macronyssidae) from Small Terrestrial Mammals in Brazil. *Exp. Appl. Acarol.* **2011**, *55*, 305–327. [CrossRef]
32. Niu, D.; Wang, R.; Zhao, Y.; Yang, R.; Hu, L. De Novo RNA-seq and Functional Annotation of *Ornithonyssus bacoti*. *Exp. Appl. Acarol.* **2018**, *75*, 191–208. [CrossRef]
33. Guo, Y.; Wang, R.; Zhao, Y.; Niu, D.; Gong, X.; Hu, L. Study on the Relationship between Microbial Composition and Living Environment in Important Medical Mites Based on Illumina Miseq Sequencing Technology. *J. Med. Entomol.* **2020**, *57*, 1049–1056. [CrossRef] [PubMed]
34. Zhao, Y.; Zhang, W.Y.; Wang, R.L.; Niu, D.L. Divergent Domains of 28S Ribosomal RNA Gene: DNA Barcodes for Molecular Classification and Identification of Mites. *Parasites Vectors* **2020**, *13*, 1–12. [CrossRef] [PubMed]
35. Liu, Z.; Guo, X.G.; Fan, R.; Zhao, C.F.; Mao, K.Y.; Zhang, Z.W.; Zhao, Y. Ecological Analysis of Gamasid Mites on the Body Surface of Norway Rats (*Rattus norvegicus*) in Yunnan Province, Southwest China. *Biologia* **2019**, *75*, 1325–1336. [CrossRef]
36. Luo, L.P.; Guo, X.G.; Qian, T.J.; Wu, D.; Men, X.Y.; Dong, W.G. Distribution of Gamasid Mites on Small Mammals in Yunnan Province, China. *Insect Sci.* **2007**, *14*, 71–78. [CrossRef]
37. Peng, P.Y.; Guo, X.G.; Jin, D.C. A New Species of *Laelaps Koch* (Acari: Laelapidae) Associated with Red Spiny Rat from Yunnan Province, China. *Pak. J. Zool.* **2018**, *50*, 1279–1283. [CrossRef]
38. Peng, P.Y.; Guo, X.G.; Song, W.Y.; Hou, P.; Zou, Y.J.; Fan, R.; HE, X.S. Analysis of Ectoparasites (Chigger Mites, Gamasid Mites, Fleas and Sucking Lice) of the Yunnan Red-Backed Vole (*Eothenomys miletus*) Sampled Throughout Its Range in Southwest China. *Med. Vet. Entomol.* **2015**, *29*, 403–415. [CrossRef] [PubMed]
39. Peng, P.Y.; Guo, X.G.; Song, W.Y.; Hou, P.; Zou, Y.J.; Fan, R. Ectoparasitic Chigger Mites on Large Oriental Vole (*Eothenomys miletus*) across Southwest, China. *Parasitol. Res.* **2016**, *115*, 623–632. [CrossRef]
40. Ritzi, C.M.; Whitaker, J.O. Ectoparasites of Small Mammals from the Newport Chemical Depot, Vermillion County, Indiana. *Northeast. Nat.* **2003**, *10*, 149–158. [CrossRef]
41. Storm, J.J.; Ritzi, C.M. Ectoparasites of Small Mammals in Western Iowa. *Northeast. Nat.* **2008**, *15*, 283–292. [CrossRef]
42. Zhan, Y.Z.; Guo, X.G.; Speakman, J.R.; Zuo, X.H.; Wu, D.; Wang, Q.H.; Yang, Z.H. Abundances and Host Relationships of Chigger Mites in Yunnan Province, China. *Med. Vet. Entomol.* **2013**, *27*, 194–202. [CrossRef]
43. Taylor, L.R. Aggregation, Variance and the Mean. *Nature* **1961**, *189*, 732–735. [CrossRef]
44. Kuno, E. Sampling and Analysis of Insect Populations. *Annu. Rev. Entomol.* **1991**, *36*, 285–304. [CrossRef]
45. Lopez, J.E. Parasite Prevalence and the Size of Host Populations: An Experimental Test. *J. Parasitol.* **2005**, *91*, 32–37. [CrossRef] [PubMed]
46. Guo, X.G.; Qian, T.J.; Meng, X.Y.; Dong, W.G.; Shi, W.X.; Wu, D. Preliminary Analysis of Chigger Communities Associated with House Rats (*Rattus flavipectus*) from Six Counties in Yunnan, China. *Syst. Appl. Acarol.* **2006**, *11*, 13–21. [CrossRef]
47. Watson, J. New Building, Old Parasite: Mesostigmatid Mites—An Ever-Present Threat to Barrier Rodent Facilities. *ILAR J.* **2008**, *49*, 303–309. [CrossRef] [PubMed]
48. Charlesworth, E.N.; Clegern, R.W. Tropical Rat Mite Dermatitis. *Arch. Dermatol.* **1977**, *113*, 937–938. [CrossRef]
49. Guo, X.G.; Dong, W.G.; Men, X.Y.; Qian, T.J.; Wu, D.; Ren, T.G.; Qin, F.; Song, W.Y.; Yang, Z.H.; Fletcher, Q.E. Species Abundance Distribution of Ectoparasites on Norway Rats (*Rattus norvegicus*) from a Localized Area in Southwest China. *J. Arthropod Borne Dis.* **2016**, *10*, 192.
50. Paramasvaran, S.; Sani, R.; Hassan, L.; Krishnasamy, M.; Jeffery, J.; Oothuman, P.; Salleh, I.; Lim, K.; Sumarni, M.; Santhana, R. Ectoparasite Fauna of Rodents and Shrews from Four Habitats in Kuala Lumpur and the States of Selangor and Negeri Sembilan, Malaysia and Its Public Health Significance. *Trop. Biomed.* **2009**, *26*, 303–311.
51. Negm, M.W.; Mohamed, A.A.; El-Gepaly, H.M.K.; Abdelaziz, S.M. Mesostigmata Mites (Acari: Parasitiformes) Associated with Birds and Their Nests from Egypt. *Turk. J. Zool.* **2018**, *42*, 722–731. [CrossRef]
52. Peng, P.Y.; Guo, X.G.; Ren, T.G.; Dong, W.G.; Song, W.Y. An Updated Distribution and Hosts: Trombiculid Mites (Acari: Trombidiformes) Associated with Small Mammals in Yunnan Province, Southwest China. *Parasitol. Res.* **2016**, *115*, 1923–1938. [CrossRef]
53. Zhang, M.; Xu, Z. The Effect of Three Gorge Project on the Small Mammals in Yangtze River of China. *Arch. Zool. Stud.* **2018**, *1*. [CrossRef]
54. Stuart, A.M.; Singleton, G.R.; Prescott, C.V. Population Ecology of the Asian House Rat (*Rattus tanezumi*) in Complex Lowland Agroecosystems in the Philippines. *Wildl. Res.* **2015**, *42*, 165–175. [CrossRef]
55. Guo, S.; Li, G.; Liu, J.; Wang, J.; Lu, L.; Liu, Q. Dispersal Route of the Asian House Rat (*Rattus tanezumi*) on Mainland China: Insights from Microsatellite and Mitochondrial DNA. *BMC Genet.* **2019**, *20*, 11. [CrossRef] [PubMed]
56. Zhang, Y.Z.; Zou, Y.; Fu, Z.F.; Plyusnin, A. Hantavirus Infections in Humans and Animals, China. *Emerg. Infect. Dis.* **2010**, *16*, 1195. [CrossRef]

57. Ying, B.; Kosoy, M.Y.; Maupin, G.O.; Tsuchiya, K.R.; Gage, K.L. Genetic and Ecologic Characteristics of *Bartonella* Communities in Rodents in Southern China. *Am. J. Trop. Med. Hyg.* **2002**, *66*, 622–627. [CrossRef]
58. Yan, X.; Zhenyu, L.; Gregg, W.P.; Dianmo, L. Invasive Species in China—An Overview. *Biodivers. Conserv.* **2001**, *10*, 1317–1341. [CrossRef]
59. Vandenbergh, J.G. Use of House Mice in Biomedical Research. *ILAR J.* **2000**, *41*, 133–135. [CrossRef]
60. Jing, M.; Yu, H.T.; Bi, X.; Lai, Y.C.; Jiang, W.; Huang, L. Phylogeography of Chinese House Mice (*Mus musculus musculus/castaneus*): Distribution, Routes of Colonization and Geographic Regions of Hybridization. *Mol. Ecol.* **2014**, *23*, 4387–4405. [CrossRef]
61. Badgley, C. Tectonics, Topography, and Mammalian Diversity. *Ecography* **2010**, *33*, 220–231. [CrossRef]
62. Rangwala, I.; Miller, J.R. Climate Change in Mountains: A Review of Elevation-Dependent Warming and Its Possible Causes. *Clim. Change* **2012**, *114*, 527–547. [CrossRef]
63. Wang, Y.X.; Ding, K.; Li, M.B.; Zhou, R.L. Spatial Distribution Modeling of Temperature Increase for the Uplifted Mountain Terrains and Its Characteristics in Southwest China. *J. Mt. Sci.* **2017**, *14*, 2270–2283. [CrossRef]
64. Jen, M.N. Agricultural Landscape of Southwestern China: A Study in Land Utilization. *Econ. Geogr.* **1948**, *24*, 157–169. [CrossRef]
65. Deng, G.F.; Wang, D.Q.; Gu, Y.M.; Meng, Y.C. *Economic Insect Fauna of China. Fasc 40, Acari, Dermanyssoidea*; Science Press: Beijing, China, 1993.
66. Habedank, B.; Betke, P. Aktuelle Nachweise der Tropischen Rattenmilbe, *Ornithonyssus bacoti* (Acari: Macronyssidae) in Wohnungen. *DVG Tag. Bekämpfung Epidemiol. Parasitosen* **2002**, *19*, 44.
67. Mullen, G.R.; O'Connor, B.M. Mites (Acari). In *Medical and Veterinary Entomology*; Elsevier: Amsterdam, The Netherlands, 2019; pp. 533–602.
68. Varma, M. Ticks and Mites (Acari). In *Medical Insects and Arachnids*; Springer: Berlin, Germany, 1993; pp. 597–658.
69. Grenfell, B.T.; Dobson, A.P.; Moffatt, H. *Ecology of Infectious Diseases in Natural Populations*; Cambridge University Press: Cambridge, UK, 1995; Volume 7.
70. Grear, J.S.; Schmitz, O.J. Effects of Grouping Behavior and Predators on the Spatial Distribution of a Forest Floor Arthropod. *Ecology* **2005**, *86*, 960–971. [CrossRef]
71. Touati, L.; Figuerola, J.; Alfarhan, A.H.; Samraoui, B. Distribution Patterns of Ectoparasites of Glossy Ibis (*Plegadis falcinellus*) Chicks. *Zoology* **2015**, *25*, 46–53. [CrossRef]
72. Olson, L.E.; Squires, J.R.; Oakleaf, R.J.; Wallace, Z.P.; Kennedy, P.L. Predicting Above-Ground Density and Distribution of Small Mammal Prey Species at Large Spatial Scales. *PLoS ONE* **2017**, *12*, e0177165. [CrossRef]
73. Nicacio, J.; Oliveira, I.D.; Uchoa, M.A.; Faccenda, O.; Abot, A.R.; Fernandes, M.G.; Garcia, F.R. Spatial Distribution and Control Levels of *Anastrepha* spp. (Diptera: Tephritidae) in Guava Orchards. *Anais Acad. Bras. Ciências* **2019**, *91*, e20180428. [CrossRef]
74. Zuo, X.H.; Guo, X.G.; Zhan, Y.Z.; Wu, D.; Yang, Z.H.; Dong, W.G.; Huang, L.Q.; Ren, T.G.; Jing, Y.G.; Wang, Q.H. Host Selection and Niche Differentiation in Sucking Lice (Insecta: Anoplura) among Small Mammals in Southwestern China. *Parasitol. Res.* **2011**, *108*, 1243–1251. [CrossRef]
75. Elton, C.S. *Animal Ecology*; University of Chicago Press: Chicago, IL, USA, 2001.
76. Dice, L.R. Measures of the Amount of Ecologic Association between Species. *Ecology* **1945**, *26*, 297–302. [CrossRef]
77. Hurlbert, S.H. A Coefficient of Interspecific Assciation. *Ecology* **1969**, *50*, 1–9. [CrossRef]
78. Blackburn, T.M.; Gaston, K.J. A Critical Assessment of the Form of the Interspecific Relationship between Abundance and Body Size in Animals. *J. Anim. Ecol.* **1997**, 233–249. [CrossRef]
79. Cole, L.C. The Measurement of Interspecific Associaton. *Ecology* **1949**, 411–424. [CrossRef]
80. Cole, L.C. The Measurement of Partial Interspecific Association. *Ecology* **1957**, *38*, 226–233. [CrossRef]

animals

MDPI

Article

Comparison of Two DNA Extraction Methods and Two PCRs for Detection of *Echinococcus multilocularis* in the Stool Samples of Naturally Infected Red Foxes

Katarzyna Skrzypek, Jacek Karamon *, Małgorzata Samorek-Pieróg, Joanna Dąbrowska, Maciej Kochanowski, Jacek Sroka, Ewa Bilska-Zając and Tomasz Cencek

Department of Parasitology and Invasive Diseases, National Veterinary Research Institute, Partyzantów Avenue 57, 24-100 Puławy, Poland; katarzyna.skrzypek@piwet.pulawy.pl (K.S.); malgorzata.samorek-pierog@piwet.pulawy.pl (M.S.-P.); Joanna.dabrowska@piwet.pulawy.pl (J.D.); maciej.kochanowski@piwet.pulawy.pl (M.K.); jacek.sroka@piwet.pulawy.pl (J.S.); ewa.bilska@piwet.pulawy.pl (E.B.-Z.); tcencek@piwet.pulawy.pl (T.C.)

* Correspondence: j.karamon@piwet.pulawy.pl; Tel.: +48-81-889-3040

Received: 9 November 2020; Accepted: 8 December 2020; Published: 11 December 2020

Simple Summary: The goal of the study was to compare the efficiency of two commercial DNA extraction kits together with two different Polymerase Chain Reaction (PCR) protocols in the detection of *Echinococcus multilocularis* in the feces of naturally infected foxes. Stool samples from red foxes were collected in a highly endemic area in Poland. Sedimentation and counting technique (SCT) was used as a reference method. From 48 samples, 35 were positive in SCT. Further investigations showed that 40.0% of samples (from those with SCT positive result) after Z—DNA extraction and 45.7% after Q—DNA extraction gave positive results in nested PCR. In multiplex PCR, positive results were obtained in 54.3% of samples after Z isolation and 48.6% of samples after Q. Additionally, one sample negative in SCT gave a positive result in PCR. The number of worms detected in the intestines had no influence on the PCR results. Both of the extraction methods showed similar efficiency in DNA isolation and dealing with inhibitors; however, they showed relatively low sensitivity.

Abstract: (1) Background: Due to the increasing distribution of *Echinococcus multilocularis* infections in final hosts, epidemiological investigations are important for recognizing the spreading pattern of this parasite and also to estimate risk infection for humans. (2) Methods: Investigations were conducted with two commercial kits dedicated for DNA extraction from feces: ZR Fecal DNA Mini Prep (Zymo Research, Freiburg, Germany) and QIAamp FAST DNA Stool Mini Kit (Qiagen, Hilden, Germany) (marked as Z and Q), together with two common PCR protocols (nested PCR and multiplex PCR). The goal was to compare their efficiency in detecting the genetic material of *E. multilocularis* in the samples of feces. Stool samples from red foxes were collected in a highly endemic area in Poland. Sedimentation and counting technique (SCT) was used as a reference method. (3) Results: From 48 samples, 35 were positive in SCT. Further investigations showed that 40.0% of samples (from those with SCT positive result) after Z-DNA extraction and 45.7% after Q-DNA extraction gave positive results in nested PCR. In multiplex PCR, positive results were obtained in 54.3% of samples after Z isolation and 48.6% of samples after Q. Additionally, one sample that resulted in being negative in SCT gave a positive result in PCR. The number of worms detected in the intestines had no influence on PCR results. (4) Conclusions: Both of the extraction methods showed similar efficiency in DNA isolation and dealing with inhibitors; however, they showed relatively low sensitivity. This was probably caused by degradation of genetic material in the field-collected samples.

Keywords: *Echinococcus multilocularis*; PCR; DNA extraction; feces

1. Introduction

Echinococcus multilocularis is a zoonotic cestode belonging to the Taenidae family. Circulation of this parasite occurs mostly in the sylvatic cycle, based on predator–prey relationship. Typical definitive hosts for this tapeworm are canids—in Europe, mainly red foxes (*Vulpes vulpes*) but also raccoons, dogs, arctic foxes and wolves. Its intermediate hosts are small rodents, mainly from the Cricetidae family [1,2]. Humans can become an incidental intermediate host; as a consequence, this infection leads to the development of alveolar echinococcosis disease caused by larval stage of *E. multilocularis*, which is one of the most dangerous parasitic zoonosis in Europe [2]. This happens through accidental ingestion of the tapeworm's eggs shed into the environment with feces of infected final hosts. It has been shown that domestic animals, i.e., dogs and cats, can be infected with this parasite and can become a potential source of human infections. Nevertheless, the red fox is still the crucial species responsible for contamination of the environment [2–5]. In light of the recent studies, the distribution pattern of *E. multilocularis* and prevalence in previously known areas have increased. Thus, it is important to conduct monitoring among populations of final hosts. The "gold standard" in detection of adult worms in definitive host is sedimentation and counting technique (SCT). However, since the SCT technique is applied postmortem, it is impractical for use in domestic animals. There are many alternative methods developed for recognizing *E. multilocularis* infection in definitive hosts, i.e., coproantigen detection by ELISA [6] or different PCR techniques designed for detection of the parasite's DNA (from eggs or copro-DNA) in animal feces [7–12]. One of the advantages of the abovementioned molecular methods is their sensitivity; however, it is possible that inhibitors present in the feces may prevent the successful PCR reaction. Consequently, reported results can be false negative. In order to avoid the negative influence of inhibitors on the amplification process, there are several methods of removing them from the samples, e.g., using the inhibitor-binding substances included in commercial kits dedicated for DNA extraction from stool or using the magnetic capture (MC) technique that binds the target DNA on the magnetic beads in a very specific way [9,10].

The aim of this study was to compare two commercial DNA extraction kits combined with two PCR techniques commonly used for detection of *E. multilocularis* DNA. We evaluate their effectiveness and ability of dealing with inhibitors present in feces.

2. Materials and Methods

The material for this study (48 stool samples) was collected from the large intestine of red foxes shot by hunters in the area of Podkarpackie province in Southeastern Poland. Intestines were kept in −80 °C for 7 days, for safety reasons, before examination. In a further stage, they were taken out to thaw overnight, and the next day, sedimentation and counting technique (SCT) was performed [13,14]. SCT results were treated as reference data for analyzing PCR results. During SCT protocol, stool samples were collected from posterior part of the rectum and then frozen at −20 °C, for further molecular analysis.

2.1. DNA Extraction

DNA extraction was carried out for all stool samples with two extraction kits:

Z—ZR Fecal DNA Mini Prep (Zymo Research, Freiburg, Germany): An isolation was conducted in accordance with the standard manufacturer's protocol.

Q—QIAamp FAST DNA Stool Mini Kit (Qiagen, Hilden, Germany): The isolation was performed in accordance with the special manufacturer's protocol for DNA isolation from large volume of stool with few modifications. In the first step, 1 g of stool sample diluted in 10 mL of InhibitEX Buffer was vortexed vigorously in 50 mL centrifuge tube containing glass beads. In the first incubation, the suspension was heated at 95 °C (5 min), according to manufacturer's recommendation for samples difficult to lyse. Second incubation was carried out at 70 °C for 20 min. In the last step, after adding ATE buffer, samples were left at room temperature for 5 min. Then, the isolates were kept in −20 °C, until further analysis.

2.2. Polymerase Chain Reactions (PCRs)

PCR was performed simultaneously with two different PCR methods for all DNA samples. Multiplex PCR [8] was accomplished for amplification of NADH dehydrogenase subunit 1 (nad1) of *E. multilocularis*, small subunit of ribosomal RNA (rrnS) of *Taenia* spp. and rrnS of *E. granulosus*. Nested PCR [7], with some modifications [15], was conducted for amplification of mitochondrial 12S ribosomal RNA (12S rRNA) of *E. multilocularis*. Each DNA sample was tested in undiluted (1/1) and tenfold diluted (1/10) variant. Moreover, each variant of DNA sample was tested in repetition. Internal control (DNA extracted from *E. multilocularis* adult worms) was added to one of two repeated samples, to make sure that no inhibition occurred. There was no inhibition if the specific band was present on the gel after electrophoresis.

2.3. Statistical Analysis

Differences in numbers of positive PCR results among two isolation methods (Z and Q) in each PCR variant were estimated by a chi-square test (or chi-square with Yates correction). The distribution of quantitative variables was tested by the Shapiro–Wilk test, and the normality hypothesis of the data was rejected. The relationship between the intensity of infection (number of *E. multilocularis* worms in the intestine assessed with SCT) and PCR results was calculated with Spearman's rank-order correlation. The differences were considered statistically significant when $p < 0.05$. The statistical analysis was performed by using Statistica 9.1 (StatSoft Inc., Tulsa, OK, USA).

3. Results

SCT examination showed the presence of *E. multilocularis* tapeworms in 35 of the 48 tested samples. The intensity of infection ranged from 1 to 75,000 tapeworms per intestine.

The PCR results (for *E. multilocularis*) were compared to the results of SCT. Out of 35 samples positive for *E. multilocularis* in SCT, 23 gave a positive result, simultaneously in nested and multiplex PCR. Nested PCR showed the presence of *E. multilocularis* genetic material in 14 (after Z isolation) and 16 (after Q isolation) stool samples in total, 40.0% and 45.7%, respectively (Table 1). In multiplex PCR, 19 samples (Z) and 17 (Q) gave positive results, 54.3% and 48.6%, respectively (Table 1). There were no significant differences between Z and Q extraction methods in any variant of PCR and dilution. Twelve SCT positive samples did not give a band specific for *E. multilocularis* DNA in any of the PCRs applied, regardless of the intensity of the infection. Additionally, one negative sample in SCT, gave a positive result in multiplex PCR (only in undiluted DNA) (Q).

Table 1. Percentages of PCR *Echinococcus multilocularis* positive results among fecal samples obtaining from *Echinococcus multilocularis* sedimentation and counting technique (SCT) positive intestines.

Method of Isolation	Percentage of Positive Results in *Echinococcus multilocularis*									
	Nested PCR					Multiplex PCR				
	Total	1/1	1/10	Only 1/1	Only 1/10	Total	1/1	1/10	Only 1/1	Only 1/10
Z	40.0	40.0	34.3	5.7	-	54.3	51.4	42.9	11.4	2.9
Q	45.7	45.7	25.7	20.0	-	48.6	45.7	28.6	22.9	2.9

1/1—results obtained in undiluted DNA variant; 1/10—results obtained in tenfold diluted DNA variant; only 1/1—positive results obtained only in undiluted variants (while diluted variants were negative); only 1/10—positive results obtained only in diluted variants (while undiluted variants were negative); Z—ZR Fecal DNA Mini Prep (Zymo Research, Freiburg, Germany); Q—QIAamp FAST DNA Stool Mini Kit (Qiagen, Hilden, Germany).

The analysis of relationship between number of *E. multilocularis* worms in the intestine and PCR results showed modern positive correlation for Z extraction ($r_s = 0.5429$ for nested PCR; $r_s = 0.4629$ for multiplex PCR) and low positive correlation for Q extraction ($r_s = 0.3436$ for nested PCR; $r_s = 0.3680$ for multiplex PCR).

Figure 1 presents percentages of positive PCR results grouped in three levels of intensity estimated with SCT: low (1–10 worms) ($n = 6$), medium (11–1000) ($n = 17$) and high (>1000) ($n = 12$) intensity. We observed that only the Q extraction allowed for positive PCR results in low number of worms in the intestine. Moderate- and high-intensity infection percentages of positive PCR results were similar in both extraction methods. The highest number of PCR positive results in each variant was detected in feces coming from intestines infected with a high number of worms (>1000).

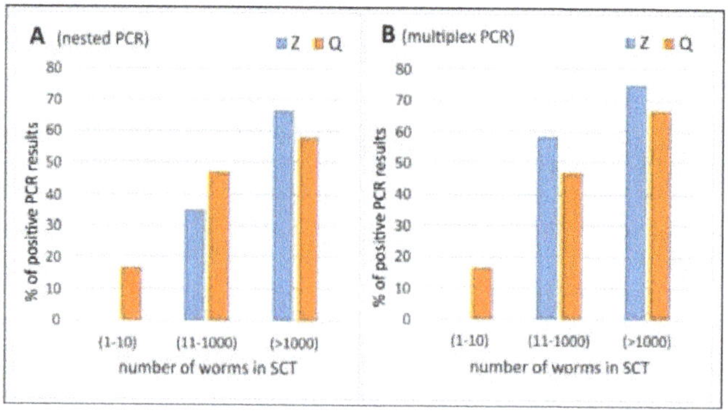

Figure 1. Percentages of *Echinococcus multilocularis* positive PCR results in samples of feces coming from intestines with different number of worms estimated with sedimentation and counting technique (SCT)—grouped in three levels of intensity: low (1–10 worms), medium (11–1000) and high (>1000) intensity. Z—ZR Fecal DNA Mini Prep (Zymo Research, Freiburg, Germany); Q—QIAamp FAST DNA Stool Mini Kit (Qiagen, Hilden, Germany); (**A**) nested PCR and (**B**) multiplex PCR.

Internal control gave negative results (both in diluted and undiluted DNA) only in one sample after Q isolation tested with nested PCR (in multiplex PCR it was positive). This sample after DNA extraction with the Z method gave positive results in both PCRs. Additionally, in nested PCR, undiluted samples with added internal control gave significantly more positive results after Z isolation (97.9%) than Q (83.3%) (Table 2)

Table 2. Results in *Echinococcus multilocularis* internal controls obtained in nested PCR and multiplex PCR.

Method of Isolation	Percentage of Positive Results in *Echinococcus multilocularis* Internal Controls									
	Nested PCR					Multiplex PCR				
	Total	1/1	1/10	Only 1/1	Only 1/10	Total	1/1	1/10	Only 1/1	Only 1/10
Z	100.0	97.9 [a]	100.0	0.0	2.1	100.0	100.0	100.0	0.0	2.1
Q	97.9	83.3 [b]	97.9	0.0	14.6	100.0	100.0	97.9	0.0	0.0

[a,b] Different letters in superscript indicate statistically significant differences among methods of isolation in individual variants ($p < 0.05$). 1/1—results obtained in undiluted DNA variant; 1/10—results obtained in tenfold diluted DNA variant; only 1/1—positive results obtained only in undiluted variants (while diluted variants were negative); only 1/10—positive results obtained only in diluted variants (while undiluted variants were negative); Z—ZR Fecal DNA Mini Prep (Zymo Research, Freiburg, Germany); Q—QIAamp FAST DNA Stool Mini Kit (Qiagen, Hilden, Germany).

Taenia spp. band in multiplex PCR gave a total of 42 out of 48 isolates (87.5%) after Z isolation and 91.7% after Q isolation (Table 3). There were no significant differences between method of DNA extraction in detection of *Taenia* spp. in any variant.

Table 3. Percentages of PCR *Taenia* spp. positive results (multiplex PCR).

Method of Isolation	Percentage of Positive Results for *Taenia* spp.				
	Multiplex PCR				
	Total	1/1	1/10	only 1/1	only 1/10
Z	87.5	87.5	77.1	10.4	-
Q	91.7	89.6	64.6	27.1	2.1

1/1—results obtained in undiluted DNA variant; 1/10—results obtained in tenfold diluted DNA variant; only 1/1—positive results obtained only in undiluted variants (while diluted variants were negative); only 1/10—positive results obtained only in diluted variants (while undiluted variants were negative); Z—ZR Fecal DNA Mini Prep (Zymo Research, Freiburg, Germany); Q—QIAamp FAST DNA Stool Mini Kit (Qiagen, Hilden, Germany).

Product specific for *E. granulosus* was not obtained in any of tested samples by multiplex PCR.

4. Discussion

The present study was undertaken to compare two widely used commercial DNA extraction kits, together with two PCR protocols, i.e., multiplex PCR [8] and nested PCR [7]. We determined the efficiency of detecting *E. multilocularis* infection in final hosts by molecular detection of copro-DNA.

Some authors reported higher sensitivity of molecular methods over widely used reference standards such as SCT or intestinal scraping technique (IST) [10,16]. Isaksson et al. [9] estimated sensitivity of magnetic capture (MC)-PCR to be 88.2%, and Dinkel et al. [7] estimated sensitivity of 89% for nested PCR. In our study, from all of the samples confirmed as positive by SCT, only 40.0–54.3% of them gave positive results in molecular tests (maximum with combination of Z extraction and multiplex PCR). A similar percentage of positive results (52%) was obtained by Maksimov et al. [17] in the stool samples from *E. multilocularis* naturally infected foxes. In Dinkel et al.'s [7] study, they showed correlation of *E. multilocularis* gravid worm number in IST and PCR sensitivity. It was on a level of 100% for samples containing 1000 and more adult tapeworms and 70% for 10 or less of immature ones. Similar dependency was observed by Maksimov et al. [17]. In their studies, PCR sensitivity was 90% for the samples taken from foxes with intense infection and only 30% for those with low infection level. In our study, we observed moderate or low positive correlation between number of worms detected in SCT and PCR results. However, it must be stressed that some of the SCT positive samples did not show positive PCR results, even though they were extracted from the stool taken from intestines of foxes with very differential number of tapeworms (from 1 to more than 1000 per sample). We obtained negative results in molecular tests both for samples with less than 10, as well as a few thousands of adult worms. Among all of the SCT negative samples, only one sample was PCR positive. In contrast, Dinkel et al. [7] used IST as a reference method. IST has been proven to be less sensitive than SCT [18,19]. Therefore, this could be a reason for the higher PCR sensitivity (compared to IST) in Dinkel's study than in presented investigation, where SCT was used as a standard. The other factor that might have affected the detection of *E. multilocularis* DNA with PCR is that all intestine samples in this study were collected by hunters in the field. Thus, the storage and transport conditions could not be controlled in detail. With conventional PCR, Maksimov et al. [17] obtained much better sensitivity results for samples experimentally spiked with tapeworm eggs in the laboratory conditions than from naturally infected foxes. Natural conditions and DNase enzymes in the feces may influence or even damage DNA [20], resulting in a negative PCR outcome even in samples positive in SCT.

PCR amplification of genetic material extracted from stool samples can be disrupted by inhibitors present in the feces, giving false negative results. There are two most common approaches for sample treatment that can be applied in order to limit inhibitors presence and lower the risk of amplification disturbance. One of these approaches is to perform extraction not directly from stool but from tapeworm eggs separated from the sample prior to extraction [21,22]. This method is useful especially for experiments with small amounts of stool samples [22]. However, it is time-consuming, laborious and carries the risk of losing the positive material. Additionally, flotation is not so highly efficient in detecting tapeworm eggs; thus, this method may eliminate some of the potentially positive samples [23]. It is also limited because there are no eggs in the stool in early stage infections with immature worms [7]. The second approach is conducting DNA extraction from the stool and then diluting the isolate [23]. Extracting genetic material directly from feces lowers the risk of losing positive material. Diluting the DNA helps to lower the concentration of inhibitors in the final sample volume, decreasing the possibility of their influence on the amplification process. It leads not only to lower the concentration of inhibitors but also a specific DNA, so it is important to use sensitive and specific PCR protocol in order to obtain satisfactory results. Additionally, the use of an extraction method suitable for feces is a key factor. In Al-Sabi et al.'s [21] study, they obtained results with sensitivity on a very low level, by using a commercial kit dedicated for tissues. However, they isolated the genetic material from stool samples with no inhibitors eliminating step. In Karamon's [23] study, two stool dedicated extraction kits gave significantly more positive results in comparison to the method dedicated for tissues. In order to determine if any DNA inhibition occurred, each sample was tested in repetition with internal control. In one variant (extraction kit Q combined with nested PCR), there were eight negative results of internal controls in undiluted DNA samples, while there was only one in diluted DNA. This confirms the need to examine additional diluted variants of samples, to avoid false negative results.

The impact of freezing-and-thawing process on the quality of DNA occurring in feces should also be presented. In this study, all of the intestines were frozen at −80 °C, for one week, due to safety reasons. Then, stool samples taken from the intestines during SCT preparation were frozen at −20 °C and kept for further analysis. Klein et al.'s [16] study shows that freeze–thawing cycles significantly increase PCR sensitivity, possibly by mechanical disruption of egg shells, which helps release DNA. In this method, the number of eggs prevalent in the feces influenced the result. On the other hand, repeated freezing and thawing degrades "free" DNA (DNA from tissue fragments or the whole worms shed with feces), which negatively affects the sensitivity of further molecular analysis when there are no or few parasite eggs in the sample [7,23]. False negative results may be also due the fact that, even if the eggs are present in stool, they may contain too little DNA for efficient amplification [24].

5. Conclusions

Our investigation showed similar effectiveness of two DNA extraction kits dedicated for stool samples in *E. multilocularis* detection. Moreover, the presented methods were found to be good at eliminating of copro-inhibitors. The study indicated the problem with relatively low sensitivity of PCRs in the examination of field samples from naturally infected animals (in comparison to microscopic examination of intestine), what was probably connected with degradation of genetic material in feces.

Author Contributions: Conceptualization, K.S., J.K. and T.C.; methodology, J.K., M.S.-P. and J.S.; software, J.K.; investigation, J.K., J.D. and M.K.; data curation, J.S. and E.B.-Z.; writing—original draft preparation, K.S., J.K. and M.S.-P.; writing—review and editing, J.K. and T.C.; visualization, M.K.; supervision, T.C. All authors have read and agreed to the published version of the manuscript.

Funding: This work was supported by funding from the European Union's Horizon 2020 Research and Innovation programme, under grant agreement number 773830: One Health European Joint Programme (MEME project; https://onehealthejp.eu/jrp-meme/).

Conflicts of Interest: The authors have no conflict of interest. The funders had no role in the design of the study; in the collection, analyses, or interpretation of data; in the writing of the manuscript, or in the decision to publish the results.

References

1. Nakao, M.; Xiao, N.; Okamoto, M.; Yanagida, T.; Sako, Y.; Ito, A. Geographic pattern of genetic variation in the fox tapeworm *Echinococcus multilocularis*. *Parasitol. Int.* **2009**, *58*, 384–389. [CrossRef] [PubMed]

2. Oksanen, A.; Siles-Lucas, M.; Karamon, J.; Possenti, A.; Conraths, F.J.; Romig, T.; Wysocki, P.; Mannocci, A.; Mipatrini, D.; La Torre, G.; et al. The geographical distribution and prevalence of *Echinococcus multilocularis* in animals in the European Union and adjacent countries: A systematic review and meta-analysis. *Parasites Vectors* **2016**, *9*, 1–23. [CrossRef] [PubMed]
3. Kapel, C.M.O.; Torgerson, P.R.; Thompson, R.C.A.; Deplazes, P. Reproductive potential of *Echinococcus multilocularis* in experimentally infected foxes, dogs, raccoon dogs and cats. *Int. J. Parasitol.* **2006**, *36*, 79–86. [CrossRef] [PubMed]
4. Romig, T. *Echinococcus multilocularis* in Europe—state of the art. *Vet. Res. Commun.* **2009**, *33*, S31–S34. [CrossRef] [PubMed]
5. Hegglin, D.; Deplazes, P. Control of *Echinococcus multilocularis*: Strategies, feasibility and cost-benefit analyses. *Int. J. Parasitol.* **2013**, *43*, 327–337. [CrossRef]
6. Deplazes, P.; Alther, P.; Tanner, I.; Thompson, R.C.A.; Eckert, J. *Echinococcus multilocularis* coproantigen detection by enzyme-linked immunosorbent assay in fox, dog, and cat populations. *J. Parasitol.* **1999**, *85*, 115–121. [CrossRef]
7. Dinkel, A.; Von Nickisch-Rosenegk, M.; Bilger, B.; Merli, M.; Lucius, R.; Romig, T. Detection of *Echinococcus multilocularis* in the definitive host: Coprodiagnosis by PCR as an alternative to necropsy. *J. Clin. Microbiol.* **1998**, *36*, 1871–1876. [CrossRef]
8. Trachsel, D.; Deplazes, P.; Mathis, A. Identification of taeniid eggs in the faeces from carnivores based on multiplex PCR using targets in mitochondrial DNA. *Parasitology* **2007**, *134*, 911–920. [CrossRef]
9. Isaksson, M.; Hagstrom, A.; Armua-Fernandez, M.T.; Wahlstrom, H.; Agren, E.O.; Miller, A.; Holmberg, A.; Lukacs, M.; Casulli, A.; Deplazes, P.; et al. A semi-automated magnetic capture probe based DNA extraction and real-time PCR method applied in the Swedish surveillance of *Echinococcus multilocularis* in red fox (*Vulpes vulpes*) faecal samples. *Parasites Vectors* **2014**, *7*, 583. [CrossRef]
10. Maas, M.; Van Roon, A.; Dam-Deisz, C.; Opsteegh, M.; Massolo, A.; Deksne, G.; Teunis, P.; Van der Giessen, J. Evaluation by latent class analysis of a magnetic capture based DNA extraction followed by real-time qPCR as a new diagnostic method for detection of *Echinococcus multilocularis* in definitive hosts. *Vet. Parasitol.* **2016**, *230*, 20–24. [CrossRef]
11. Dinkel, A.; Kern, S.; Brinker, A.; Oehme, R.; Vaniscotte, A.; Giraudoux, P.; Mackenstedt, U.; Romig, T. A real-time multiplex-nested PCR system for coprological diagnosis of *Echinococcus multilocularis* and host species. *Parasitol. Res.* **2011**, *109*, 493–498. [CrossRef] [PubMed]
12. Knapp, J.; Umhang, G.; Poulle, M.-L.; Millon, L. Development of a Real-Time PCR for a Sensitive One-Step Coprodiagnosis Allowing both the Identification of Carnivore Feces and the Detection of *Toxocara* spp. and *Echinococcus multilocularis*. *Appl. Environ. Microbiol.* **2016**, *82*, 2950–2958. [CrossRef] [PubMed]
13. Hofer, S.; Gloor, S.; Muller, U.; Mathis, A.; Hegglin, D.; Deplazes, P. High prevalence of *Echinococcus multilocularis* in urban red foxes (*Vulpes vulpes*) and voles (*Arvicola terrestris*) in the city of Zurich, Switzerland. *Parasitology* **2000**, *120*, 135–142. [CrossRef] [PubMed]
14. OIE. Echinococcosis/hydatidosis (infection with *Echinococcus granulosus* and *E. multilocularis*). In *Manual of Diagnostic Tests and Vaccines for Terrestrial Animals*, 6th ed.; Office International des Epizooties: Paris, France, 2008; Volume 1, pp. 1–15.
15. Karamon, J.; Sroka, J.; Cencek, T. The first detection of *Echinococcus multilocularis* in slaughtered pigs in Poland. *Vet. Parasitol.* **2012**, *185*, 327–329. [CrossRef] [PubMed]
16. Klein, C.; Liccioli, S.; Massolo, A. Egg intensity and freeze-thawing of fecal samples affect sensitivity of *Echinococcus multilocularis* detection by PCR. *Parasitol Res* **2014**, *113*, 3867–3873. [CrossRef] [PubMed]
17. Maksimov, P.; Schares, G.; Press, S.; Frohlich, A.; Basso, W.; Herzig, M.; Conraths, F.J. Comparison of different commercial DNA extraction kits and PCR protocols for the detection of *Echinococcus multilocularis* eggs in faecal samples from foxes. *Vet. Parasitol.* **2017**, *237*, 83–93. [CrossRef]
18. Karamon, J.; Sroka, J.; Cencek, T.; Kochanowski, M.; Dąbrowska, J. Efficacy of intestinal scraping technique in the detection of *Echinococcus multilocularis*—Estimation of the limit of the detection and comparison with sedimentation and counting technique. *Bull. Vet. Inst. Pulawy* **2012**, *56*, 535–538. [CrossRef]
19. EFSA. The EU summary Report on Trends and Sources of Zoonoses, Zoonotic Agents and Food-borne Ourbreaks in 2013. *EFSA J.* **2015**, *13*, 3991.

20. Irie, T.; Ito, T.; Kouguchi, H.; Yamano, K.; Uraguchi, K.; Yagi, K.; Nonaka, N. Diagnosis of canine *Echinococcus multilocularis* infections by copro-DNA tests: Comparison of DNA extraction techniques and evaluation of diagnostic deworming. *Parasitol. Res.* **2017**, *116*, 2139–2144. [CrossRef]
21. Al-Sabi, M.N.S.; Kapel, C.M.O.; Deplazes, P.; Mathis, A. Comparative copro-diagnosis of *Echinococcus multilocularis* in experimentally infected foxes. *Parasitol. Res.* **2007**, *101*, 731–736. [CrossRef]
22. Mathis, A.; Deplazes, P.; Eckert, J. An improved test system for PCR-based specific detection of *Echinococcus multilocularis* eggs. *J. Helminthol.* **1996**, *70*, 219–222. [CrossRef] [PubMed]
23. Karamon, J. Detection of *Echinococcus multilocularis* in faeces by nested PCR with the use of diluted DNA samples. *Pol. J. Vet. Sci.* **2014**, *17*, 79–83. [CrossRef] [PubMed]
24. Oines, O.; Isaksson, M.; Hagstrom, A.; Tavornpanich, S.; Davidson, R.K. Laboratory assessment of sensitive molecular tools for detection of low levels of *Echinococcus multilocularis*-eggs in fox (*Vulpes vulpes*) faeces. *Parasites Vectors* **2014**, *7*, 246. [CrossRef] [PubMed]

Publisher's Note: MDPI stays neutral with regard to jurisdictional claims in published maps and institutional affiliations.

animals MDPI

Article

Long-Term Determinants of the Seroprevalence of *Toxoplasma gondii* in a Wild Ungulate Community

Patricia Barroso [1,*], Ignacio García-Bocanegra [2], Pelayo Acevedo [1], Pablo Palencia [1], Francisco Carro [3], Saúl Jiménez-Ruiz [1,2], Sonia Almería [4], Jitender P. Dubey [5], David Cano-Terriza [2] and Joaquín Vicente [1,6]

1 Grupo Sanidad y Biotecnología (SaBio), Instituto de Investigación en Recursos Cinegéticos (IREC) CSIC-UCLM-JCCM, 13071 Ciudad Real, Spain; pelayo.acevedo@uclm.es (P.A.); pablo.palencia@uclm.es (P.P.); saul.jimenez.ruiz@gmail.com (S.J.-R.); joaquin.vicente@uclm.es (J.V.)
2 Grupo de Investigación en Sanidad Animal y Zoonosis (GISAZ), Departamento de Sanidad Animal, Universidad de Córdoba, 14014 Córdoba, Spain; v62garbo@uco.es (I.G.-B.); davidcanovet@gmail.com (D.C.-T.)
3 Estación Biológica Doñana, CSIC, 41092 Sevilla, Spain; pcarro@ebd.csic.es
4 Division of Virulence Assessment, Office of Applied Research and Safety Assessment (OARSA), Center for Food Safety and Nutrition (CFSAN), Department of Health and Human Services, Food and Drug Administration, Laurel, MD 20708, USA; Maria.Almeria@fda.hhs.gov
5 Animal Parasitic Disease Laboratory, Beltsville Agricultural Research Center, Agricultural Research Service, United States Department of Agriculture, Building 1001, BARC-East, Beltsville, MD 20705-2350, USA; jitender.dubey@usda.gov
6 Escuela Técnica Superior de Ingenieros Agrónomos, UCLM, 13071 Ciudad Real, Spain
* Correspondence: pbarrososgg@gmail.com

Received: 17 November 2020; Accepted: 8 December 2020; Published: 9 December 2020

Simple Summary: *Toxoplasma gondii* is a zoonotic intracellular parasite which infects a wide range of warm-blooded animals. Long-term studies provide the necessary perspective required to understand those processes which took place over many years in order to address epidemiology and ecology in complex host communities. This study is focused on evaluating what the main long-term determinants of the seroprevalence of *T. gondii* are in the wild ungulate community from Doñana National Park (southwestern Spain). With this purpose, we assayed sera from 1573 wild ungulates (wild boar, red deer, and fallow deer), collected for 13 years (from 2005 to 2018). We found high seroprevalence values of *T. gondii* (% ± CI 95%; wild boar 39 ± 3.3; red deer 30.7 ± 4.4; and fallow deer 29.7 ± 4.2. Several factors operating in the medium and long-term (individual, environmental, population and stochastic) explained the risk of *T. gondii* in wild boar and deer, some of them operating at the community level.

Abstract: *Toxoplasma gondii* is an obligate intracellular protozoan which infects warm-blooded vertebrates, including humans, worldwide. In the present study, the epidemiology of *T. gondii* was studied in the wild ungulate host community (wild boar, red deer, and fallow deer) of Doñana National Park (DNP, south-western Spain) for 13 years (2005–2018). We assessed several variables which potentially operate in the medium and long-term (environmental features, population, and stochastic factors). Overall, the wild ungulate host community of DNP had high seroprevalence values of *T. gondii* (STG; % ± confidence interval (CI) 95%; wild boar (*Sus scrofa*) 39 ± 3.3, $n = 698$; red deer (*Cervus elaphus*) 30.7 ± 4.4, $n = 423$; fallow deer (*Dama dama*) 29.7 ± 4.2, $n = 452$). The complex interplay of hosts and ecological/epidemiological niches, together with the optimal climatic conditions for the survival of oocysts that converge in this area may favor the spread of the parasite in its host community. The temporal evolution of STG oscillated considerably, mostly in deer species. The relationships shown by statistical models indicated that several factors determined species patterns. Concomitance of effects among species, indicated that relevant drivers of risk operated at the community level. Our focus, addressing factors operating at broad temporal scale, allows showing their impacts on the epidemiology of *T. gondii* and its trends. This approach is key to understanding the epidemiology and ecology to *T. gondii* infection in wild host communities in a context where the decline in seroprevalence leads to loss of immunity in humans.

Keywords: parasite; long-term study; protozoan; shared infections; zoonoses; wildlife-livestock interface

1. Introduction

Toxoplasma gondii is a zoonotic obligate intracellular protozoan which infects warm-blooded vertebrates [1]. It has an indirect life cycle where wild and domestic felids are the definitive hosts, excreting oocyst in feces. Humans, as well as many mammal and bird species, serve as intermediate hosts of *T. gondii* and can become infected by vertical transmission, the fecal-oral route, through the ingestion of water or food contaminated with sporulated *T. gondii* oocysts, or through the consumption of tissues from animals infected with encysted bradyzoites [1,2].

T. gondii has been detected in wildlife and livestock worldwide [1]. Previous Spanish studies revealed a widespread distribution of this parasite in both wild and domestic ungulates, showing significant differences in the presence of *T. gondii* among geographic areas [3–7]. In Mediterranean ecosystems in southern Spain, antibodies against *T. gondii* have been detected in wild ungulates including wild boar (*Sus scrofa*), red deer (*Cervus elaphus*), fallow deer (*Dama dama*), roe deer (*Capreolus capreolus*), Barbary sheep (*Ammotragus lervia*), mouflon (*Ovis aries musimon*) and Iberian ibex (*Capra pyrenaica*). In these studies, seroprevalences of 40.2%, 15.6% and 10.5% were reached in wild boar, fallow deer and red deer, respectively [3,8], being lower or calculated from a few samples in the other species. Regarding livestock, serosurveys in this region revealed rate levels ranging between 18.6–83.3% and 16.2–24.3% in domestic ungulates [5,8,9] and pigs, respectively [7,10,11]. Higher seroprevalences from this area have been reported in wild carnivores, especially in the Iberian lynx (*Lynx pardinus*), reaching rates of 81.5% [12].

Host-pathogen dynamics are subjected to several processes which operate over broad temporal scales; however, little attention has been paid to *T. gondii*, and particularly, in intermediate host communities at the wildlife-livestock interface [4,8]. Wide temporal data series are essential to address epidemiology and ecology in complex host communities with the necessary perspective required to understand processes taking place over many years [13–16]. In Doñana National Park (DNP, South West Spain), the wild ungulate community (including wild boar, red deer and fallow deer) occurs sympatrically with free-ranging cattle and horses, and one of the most important meta-population of the endangered Iberian lynx [17]. Studies on *T. gondii* in DNP has been exclusively conducted in felid populations with conservational purposes, showing a widespread infection in the area and reporting seroprevalence rates up to 60% [12,18,19].

The multiple transmission routes and capacity of *T. gondii* to find niches into the hosts studied provided an excellent scenario to improve our understanding of the transmission dynamics of this pathogen. While *T. gondii* has normally been considered an excellent model to study host-pathogen interactions, we also showed that it may also be used to address the study of population, community and environmental factors. The present long-term study illustrates the interplay of factors, particularly factors operating at broad temporal scale that may contribute to the spread and maintenance of a pathogen over host communities. In this context, we present data on serosurveillance of *T. gondii* in wild ungulates (wild boar, red deer, and fallow deer) from DNP for a 13-years, with the specific aims of: (I) evaluating the factors (individual, populational and environmental) modulating the seroprevalence of *T. gondii* (STG), and (II) assessing the factors operating in the long-term (population and stochastic) in order to explain the temporal trend of STG in the intermediate host ungulate community from 2005 to 2018.

2. Materials and Methods

2.1. Study Area

This study was conducted in DNP (54,000 ha), one of the most relevant biodiversity reserves in Europe, located on the Atlantic coast of southwestern Spain (37°09′ N, 6°30′ W). Human access to the park is restricted and agriculture and hunting are prohibited inside the park; cattle and horse breeding are allowed and are, mainly focused on autochthonous and traditional breeds [20].

The habitat consists of a greater proportion of sand dune habitat and marshland, combined with pine forest and Mediterranean scrubland (see Figure 1 and [21] for a more detailed description). Between the scrublands and the edge of the marshland, there is a narrow north-south longitudinal strip of humid ecotone of high ecological richness.

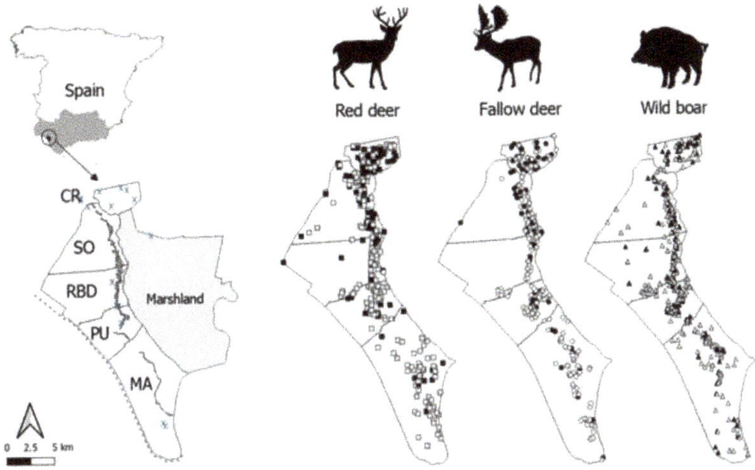

Figure 1. Map of the study area, Doñana National Park. The sampling areas (cattle management units: Coto del Rey (CR), Sotos (SO), Doñana Biological Reserve (RBD), Puntal (PU) and Marismillas (MA)) are delimited and the ecotone and small human settlements are displayed by a dark band and blue "X", respectively. Red deer (squares), fallow deer (circles), and wild boar (triangles) sampled are shown. Black and white symbols mean animals positive and negative for antibodies against *Toxoplasma gondii*, respectively.

DNP has a dry sub-humid Mediterranean climate with strong seasonality, especially in terms of water availability to animals and vegetation. The average annual temperature is 17 °C, and the mean annual precipitation is 550 mm, with high intra and interannual fluctuation (170 to 1000 mm), which determine the dynamics of the marshlands [22]. During the wet seasons (winter and spring) the marshlands may flood, so ungulates concentrate and browse in the remaining uncovered scrublands. In late summer and autumn, the hardest season for ungulates due to the shortage of resources and the seasonal drought, an aggregation of wild and domestic ungulates on the ecotone and around water points occurs [21].

The territory of DNP included in this study is divided into five cattle management areas from north to south: Coto del Rey (CR), Sotos (SO), Doñana Biological Reserve (RBD), Puntal (PU) and Marismillas (MA). Free-ranging livestock is distributed through the entire park, except in the northernmost area (CR). In this area, despite the existence of a low number of horses since the last year, cattle husbandry is prohibited since 2002 as a conservation measure for the endangered Iberian lynx. A meta-population of 94 individuals of Iberian lynx currently inhabit DNP and the surrounding areas [17]. The remaining community of carnivores is comprised by red fox (*Vulpes vulpes*), Eurasian badger (*Meles meles*),

Eurasian otter (*Lutra lutra*), polecat (*Mustela putorius*), European genet (*Genetta genetta*), Egyptian mongoose (*Herpestes ichneumon*), and occasionally, wild cat (*Felis silvestris silvestris*), whose presence is very scarce, probably due to the presence of a larger predator such as the Iberian lynx [23]. Furthermore, domestic carnivores including stray cats (*Felis silvestris catus*) and dogs (*Canis lupus familiaris*) are also occasionally present throughout DNP, although a population control plan of stray dogs and cats has been carried out in DNP since 2007 [24].

Finally, population control (by culling) of the wild ungulate population is practiced exclusively by park rangers as part of the park management scheme, and it is also used to carry out a health-monitoring program [14].

2.2. Animal Sampling

From October–January of 2005 to 2018 (sampling seasons 2005–2006 to 2017–2018), 423 red deer, 452 fallow deer and 698 wild boar were randomly (in terms of sex, age and health status) sampled in the population control context performed by park rangers and necropsied as part of the DNP health-monitoring program (approved by the Research Commission of DNP in accordance with management rules established by the Autonomous Government of Andalusia). Table S1 displays the sample size by species, sampling site and season, as well as the seroprevalences found. For each individual, the geographical location of the sighting was recorded through a portable GPS (Garmin Ltd., Olathe, KS, USA).

The sampling was performed according to European (EC Directive 86/609/EEC; [25]) and Spanish laws (RD 223/1988; [26]), current guidelines for the ethical use of animals in research [27], the Animal Experiment Committee of Castilla-La Mancha University and the Spanish Ethics Committee (PR-2015-03-08). Necropsies and sample collection were undertaken in the field by qualified veterinarians. During the examination, blood samples were collected into sterile plastic tubes (Vacutainer®, Becton-Dickinson, NJ, USA) from the heart, thoracic cavity, or preferably by endocranial venous sinuses puncture [28].

2.3. Serological Testing

Sera were obtained after centrifugation at 40× *g* for 5 min and stored at −20 °C until assayed for antibodies. Antibodies to *T. gondii* were tested using the modified agglutination test (MAT) as previously described [29]. This technique has been employed broadly for the diagnosis of antibodies against *T. gondii* in both domestic and wildlife species [1]. Two recent large studies in wild pigs and white-tailed deer in the USA added evidence for the validity of serological analysis by MAT in those species since viable *T. gondii* was isolated from a large number of seropositive animals and the rate of isolating viable parasites was positively associated with MAT titers in those studies [30–33]. Each serum sample was tested at 1:25 and 1:50 dilutions, including positive and negative controls in each test. Sera with a titer of 1:25 or higher were considered positive and those with doubtful or positive results were re-tested [12,34,35].

2.4. Data Collection

2.4.1. Individual Factors

The sex and age of the animals were determined, classifying them into three age classes on the basis of dentition eruption patterns [36]: calves (<1-year-old), juveniles (1–2 years) and adults (≥3 years) for deer species, and piglets (<6 months), juveniles (0.5–2 years) and adults (>2 years) for wild boar.

Considering the well-known debilitating effect of tuberculosis (TB) progression on immune response [37], we assessed the potential effect of TB severity on the seroprevalence against *T. gondii*. For this purpose, the presence of concomitant tuberculosis-like lesions (TBL) was used as a proxy to infection by the *Mycobacterium tuberculosis* Complex (MTC), since it provides a relatively accurate diagnosis without the need for expensive laboratory confirmation [14,15]. The presence of TBL was recorded by macroscopic inspection of the head, thoracic and mesenteric lymph nodes as well as abdominal and thoracic organs in the laboratory (see [38] for a detailed methodology). When TBL are identified in at least two of the three anatomical compartments examined (head, thorax, and abdomen) we considered the TBL as generalized, indicative of a more severe and evolved infection [38]. According to the generalized TBL status, wild ungulates were grouped in two classes: those without TBL or showing localized TBL in a single anatomical compartment, and a second class including animals with presence of generalized TBL.

2.4.2. Environmental Factors

As for environmental information, several variables were included in our analysis to assess their effect on STG because of their importance to ungulate behavior, distribution, and transmission of pathogens in DNP and South Spain [15,21,39]. A grid of one hectare of surface was built, generating territorial units in which the proportional cover of dense scrub, low-clear shrubland, herbaceous grassland, woodland, bare land and watercourse vegetation were calculated for each territorial unit (see [40]). This grid was merged with the geographical location of the animals through a point sampling tool with QGIS version 2.12.1 [41]. Landcover data was obtained from Andalusia Environmental Information [42].

Given the effects reportedly associated with urban areas [43], the coast [44] and surface water on the infection risk of *T. gondii* [45], the effect of the nearest location of animals to these areas was assessed. For that purpose, we calculated the straight-line distance (m) from the exact location of each animal sampled to the nearest: urban area (DURB), small human settlements (DHS), coast line (DCOAST), water point (DWAT) and marsh-shrub ecotone (DE) (see [14,21,46]).

2.4.3. Population Factors

To estimate the population density of wild ungulates we applied distance sampling methodology [47]. Every year during September, and two hours before sunset, we sampled twice 7 line transects of 10–15 km each one, distributed throughout the study area. Additionally, for wild boar we repeated the transect one hour after sunset in order to increase the sample size. We carried out the surveys during September because it is the month of maximum detectability for these species [48], and to obtain density results just before the health-monitoring program. Moreover, during September the marshland was dry, and it allowed us to sample all the habitats in DNP. The surveys were carried out from a vehicle (average speed was 10 km/h), and the perpendicular distance between animals and transect was recorded with a telemeter (Garmin Ltd., Olathe, KS, USA). The analysis were carried out using Distance Sampling 6.2 software [49] by considering stratification. We defined three strata according to its abundance and visibility: shrubland, marshland and ecotone. The data of all the years (2005–2018) were considered to estimate a detection function for each stratum, and we considered the data of each strata, sampling season and livestock management area to estimate the encounter rate and mean group size. Data were right-truncated when the probability of detection was lower than 0.15 [47]. Half-normal, uniform and hazard rate models for the detection function were fitted against the data using cosine, hermite polynomial, and simple polynomial adjustment terms, fitted sequentially. The selection of the best model was based on the Akaike's Information Criterion (AIC) [50].

The abundance of the diverse community of carnivores from DNP was monitored by means of track surveys along prefixed transects on sandy substrate according monitoring team program (ESPN-EBD-CSIC). Tracks left on moist sand over a 24 h period were tracked in transects of 1.5 m width and 2 km length, from dawn to midday and were expressed as Kilometric Abundance Index (KAI) of footprints. The surveys include 12 different transects distributed across the DNP which were repeated during three consecutive days, being cleaned daily.

As for livestock, we calculated the cattle and horse stocks per square kilometer for each sampling site and season.

2.4.4. Stochastic Factors

Meteorological information (average rainfall and temperature) was collected from the meteorology station located at RBD for each sampling season [51]. In Mediterranean environments, rainfall and temperature have potential relevance to the dynamics of ungulate populations, as well as effects on the susceptibility or exposure to pathogens [15]. Specifically for *T. gondii*, both factors are key for the survival of oocyst in the environment [52]. Therefore, they were considered here for their potential effect in *T. gondii* epidemiology.

2.5. Risk Factor Analysis

Initially, collinearity between environmental and population variables was explored [53]. Given the high collinearity observed between environmental variables and with the purpose of simplifying the environmental information, a principal component analysis (PCA) was performed, obtaining two uncorrelated environmental factors: closed habitats, in which dense scrub and woodlands predominates, and watercourse habitats in which watercourse vegetation predominates.

Generalized linear mixed models (GzLMMs; binomial family) were used to assess the effect of the range of explanatory variables on the individual serological status against *T. gondii* (negative/positive). The statistical differences in STG among sampling areas (CR, SO, RBD, PU and MA) were evaluated in a first exploratory approach, the purpose of which was showing spatial differences in the serological status against *T. gondii*. A GzLMM for each species (red deer, fallow deer, and wild boar) was designed. In these models, serological status against *T. gondii* was the response variable; the sex, age class, and the sampling area were the explanatory variables. The sampling season and month were fitted in the model as random-effect factors.

Concerning the final model, it included sampling area and season as random-effect factors, since the main aim of this study was to generalize the effect of the variables included on the serological status against *T. gondii* regardless of the sampling area. Models were also performed separately for each species (red deer, fallow deer, and wild boar). The explanatory variables included individual, environmental, population and stochastic factors. Individual factors encompassed sex, age class, and tuberculosis status. Environmental factors comprised DWAT, DE, DCOAST, DHS, DURB, closed habitats, and watercourse habitats. Regarding populational factors, the population densities of wild (fallow deer, red deer, and wild boar) and domestic ungulates (cattle and horses), as well as the abundance (KAI of footprints) of wild carnivores (all together genet, Eurasian badger, red fox, and Egyptian mongoose, and separately, the abundance of Iberian lynx) were included. Finally, the stochastic factors were the previous seasons' rainfall and temperature. The two-way interactions between individual-stochastic factors separately (sex-age, and previous season's rainfall-temperature) and all together (sex-rainfall and age-rainfall), as well as between population-individual factors (density-age), and population-stochastic factors (previous season´s rainfall-density) were also included in the models. For the GzLMMs, a binomial error and a logit link function were used. Stepwise selection processes for the final models were performed on the basis of the AIC [50] (Table S2). Furthermore, the assumptions of binomial GzLMMs were met in all the best models selected [53]. The predicted probabilities of serological response to *T. gondii* obtained from these models were used to represent the results. Finally, cross-correlations and autocorrelations between STG and its predicted response probability between the different species were carried out to explore similarities of temporal patterns [54].

The statistical analyses were performed using R-studio software version 4.0.2 [55]. All models were performed using the R package glmer [56]. Significant *p*-values were set at 0.05.

3. Results

3.1. General

The STG (MAT ≥ 1:25; % ± confidence interval (CI) 95%) in wild boar was 39.0 ± 3.3 (n = 698), followed by red deer 30.7 ± 4.4 (n = 423), and fallow deer 29.7 ± 4.2 (n = 452). Among the seropositive animals, titers of 1:25 were detected in 34.5% wild boar, 27.7% red deer, and 55.2% fallow deer, whereas titers ≥ 1:50 were found in 72.3% red deer, 65.5% wild boar, and 44.8% fallow deer. We observed increasing age trends in STG in all wild ungulate species, except for wild boar females (Figure 2a) since, interestingly, piglets already showed high STG. With respect to gender, males tended to present higher STG than females in deer species (32.1–28.8% and 32–26.4% for red deer and fallow deer, respectively), whereas the opposite was observed in wild boar (STG = 26.2% for males, STG = 29.7% for females; Figure 2a; see statistical comparisons below).

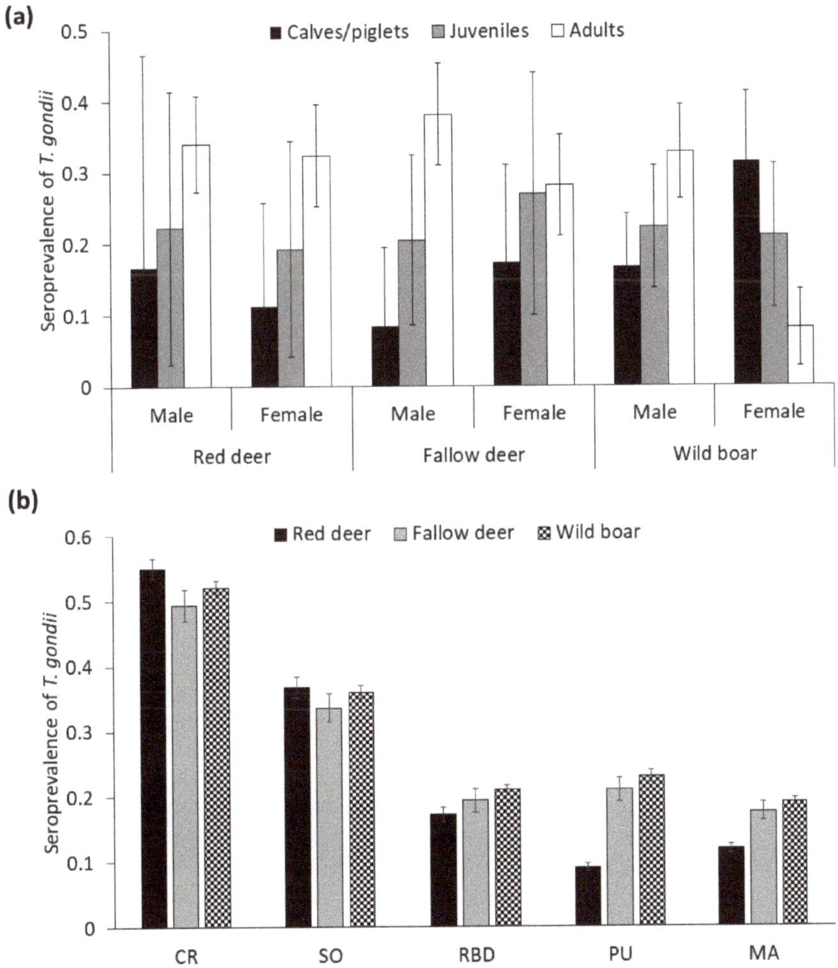

Figure 2. (**a**) Seroprevalence (±CI 95%) of *Toxoplasma gondii* depending on age class and sex in red deer, fallow deer and wild boar (**b**) Seroprevalence (±CI 95%) of *T. gondii* obtained from selected generalized linear mixed models (GzLMMs) for the species studied depending on the sampling area, from north to south areas (see Figure 1 for a map of the areas with their full names).

Contrasted STG were apparent among areas, which was consistent across species. In this sense, seroprevalence decreased from north to south, more markedly in red deer (Figure 2b). The temporal evolution of STG, and trends in the estimated density/abundance of each different species are summarized in Figure 3a,b, respectively. In this regard, the STG exhibited strong annual fluctuations, mostly in deer species (Figure 3a). Actually, it is noteworthy the significant decrease of STG in these species since the season 2013–2014. No autocorrelations or cross-correlations were observed.

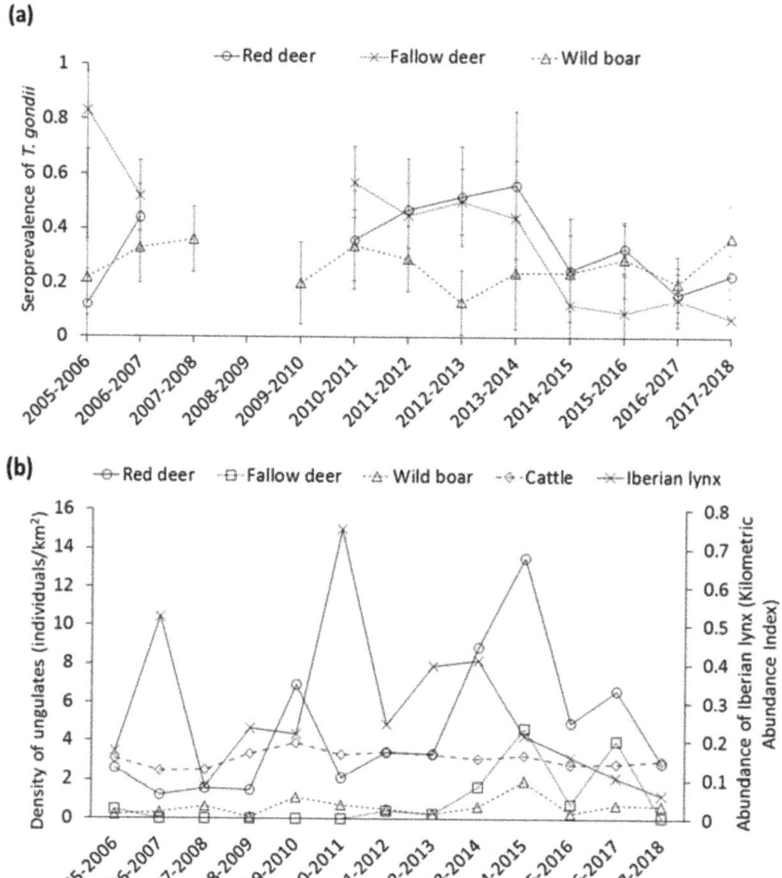

Figure 3. Temporal trend of the (**a**) seroprevalence of *Toxoplasma gondii* (±CI 95%), and (**b**) population density of red deer, fallow deer, wild boar, and cattle (individuals/km^2), and Kilometric Abundance Index of Iberian lynx.

3.2. Factors Determining the Seroprevalence of *T. gondii*

There were statistically significant differences in the STG between sampling areas for all the wild ungulates species (red deer, $F = 13.4$, df = 410, $p \leq 0.01$; fallow deer, $F = 4.5$, df = 436, $p \leq 0.01$; and wild boar $F = 10.3$, df = 682, $p \leq 0.01$), confirming the north to south spatial decreasing gradient (Figure 2b).

The results of the GzLMMs on the status against *T. gondii*, incorporating broader environmental and populational information are shown in Table 1. The conditional R^2 obtained from these models were 0.37, 0.53 and 0.25, for red deer, fallow deer, and wild boar, respectively.

Table 1. Results of the GzLMMs of risk factors associated with seroprevalence of *Toxoplasma gondii* in each species studied related to sex, age class, straight-line distance to the nearest coastline (DCOAST), previous season´s rainfall and temperature, annual density of wild ungulates and livestock (horses and cattle), abundance of carnivores and Iberian lynx (KAI), presence of generalized tuberculosis-like lesions (TBL), cover level of closed habitat, straight-line distance to the nearest small human settlements (DHS), and two-ways interactions among them. The model was fitted using sampling season and sampling area as random factors. Parameter estimates for the level of fixed factors were calculated using a reference value of 0 for the male level in the variable sex, calves, and piglets for the variable age in deer species and wild boar, respectively, and negative for the variable presence of generalized TBL.

Variables	Red Deer			Fallow Deer			Wild Boar		
	F df (x,y)	Estimate ± SD	*p*	*F* df (x,y)	Estimate ± SD	*p*	*F* df (x,y)	Estimate ± SD	*p*
Sex [1]	0.44 (1, 420)	Female: −3.01 ± 0.99	<0.01				0.22 (1, 695)	Female: 0.89 ± 0.38	0.03
Age [2]	1.46 (2, 420)	Juveniles: 0.01 ± 0.81 Adults: 1.03 ± 0.70	0.04				2.86 (2, 695)	Juveniles: −3.76 ± 1.24 Adults: 0.85 ± 0.76	<0.01
Presence of generalized TBL [3]	2.95 (1, 420)	Positive: 0.59 ± 0.31	0.05	14.47 (1, 446)	Positive: 1.86 ± 0.48	<0.01			
DCOAST	25.19 (1, 420)	0.0002 ± 0.00002	<0.01	5.69 (1, 446)	0.00007 ± 0.00003	0.01	26.70 (1, 695)	0.0001 ± 0.00002	<0.01
DE	17.67 (1, 420)	0.0002 ± 0.00009	0.07						
DHS							2.93 (1, 695)	−0.0001 ± 0.00005	<0.01
Closed habitat				5.75 (1, 446)	−0.31 ± 0.15	0.04			
Previous season´s rainfall	3.24 (1, 420)	0.01 ± 0.001	<0.01	0.07 (1, 446)	−0.05 ± 0.01	<0.01	2.40 (1, 695)	−0.0006 ± 0.001	0.15
Previous season´s temperature	3.23 (1, 420)	−0.68 ± 0.32	0.04	3.88 (1, 446)	−1.61 ± 0.57	<0.01			
Red deer density	0.43 (1, 420)	−0.11 ± 0.04	<0.01				1.27 (1, 695)	−0.05 ± 0.03	0.07
Fallow deer density	0.43 (1, 420)	0.11 ± 0.06	0.04	6.03 (1, 446)	−0.42 ± 0.18	0.02	7.07 (1, 695)	−0.16 ± 0.07	0.02

Table 1. *Cont.*

Variables	Red Deer			Fallow Deer			Wild Boar		
	F df (x,y)	Estimate ± SD	*p*	*F* df (x,y)	Estimate ± SD	*p*	*F* df (x,y)	Estimate ± SD	*p*
Wild boar density							4.1 (1, 695)	0.34 ± 0.16	0.04
Horse density	1.14 (1, 420)	0.08 ± 0.05	0.10						
Iberian lynx abundance				5.81 (1, 446)	3.56 ± 1.60	0.03	0 (1, 695)	−0.86 ± 0.45	0.06
Carnivores abundance	13.88 (1, 420)	0.35 ± 0.09	<0.01						
Sex * Rainfall	10.18 (1, 420)	Rainfall * Female: −0.01 ± 0.002	<0.01						
Sex * Age							2.34 (2, 695)	Female * juveniles: −0.97 ± 0.58 Female * ≥ adults: −1.01 ± 0.45	0.07
Temperature * Rainfall				1.95 (1, 446)	0.003 ± 0.0005	<0.01			
Rainfall * Age [2]							7.19 (2, 695)	Rainfall * juveniles: 0.01 ± 0.002 Rainfall * adults: 0.0003 ± 0.001	<0.01

[1] Reference value for sex: male, [2] Reference value for age: calves (deer species) or piglets (wild boar), [3] Reference value for presence of generalized tuberculosis-like lesions (TBL): negative, "*" represents interactions among explanatory variables.

3.2.1. Individual Factors

The sex and age classes were statistically significant factors in the models on red deer and wild boar. However, no sex or age-related differences were found in fallow deer. Regarding red deer, females had lower STG than males, and irrespective of sex, the pattern increased with the age. Concerning wild boar, different sex-related age patterns were shown, increasing for males but not for females (the sex by age interaction was marginally significant).

Regarding TB status, the prevalence of TBL (% ± CI 95%) for wild boar, red deer and fallow deer were 77.4 ± 3.1, 42.5 ± 4.7, and 16.4 ± 3.7, respectively. Wild boar had the highest prevalence of generalized TBL (% ± CI 95%; 27.73 ± 3.5), followed by red deer (17.7 ± 4.2) and fallow deer (8.11 ± 2.9). The STG was higher in red deer and fallow deer presenting generalized TBL (Figure 4) compared to generalized TBL-free individuals (TBL-free plus not generalized TBL positive). As for the wild boar, a complementary model was performed with the purpose of exploring the effect of the presence of TBL (positive or negative), since no effect of the presence of generalized TBL was observed. In this model, wild boar presenting TBL showed higher STG than negative individuals ($F = 8.96$, df = 695, $p = 0.05$).

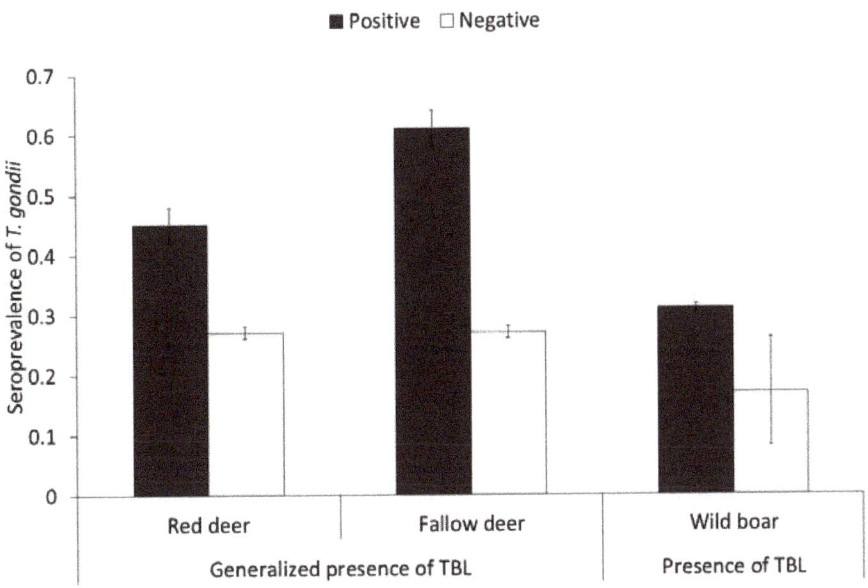

Figure 4. Seroprevalence (±CI 95%) of *Toxoplasma gondii* depending on the tuberculosis status in red deer and fallow deer (interpreted as positive animals with generalized presence of tuberculosis-like lesions (TBL), and in wild boar (interpreted as positive animals with presence of TBL).

3.2.2. Environmental Factors

The further to the coastline, the higher the STG was (see e.g., Figure 5a for red deer) in all the species. Moreover, the closer to small human settlements, the higher the STG for wild boar was (Figure 5b). The increased availability of closed habitat significantly associated with lower STG in fallow deer (Figure 5c).

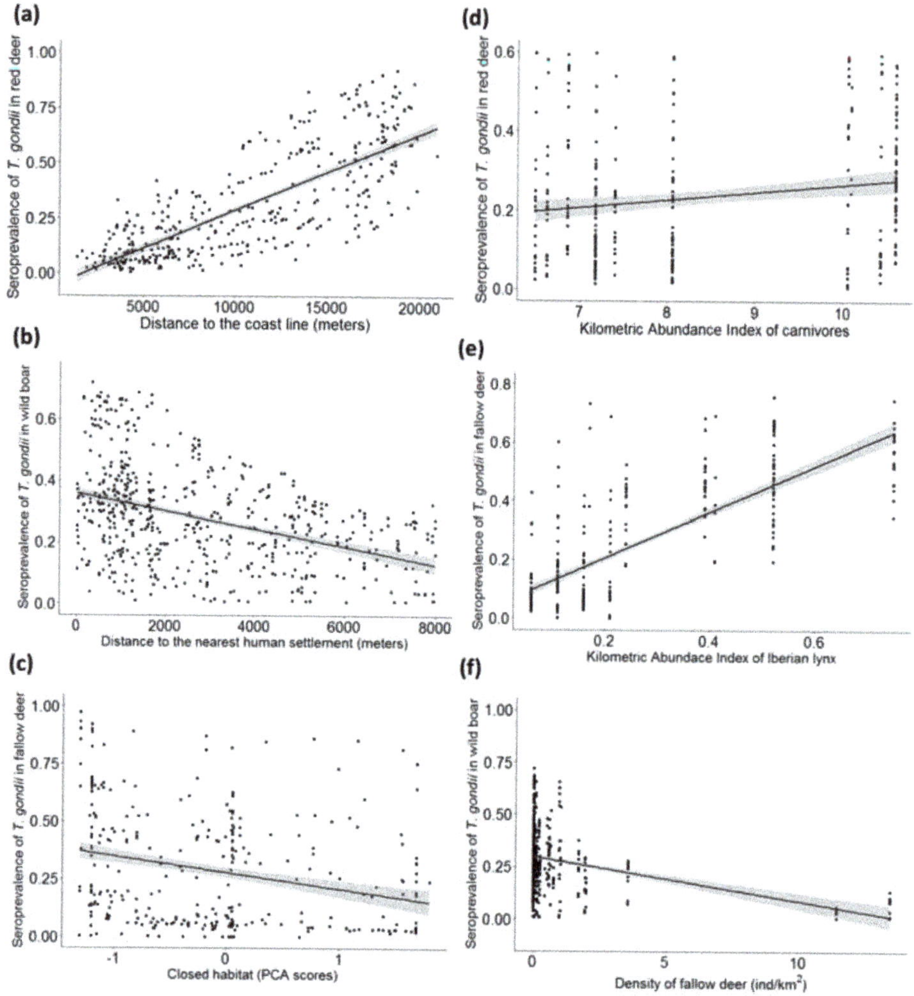

Figure 5. Seroprevalence (±CI 95%, represented by the shaded band) of *Toxoplasma gondii* obtained from selected generalized linear mixed models (GzLMMs) in (**a**) red deer depending on the distance to the coast line (m), (**b**) wild boar depending on the distance to the nearest human settlement (m), (**c**) fallow deer depending on the cover level of closed habitats, measured according to the principal component analysis (PCA) scores from axis 1, (**d**) red deer depending on the Kilometric Abundance Index of carnivores species (KAI), (**e**) fallow deer depending on the Kilometric Abundance Index of Iberian lynx (KAI), and (**f**) wild boar depending on the density of fallow deer (individuals/km^2).

3.2.3. Population Factors

The abundance of carnivores significantly and positively associated with the exposure to *T. gondii* in red deer (Figure 5d), and similarly, the Iberian lynx abundance positively associated with the seropositivity to this parasite in fallow deer (Figure 5e). The fallow deer density negatively associated with the STG in wild boar (Figure 5f) and fallow deer, but positively in the case of red deer. As for red deer, a negative association was found between STG and density. In contrast, wild boar showed higher STG at higher densities.

3.2.4. Stochastic Factors

To represent our results, and considering the mean values obtained, we established the following categories of rainfall and temperature for displaying results: low rainfall (≤521.10 mm), high rainfall (>521.10 mm), low temperature (≤17.5 °C) and high temperature (>17.5 °C). Lower annual temperature was associated with higher STG in red deer. Furthermore, higher annual rainfall was associated with higher seropositivity to *T. gondii* in red deer. Regarding fallow deer, the interaction between rainfall and temperature was significant: overall, there was a trend to higher STG in cold years, and this pattern was more marked in dry years (see Figure 6a). Rainy years were statistically associated with higher STG in male red deer, but not in females (significant rainfall by sex interaction, Figure 6b). Concerning the wild boar, the rainfall was positively associated with the STG in juveniles, but this effect was not shown in other age classes (significant annual rainfall by age interaction, Figure 6c).

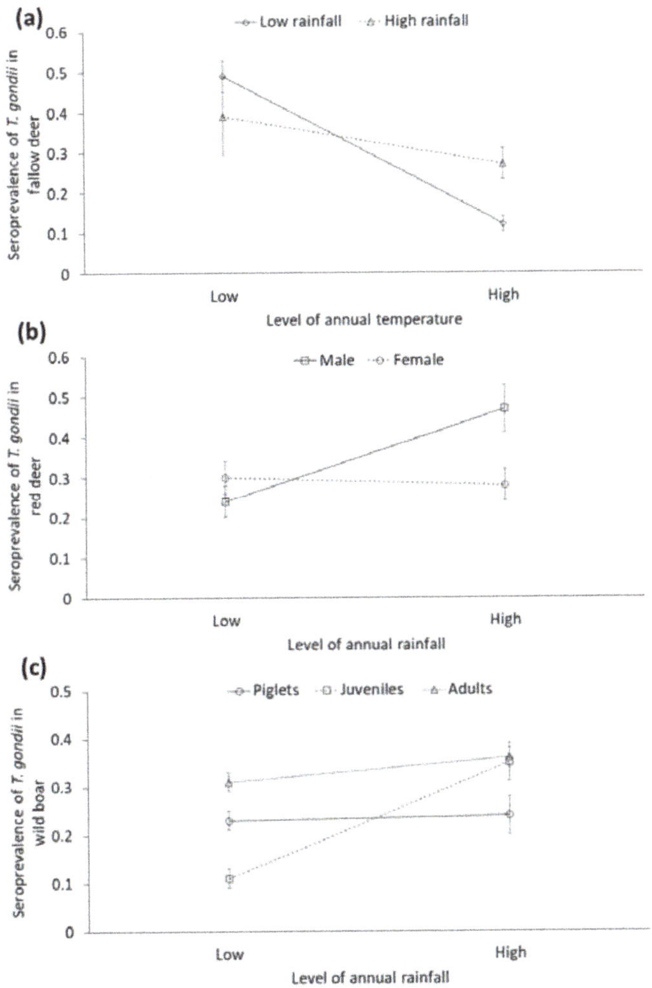

Figure 6. Seroprevalence (±CI 95%) of *Toxoplasma gondii* obtained from selected generalized linear mixed models (GzLMMs) in (**a**) fallow deer depending on the interaction between annual rainfall (mm) and temperature (°C), (**b**) red deer depending on the interaction between annual rainfall (mm) and sex, and (**c**) wild boar depending on the interaction between annual rainfall (mm) and age.

4. Discussion

4.1. General Patterns of the Seroprevalence of *T. gondii*

The STG reported in the present study oscillated considerably (from 29.7 to 39%) between the three species tested and sharing the same environment. This may be caused by differences in the susceptibility, the feeding behavior, or the habitat use of those species determining the exposure [4,52]. The seroprevalence detected in wild boar (39%) concurs with studies conducted in Europe [3,57–59]. However, most studies from European countries, STG in wild boar ranged from 6 to 25% [6,60,61]. In this regard, trophic relationships by predation and/or scavenging of a wide range of warm-blooded animals of the DNP may operate.

Concerning deer, the STG obtained (30.7% and 29.7%, for red deer and fallow deer, respectively) are also in accordance with those reported in the literature over Europe in general [62,63] and Spain in particular [8,34], ranging from 10.5 to 48%. Specifically for fallow deer, STG (29.7%) were among the highest reported in European studies that were mainly focused on Spain [8,34,63]. It was only exceeded by the rates obtained by Calero-Bernal et al. [64] in south-central Spain (reaching the 48%). The higher rate of movement between areas reported for fallow deer in DNP may imply higher exposure to *T. gondii*, explaining the high STG observed in this species [39].

Overall, the wild ungulate host community of DNP showed higher STG compared with those reported in the literature of the European and Iberian contexts [34,61,62]. Mechanisms determining seroprevalence in different host species of the studied community are related to the life cycle of *T. gondii*, which involves both an environmental and a trophic transmission route (i.e., trophic relationships among potential hosts of the community; [65]). Terrestrial herbivores should have the lowest *T. gondii* exposure, only through the ingestion of oocyst-contaminated vegetation, soil and/or drinking water. In DNP, the environmental presence of oocyst excreted by felids may be playing a major role (see below). The high biodiversity inhabiting DNP, which provides a wide range of hosts and ecological/epidemiological niches, and the optimal climatic conditions for the survival of the oocysts may favor the spread of the parasite in the DNP host community.

The specific role of the different factors in a long-term perspective is also detailed further in the discussion. Interestingly, while the STG exhibited strong annual fluctuations, mostly in deer species, it was more stable in wild boar. In populations from The Netherlands, seroprevalence in wild boar similarly established at around 35% [66]. Whereas authors stated that the actual mechanisms behind the stabilization requires further investigation, an epidemiological SIS-model that included a reversion to susceptible after infection (with loss of antibodies that may have been preceded by a loss of tissue cysts), fitted the data much better.

The north to south spatial gradient observed is similar to that exhibited by the prevalence of other shared pathogens tested in the wild ungulate community of DNP in previous studies [14,67]. This pattern may relate to spatial variation in the contamination of the environment by *T. gondii* oocyst. The main large human settlements around DNP are concentrated in the northern part of the park, with a subsequent higher presence of peri-domestic cat populations [68], which may contaminate the environment with oocysts. Iberian lynx populations also show a north to south decreasing pattern in DNP [69], contributing to a lesser extent to this contamination. *T. gondii* oocysts were found in feces of 17% of cats sharing a habitat with Iberian lynx [19]. In this regard, feral cats are the more likely reservoir host of parasites affecting the Iberian lynx and wildlife species in general, especially in areas where feral cats are abundant and widespread such as DNP surroundings [70,71].

4.2. Individual Factors

In wild boar, overall, females had significantly higher STG than males. This result is in accordance with previous studies in this species [3,57–59]. However, the age pattern observed was opposite to that of males. Several authors have reported that no statistically significant effect of age on STG in wild boar was observed [3,72,73], whereas only one study found a significantly higher prevalence in adult wild boar [59]. Nevertheless, we must consider sex by age pattern to understand the differences. The increased exposure to *T. gondii* through life, together with the high persistence of antibodies against *T. gondii*, could explain the age pattern found in the STG in males. Even so, in females, the decline in the STG rather than indicating a decrease of exposure to the parasite, may be indicative of a subtle equilibrium of chronic infection and reduced specific humoral response that is not detected. Ecological and evolutive aspects determining differences in exposure may be behind this pattern. However, further research is required. Finally, piglets exhibiting high seroprevalences could be explained by maternal-derived antibodies, whose titers depend on those shown by sows, according to previous studies [10].

As for the red deer, the STG was significantly higher in adult individuals, which has been previously reported in many studies on *T. gondii* [8,57,58,73]. An apparent similar trend, but not significant, was observed in fallow deer (Figure 2a). The increased exposure to *T. gondii* along life together with the high persistence of antibodies against *T. gondii* could explain the age pattern found in deer species.

Concerning concomitant TB infection, overall, the positive TB status of the animals significantly associated with STG in all the species studied. For deer species, the generalized presence of TBL was relevant, as well as the presence of TBL for wild boar. There are several studies on TB-*T. gondii* co-infection in humans [74,75] but not in animals. The relationship between TB and STG observed may be mediated by exposure over time (age-related) and environment. In the latter case, the conditions favoring the persistence of MTC and *T. gondii* oocyst in the environment are similar (see Sections 4.3 and 4.5).

4.3. *Environmental Factors*

The distance to the coastline significantly positively associated with STG for all wild ungulates. Despite the usually reported contamination of seawater with *T. gondii* [44], less favorable conditions to the survival of oocysts may occur in the surroundings of the coastline, according to previous studies developed in northern Spain [4]. These conditions are mainly the high temperatures reached in the sandy soils which favor the desiccation of the oocysts. The availability of closed habitat (more covered by vegetation) negatively associated with STG in fallow deer (see Figure 5c). This species typically uses and occupies meadows in the park, and individuals sampled in more densely covered areas may have experienced lower exposure to *T. gondii*. By contrast, individuals of the other ungulate species combine the use of both types of habitats which may determine the absence of this effect.

The closer to small human settlements the higher the STG was for wild boar (see Figure 5b). This result is in accordance with previous studies where human settlements have become areas of epidemiological relevance for *T. gondii* infection, mainly mediated by the presence of peri-domestic species [4,45]. There are several human dwellings inside the park, around which peri-domestic cats could settle and consequently contaminate the surroundings of these areas with oocysts. Moreover, wild boar could become infected through the consumption of food scraps from garbage located in these small settlements.

4.4. *Population Factors*

The abundances of Iberian lynx and other carnivores were statistically positively associated with STG in deer species. The presence and abundance of felids have been considered a relevant risk factor associated with *T. gondii* in livestock and wildlife worldwide [1,7–10]. It has been reported that Iberian lynx could prey on fallow deer and less frequently on juveniles of red deer during seasons of rabbit´s scarceness in DNP, especially in winter and autumn, reaching 5–10% of the biomass in the Iberian lynx diet [76]. This leads to increased environmental contamination with *T. gondii* in areas with presence of wild ungulates. Furthermore, *T. gondii* infected red deer carcasses could pose a potential risk of *T. gondii* infection for carnivores species via scavenging and may therefore play a role as an amplifier of infection in the community [63]. This allows *T. gondii* to finish its life cycle, perpetuating its maintenance in the

DNP host community. Little is known about the abundance of small mammals, as well as their STG in DNP. Further studies should focus on investigating their role in the maintenance and spread of this parasite in DNP.

STG significantly increases with density in wild boar. Several authors have described *T. gondii* as a density-dependent parasite for swine [3,45]. Density, together with ecological and behavioral factors typical of this species, could determine increased exposure by wild boar. Scavenging (including cannibalism) may be increased in high density situations when resources are scarce. In the dry season, both the availability of carcasses and exacerbated cannibalism behavior [77], but also ingestion of rodents and birds occur [78]. The negative relationship found both in red deer and fallow deer between density and the risk to test positive can be explained by high recruitment of susceptible individuals (non-infected offspring) associated with high density years. Moreover, unlike for wild boar, this negative association indicates that no density-dependent effect in *T. gondii* infection occurs in deer species. These species become infected only through the ingestion of water or food contaminated with sporulated *T. gondii* oocysts, so no direct transmission route exists as in the case of wild boar, which possesses scavenging habits. However, the positive relationship showed between STG in red deer and fallow deer density may be mediated by an increased susceptibility and/or exposure at high densities of ungulates due to the competition for scarce resources, but the exact mechanism deserves further research.

4.5. Stochastic Factors

Temperature was a significant factor for red deer, displaying the lower STG during the following seasons to the warmest ones. Furthermore, rainfall significantly interacted with temperature to explain STG in fallow deer, so that the effect of the previous season´s temperature on STG was more marked when the previous season was dry. Previous driest and warmest seasons markedly associated with a lower STG (see Figure 6a). Drought together with warm temperatures leads to higher rates of evaporation and the subsequent desiccation, limiting the survival and sporulation of the oocysts in the environment. As consequence, the exposure of herbivores to infective *T. gondii* oocyst decreases [34,35,52].

Additionally, the interaction between rainfall and individual factors was significantly associated with the STG in red deer and wild boar. Concerning red deer, higher previous annual rainfall was related to higher STG in stags, but not in females. This effect of rainy years on the prevalence of pathogens exhibited by males with respect to females have been shown in previous studies on TB in DNP, and may relate to increased exposure and/or susceptibility mediated by sexual behavior and life history traits [14]. An immunosuppressant effect of the intense rut that typically occurs in rainy years has been suggested. Intense rut implies greatest investments by red deer stags in terms of reproductive effort (testosterone metabolite levels and sexual signals), and the conflict between the immune response and the reproductive effort this species is well known [79,80]. *T. gondii* tissue cysts in many organs, including viscera, are believed to persist for the lifetime of the host. In addition, deer rutting typically occurs in the ecotone, which provides excellent wet conditions for oocysts to persist. Despite the same risk derived from the reproductive efforts during the rut exist for fallow deer, no sex-dependent effect of rainfall was observed in STG in this species, which is not surprising since rutting in fallow deer takes place later in Autumn, when rainfall conditions normally are less determinant.

As for wild boar, the positive effect of the rainfall on STG was more markedly in juveniles than adults and piglets. The early dispersal behavior of young males from the natal area may leads to higher exposure to *T. gondii* [81]. This, together with the increased survival of oocysts would give rise to higher infection rates in this age group [35,52].

5. Conclusions

This study provides evidence that factors behind the risk of *T. gondii* infection in wild boar, red deer and fallow deer are related to environmental and trophic transmission routes, so as to individual, population and species characteristics. We provided evidence for most of these relationships (e.g., climate or population mediated) and trends. Concomitant pattern among species, indicated that overall, drivers of risk also operated at the community level. However, this research raised several questions that deserve further research. Approximately one-third of the human world's population is chronically infected while seroprevalence tends to decrease since the early 1960s in many countries [82]. As this decline in seroprevalence leads to loss of immunity, it becomes more relevant for the identification of the epidemiological role of wild host and the understanding of the epidemiology and ecology of *T. gondii* infection in wild host communities and at their interfaces with livestock and human. Thus, game meat, in particular venison, consumption should not be neglected as a public health risk for humans, with the subsequent impact to the public health [31]. For these purposes, addressing host population, community and environmental factors at broad temporal scale is key.

Supplementary Materials: The following are available online at http://www.mdpi.com/2076-2615/10/12/2349/s1, Table S1: Sample size and seroprevalence of *Toxoplasma gondii* by species, season and sampling site in wild ungulates; Table S2: Summary of the stepwise model selection procedure, based on the AIC, used to explain the serological status against *Toxoplasma gondii*.

Author Contributions: Conceptualization, J.V., I.G.-B. and P.A.; Methodology, J.V., I.G.-B., P.A., and F.C.; Validation, J.V., I.G.-B., P.A., and F.C.; Formal Analysis, J.V., P.B., F.C. and P.P.; Investigation, P.B., P.P., F.C., S.J.-R., and D.C.-T., Resources, J.V. and P.A.; Data Curation, P.B., F.C. and P.P.; Writing—Original Draft Preparation, P.B. and J.V.; Writing—Review and Editing, P.B., J.V., P.A., P.P., F.C., S.J.-R., S.A., J.P.D. and D.C.-T.; Visualization, P.B.; Supervision, J.V., I.G.-B. and P.A.; Project Administration, J.V. and I.G.-B.; Funding Acquisition, J.V. and P.A. All authors have read and agreed to the published version of the manuscript.

Funding: This research was funded by the Ministerio de Economía y Competitividad (MINECO; AEI/FEDER, UE; AGL2016-76358-R). S.J. is co-supported by the UCLM and the European Social Fund (2018/12504). P.P. received support from the MINECO (FPU/16/00039).

Acknowledgments: The authors would like to thank the EBD-CSIC monitoring team and their colleagues at the IREC (UCLM, Spain), Doñana National Park and Doñana ICTS-RBD, which provided logistic and technical support and the data essential for carrying out this work, and to all those who participated in the fieldwork and data collection of the wild ungulates, with special thanks to the park ranger Jose Antonio Muriel, and María de los Ángeles Risalde for her kind help with this study.

Conflicts of Interest: The authors declare no conflict of interest. The funders had no role in the design of the study; in the collection, analyses, or interpretation of data; in the writing of the manuscript, or in the decision to publish the results.

References

1. Dubey, J.P. *Toxoplasmosis of Animals and Humans*, 2nd ed.; CRC Press: Boca Raton, FL, USA, 2010.
2. Hill, D.E.; Dubey, J.P. Toxoplasmosis. *US Geol. Surv. Circ.* **2014**, *1389*, 84. [CrossRef]
3. Gauss, C.; Dubey, J.; Vidal, D.; Fons, F.R.; Vicente, J.; Marco, I.; Lavín, S.; Gortazar, C.; Almeria, S.; Vidal, D. Seroprevalence of Toxoplasma gondii in wild pigs (Sus scrofa) from Spain. *Vet. Parasitol.* **2005**, *131*, 151–156. [CrossRef] [PubMed]
4. Panadero, R.; Painceira, A.; López, C.; Vázquez, L.; Paz, A.; Díaz, P.; DaCal, V.; Cienfuegos, S.; Fernández, G.; Lago, N.; et al. Seroprevalence of Toxoplasma gondii and Neospora caninum in wild and domestic ruminants sharing pastures in Galicia (Northwest Spain). *Res. Vet. Sci.* **2010**, *88*, 111–115. [CrossRef]
5. García-Bocanegra, I.; Cabezón, O.; Hernández, E.; Martínez-Cruz, M.S.; Martínez-Moreno, Á.; Martínez-Moreno, J. Toxoplasma gondii in Ruminant Species (Cattle, Sheep, and Goats) from Southern Spain. *J. Parasitol.* **2013**, *99*, 438–440. [CrossRef] [PubMed]
6. Calero-Bernal, R.; Pérez-Martín, J.E.; Reina, D.; Serrano, F.J.; Frontera, E.; Fuentes, I.; Dubey, J.P. Detection of Zoonotic Protozoa Toxoplasma gondii and Sarcocystis suihominis in Wild Boars from Spain. *Zoonoses Public Health* **2016**, *63*, 346–350. [CrossRef] [PubMed]
7. Castillo-Cuenca, J.C.; Díaz-Cao, J.M.; Martínez-Moreno, Á.; Cano-Terriza, D.; Jiménez-Ruiz, S.; Almería, S.; García-Bocanegra, I. Seroepidemiology of Toxoplasma gondii in extensively raised Iberian pigs in Spain. *Prev. Vet. Med.* **2020**, *175*, 104854. [CrossRef] [PubMed]

8. Almeria, S.; Cabezón, O.; Paniagua, J.; Cano-Terriza, D.; Jiménez-Ruiz, S.; Arenas-Montes, A.; Dubey, J.P.; García-Bocanegra, I. Toxoplasma gondii in sympatric domestic and wild ungulates in the Mediterranean ecosystem. *Parasitol. Res.* **2018**, *117*, 665–671. [CrossRef] [PubMed]
9. Jiménez-Martín, D.; García-Bocanegra, I.; Almería, S.; Castro-Scholten, S.; Dubey, J.P.; Amaro-López, M.A.; Cano-Terriza, D. Epidemiological surveillance of Toxoplasma gondii in small ruminants in southern Spain. *Prev. Vet. Med.* **2020**, *183*. [CrossRef]
10. García-Bocanegra, I.; Simon-Grifé, M.; Dubey, J.P.; Casal, J.; Martín, G.E.; Cabezón, O.; Perea, A.; Almería, S. Seroprevalence and risk factors associated with Toxoplasma gondii in domestic pigs from Spain. *Parasitol. Int.* **2010**, *59*, 421–426. [CrossRef]
11. García-Bocanegra, I.; Dubey, J.; Simon-Grifé, M.; Cabezón, O.; Casal, J.; Allepuz, A.; Napp, S.; Almeria, S. Seroprevalence and risk factors associated with Toxoplasma gondii infection in pig farms from Catalonia, north-eastern Spain. *Res. Vet. Sci.* **2010**, *89*, 85–87. [CrossRef]
12. Sobrino, R.; Cabezón, O.; Millán, J.; Pabón, M.; Arnal, M.; Luco, D.; Gortazar, C.; Dubey, J.P.; Almeria, S. Seroprevalence of Toxoplasma gondii antibodies in wild carnivores from Spain. *Vet. Parasitol.* **2007**, *148*, 187–192. [CrossRef] [PubMed]
13. McDonald, J.; Robertson, A.; Silk, M.J. Wildlife disease ecology from the individual to the population: Insights from a long-term study of a naturally infected European badger population. *J. Anim. Ecol.* **2017**, *87*, 101–112. [CrossRef] [PubMed]
14. Barroso, P.; Barasona, J.Á.; Acevedo, P.; Palencia, P.; Carro, F.; Negro, J.J.; Torres, M.J.; Gortazar, C.; Soriguer, R.C.; Vicente, J. Long-Term Determinants of Tuberculosis in the Ungulate Host Community of Doñana National Park. *Pathogens* **2020**, *9*, 445. [CrossRef] [PubMed]
15. Vicente, J.; Barasona, J.A.; Acevedo, P.; Ruiz-Fons, J.F.; Boadella, M.; Diez-Delgado, I.; Beltran-Beck, B.; González-Barrio, D.; Queirós, J.; Montoro, V.; et al. Temporal Trend of Tuberculosis in Wild Ungulates from Mediterranean Spain. *Transbound. Emerg. Dis.* **2013**, *60*, 92–103. [CrossRef] [PubMed]
16. Barroso, P.; Acevedo, P.; Vicente, J. The importance of long-term studies on wildlife diseases and their interfaces with humans and domestic animals: A review. *Transbound. Emerg. Dis.* **2020**, 1–15. [CrossRef] [PubMed]
17. Life + IBERLINCE. Life + IBERLINCE, Recuperación de la Distribución Histórica del Lince Ibérico (Lynx pardinus) en España y Portugal. (LIFE10NAT/ES/570). Available online: http://www.iberlince.eu/index.php/esp/lince-iberico-esp/ (accessed on 18 June 2020).
18. García-Bocanegra, I.; Dubey, J.; Martinez, F.; Vargas, A.; Cabezón, O.; Zorrilla, I.; Arenas, A.; Almeria, S. Factors affecting seroprevalence of Toxoplasma gondii in the endangered Iberian lynx (Lynx pardinus). *Vet. Parasitol.* **2010**, *167*, 36–42. [CrossRef]
19. Millán, J.; Candela, M.G.; Palomares, F.; Cubero, M.J.; Rodriguez, A.W.; Barral, M.; De La Fuente, J.; Almeria, S.; León-Vizcaíno, L. Disease threats to the endangered Iberian lynx (Lynx pardinus). *Vet. J.* **2009**, *182*, 114–124. [CrossRef]
20. Fernández-Delgado, C. Conservation management of a European natural area: Doñana National Park, Spain. In *Principles of Conservation Biology*, 2nd ed.; Meffe, G.K., Carroll, C.R., Eds.; Sinauer Associates: Sunderland, MA, USA, 1997; pp. 458–467.
21. Barasona, J.Á.; Latham, M.C.; Acevedo, P.; Armenteros, J.A.; Latham, A.D.M.; Gortazar, C.; Carro, F.; Soriguer, R.C.; Vicente, J. Spatiotemporal interactions between wild boar and cattle: Implications for cross-species disease transmission. *Vet. Res.* **2014**, *45*, 1–11. [CrossRef]
22. Díaz-Delgado, R.; Aragonés, D.; Afán, I.; Bustamante, J. Long-Term Monitoring of the Flooding Regime and Hydroperiod of Doñana Marshes with Landsat Time Series (1974–2014). *Remote. Sens.* **2016**, *8*, 775. [CrossRef]
23. Valverde, J.A. *Estructura de una Comunidad de Vertebrados Terrestres*; Monografías de la Estación Biológica de Doñana; CSIC: Madrid, Spain, 1967.
24. Junta de Andalucia. *Memoria de Actividades y Resultados del Espacio Natural de Doñana*; Consejeria de Medio Ambiente y Ordenación del Territorio: Andalucia, Spain, 2013.
25. EEC. Council Directive 86/609/EEC of 24 November 1986 on the approximation of laws, regulations and administrative provisions of the Member States regarding the protection of animals used for experimental and other scientific purposes. *Off. J. Eur. Communities* **1986**, *358*, 1–29.

26. MAPA. *Real Decreto 223/1988, de 14 de Marzo, Sobre Protección de los Animales Utilizados Para Experimentación y otros Fines Científicos*; Ministerio de Agricultura, Pesca y Alimentación: Madrid, Spain, 1988; pp. 8509–8512.
27. Buchanan, K.; Burt de Perera, T.; Carere, C.; Carter, T.; Hailey, A.; Hubrecht, R.; Jennings, D.; Metcalfe, N.; Pitcher, T.; Peron, F.; et al. ASAB Guidelines for the treatment of animals in behavioural research and teaching. *Anim. Behav.* **2012**, *83*, 301–309.
28. Jiménez-Ruiz, S.; Arenas-Montes, A.; Cano-Terriza, D.; Paniagua, J.; Pujols, J.; Miró, F.; Fernández-Aguilar, X.; González, M.Á.; Franco, J.J.; García-Bocanegra, I. Blood extraction method by endocranial venous sinuses puncture in hunted wild ruminants. *Eur. J. Wildl. Res.* **2016**, *62*, 775–780. [CrossRef]
29. Dubey, J.P.; Desmonts, G. Serological responses of equids fed Toxoplasma gondii oocysts. *Equine Vet. J.* **1987**, *19*, 337–339. [CrossRef]
30. Dubey, J.P.; Cerqueira-Cézar, C.K.; Murata, F.H.A.; Verma, S.K.; Kwok, O.C.H.; Pedersen, K.; Rosenthal, B.M.; Su, C. White-tailed deer (Odocoileus virginianus) are a reservoir of a diversity of Toxoplasma gondii strains in the USA and pose a risk to consumers of undercooked venison. *Parasitology* **2020**, *147*, 775–781. [CrossRef] [PubMed]
31. Dubey, J.P.; Murata, F.; Cezar, C.; Kwok, O. Epidemiologic and public health significance of Toxoplasma gondii infections in venison: 2009–2020. *J. Parasitol.* **2021**, in press.
32. Dubey, J.P.; Cerqueira-Cézar, C.K.; Murata, F.H.A.; Verma, S.K.; Kwok, O.C.H.; Pedersen, K.; Rosenthal, B.M.; Su, C. Genotyping of viable Toxoplasma gondii from the first national survey of feral swine revealed evidence for sylvatic transmission cycle, and presence of highly virulent parasite genotypes. *Parasitology* **2019**, *147*, 295–302. [CrossRef] [PubMed]
33. Dubey, J.P.; Cerqueira-Cézar, C.K.; Murata, F.H.; Kwok, O.C.; Hill, D.; Yang, Y.R.; Su, C. WITHDRAWN: All about Toxoplasma gondii infections in pigs: The past decade. *Vet. Parasitol.* **2020**, *288*, 109182. [CrossRef]
34. Gauss, C.; Dubey, J.P.; Vidal, D.; Cabezón, O.; Ruiz-Fons, F.; Vicente, J.; Marco, I.; Lavin, S.; Gortazar, C.; Almería, S. Prevalence of Toxoplasma gondii antibodies in red deer (Cervus elaphus) and other wild ruminants from Spain. *Vet. Parasitol.* **2006**, *136*, 193–200. [CrossRef]
35. Gamarra, J.; Cabezón, O.; Pabón, M.; Arnal, M.; Luco, D.; Dubey, J.; Gortázar, C.; Almeria, S. Prevalence of antibodies against Toxoplasma gondii in roe deer from Spain. *Vet. Parasitol.* **2008**, *153*, 152–156. [CrossRef]
36. Saenz de Buruaga, M.; Lucio-Calero, A.; Purroy, F.J. *Reconocimiento de Sexo y Edad en Especies Cinegéticas*, 1st ed.; Edilesa: Madrid, Spain, 2001; ISBN 9788480123716.
37. Byrne, A.; Graham, J.; McConville, J.; Milne, G.; Guelbenzu-Gonzalo, M.; McDowell, S. Liver fluke (Fasciola hepatica) co-infection with bovine tuberculosis in cattle: A prospective herd-level assessment of herd bTB risk in dairy enterprises. *Transbound. Emerg. Dis.* **2019**, *66*, 1727–1736. [CrossRef]
38. Vicente, J.; Höfle, U.; Garrido, J.; Fernández-De-Mera, I.G.; Juste, R.; Barral, M.; Gortazar, C. Wild boar and red deer display high prevalences of tuberculosis-like lesions in Spain. *Vet. Res.* **2006**, *37*, 107–119. [CrossRef] [PubMed]
39. Braza, F.; Alvarez, F. Habitat use by Red Deer and Fallow Deer in Doñana National Park. *Misc. Zool.* **1987**, *11*, 363–367.
40. Barasona, J.Á.; Mulero-Pázmány, M.; Acevedo, P.; Negro, J.J.; Torres, M.J.; Gortázar, C.; Vicente, J. Unmanned Aircraft Systems for Studying Spatial Abundance of Ungulates: Relevance to Spatial Epidemiology. *PLoS ONE* **2014**, *9*, e115608. [CrossRef] [PubMed]
41. QGIS Development Team. Quantum GIS Geographic Information System. Open Source Geospatial Foundation Project. Available online: https://qgis.org/es/site/ (accessed on 16 May 2020).
42. REDIAM Consejería de Medio Ambiente y Ordenación del Territorio. Andalucía, España: Red de Información Ambiental de Andalucía, REDIAM. Available online: http://www.juntadeandalucia.es/medioambiente/site/rediam (accessed on 20 September 2007).
43. Miller, M.; Gardner, I.; Kreuder, C.; Paradies, D.; Worcester, K.; Jessup, D.; Dodd, E.; Harris, M.; Ames, J.; Packham, A.; et al. Coastal freshwater runoff is a risk factor for Toxoplasma gondii infection of southern sea otters (Enhydra lutris nereis). *Int. J. Parasitol.* **2002**, *32*, 997–1006. [CrossRef]
44. Dubey, J.P. Toxoplasmosis—A waterborne zoonosis. *Vet. Parasitol.* **2004**, *126*, 57–72. [CrossRef]
45. Villari, S.; Vesco, G.; Petersen, E.; Crispo, A.; Buffolano, W. Risk factors for toxoplasmosis in pigs bred in Sicily, Southern Italy. *Vet. Parasitol.* **2009**, *161*, 1–8. [CrossRef]

46. Cabal, A.; Vicente, J.; Alvarez, J.; Barasona, J.A.; Boadella, M.; Dominguez, L.; Gortazar, C. Human influence and biotic homogenization drive the distribution of Escherichia coli virulence genes in natural habitats. *Microbiologyopen* **2017**, *6*. [CrossRef]
47. Buckland, S.; Anderson, D.; Burnham, K.; Laake, J.; Borchers, D.; Thomas, L. *Introduction to Distance Sampling: Estimating Abundance of Biological Populations*; Oxford University Press: Oxford, UK, 2001.
48. Acevedo, P.; Fons, F.R.; Vicente, J.; Reyes-García, A.R.; Alzaga, V.; Gortazar, C. Estimating red deer abundance in a wide range of management situations in Mediterranean habitats. *J. Zool.* **2008**, *276*, 37–47. [CrossRef]
49. Thomas, L.; Buckland, S.T.; Rexstad, E.A.; Laake, J.L.; Strindberg, S.; Hedley, S.L.; Bishop, J.R.B.; Marques, T.A.; Burnham, K.P. Distance software: Design and analysis of distance sampling surveys for estimating population size. *J. Appl. Ecol.* **2010**, *47*, 5–14. [CrossRef]
50. Akaike, H. A new look at the statistical model identification. *IEEE Trans. Autom. Control* **1974**, *19*, 716–723. [CrossRef]
51. EBD-CSIC Equipo de Seguimiento del Espacio Nacional de Doñana. Available online: http://icts.ebd.csic.es/datos-meteorologicos (accessed on 5 July 2019).
52. Dubey, J.P.; Beattie, C.P. *Toxoplasmosis of Animals and Man*; CRC Press: Boca Raton, FL, USA, 1988.
53. Zuur, A.F.; Ieno, E.N.; Walker, N.J.; Saveliev, A.A.; Smith, G.M. GLM and GAM for Count Data. In *Environmental Forum*; Springer: New York, NY, USA, 2009; Volume 21, pp. 209–243. ISBN 9780387874579.
54. Göloglu, F.; Pott, A. Results on the Crosscorrelation and Autocorrelation of Sequences. In Proceedings of the Sequences and Their Applications-SETA 2008 5th International Conference, Lexington, KY, USA, 14–18 September 2008; Springer: Berlin/Heidelberg, Germany, 2008; pp. 95–105.
55. R Core Team. *R Core Team R: A Language and Environment for Statistical Computing*; R Core Team: Vienna, Austria, 2020.
56. Bates, D.; Mächler, M.; Bolker, B.; Walker, S. Fitting linear mixed-effects models using lme4. *arXiv* **2014**, arXiv:1406.5823.
57. Jokelainen, P.; Näreaho, A.; Hälli, O.; Heinonen, M.; Sukura, A. Farmed wild boars exposed to *Toxoplasma gondii* and *Trichinella* spp. *Vet. Parasitol.* **2012**, *187*, 323–327. [CrossRef] [PubMed]
58. Reiterová, K.; Špilovská, S.; Blaňarová, L.; Derdáková, M.; Čobádiová, A.; Hisira, V. Wild boar (Sus scrofa)—Reservoir host of Toxoplasma gondii, Neospora caninum and Anaplasma phagocytophilum in Slovakia. *Acta Parasitol.* **2016**, *61*, 255–260. [CrossRef] [PubMed]
59. Fons, F.R.; Vicente, J.; Vidal, D.; Höfle, U.; Villanúa, D.; Gauss, C.; Segalés, J.; Almeria, S.; Montoro, V.; Gortazar, C.; et al. Seroprevalence of six reproductive pathogens in European wild boar (Sus scrofa) from Spain: The effect on wild boar female reproductive performance. *Theriogenology* **2006**, *65*, 731–743. [CrossRef]
60. Coelho, C.; Vieira-Pinto, M.; Faria, A.S.; Vale-Gonçalves, H.M.; Veloso, O.; Paiva-Cardoso, M.D.N.; Mesquita, J.R.; Lopes, A.P. Serological evidence of Toxoplasma gondii in hunted wild boar from Portugal. *Vet. Parasitol.* **2014**, *202*, 310–312. [CrossRef]
61. Beral, M.; Rossi, S.; Aubert, D.; Gasqui, P.; Terrier, M.-E.; Klein, F.; Villena, I.; Abrial, D.; Gilot-Fromont, E.; Richomme, C.; et al. Environmental Factors Associated with the Seroprevalence of Toxoplasma gondii in Wild Boars (Sus scrofa), France. *EcoHealth* **2012**, *9*, 303–309. [CrossRef]
62. Rocchigiani, G.; Nardoni, S.; D'Ascenzi, C.; Nicoloso, S.; Picciolli, F.; Papini, R.; Mancianti, F. Seroprevalence of Toxoplasma gondii and Neospora caninum in red deer from Central Italy. *Ann. Agric. Environ. Med.* **2016**, *23*, 699–701. [CrossRef]
63. Bártová, E.; Sedlák, K.; Pavlik, I.; Literak, I. Prevalence of Neospora caninum and Toxoplasma gondii Antibodies in Wild Ruminants from the Countryside or Captivity in the Czech Republic. *J. Parasitol.* **2007**, *93*, 1216–1218. [CrossRef]
64. Calero-Bernal, R.; Saugar, J.M.; Frontera, E.; Pérez-Martín, J.E.; Habela, M.A.; Serrano, F.J.; Reina, D.; Fuentes, I. Prevalence and genotype identification of Toxoplasma gondii in wild animals from southwestern Spain. *J. Wildl. Dis.* **2015**, *51*, 233–238. [CrossRef]
65. Wilson, A.G.; Lapen, D.R.; Mitchell, G.W.; Provencher, J.F.; Wilson, S. Interaction of diet and habitat predicts Toxoplasma gondii infection rates in wild birds at a global scale. *Glob. Ecol. Biogeogr.* **2020**, *29*, 1189–1198. [CrossRef]
66. Opsteegh, M.; Swart, A.; Fonville, M.; Dekkers, L.; Van Der Giessen, J. Age-Related Toxoplasma gondii Seroprevalence in Dutch Wild Boar Inconsistent with Lifelong Persistence of Antibodies. *PLoS ONE* **2011**, *6*, e16240. [CrossRef] [PubMed]

67. Gortazar, C.; Torres, M.J.; Vicente, J.; Acevedo, P.; Reglero, M.; De La Fuente, J.; Negro, J.J.; Aznar-Martín, J. Bovine Tuberculosis in Doñana Biosphere Reserve: The Role of Wild Ungulates as Disease Reservoirs in the Last Iberian Lynx Strongholds. *PLOS ONE* **2008**, *3*, e2776. [CrossRef] [PubMed]
68. Jokelainen, P.; Näreaho, A.; Knaapi, S.; Oksanen, A.; Rikula, U.; Sukura, A. Toxoplasma gondii in wild cervids and sheep in Finland: North-south gradient in seroprevalence. *Vet. Parasitol.* **2010**, *171*, 331–336. [CrossRef]
69. Garrote, G.; De Ayala, R.P.; Pereira, P.; Robles, F.; Guzman, N.; García, F.J.; Iglesias, M.C.; Hervás, J.; Fajardo, I.; Simón, M.; et al. Estimation of the Iberian lynx (Lynx pardinus) population in the Doñana area, SW Spain, using capture–recapture analysis of camera-trapping data. *Eur. J. Wildl. Res.* **2011**, *57*, 355–362. [CrossRef]
70. Millán, J.; Casanova, J.C.; Millán, J.; Casanova, J.C. Helminth parasites of the endangered Iberian lynx (Lynx pardinus) and sympatric carnivores. *J. Helminthol.* **2007**, *81*, 377–380. [CrossRef] [PubMed]
71. Vicente, J.; Palomares, F.; De Ibañez, A.R.; Ortiz, J.; De Ibañez, R.R. Epidemiology of Ancylostoma spp. in the endangered Iberian lynx (Lynx pardinus) in the Doñana National Park, south-west Spain. *J. Helminthol.* **2004**, *78*, 179–183. [CrossRef]
72. Dubey, J.P. The history and life cycle of Toxoplasma gondii. In *Toxoplasma Gondii*; Elsevier: Amsterdam, The Netherlands, 2020; pp. 1–19.
73. Antolová, D.; Reiterová, K.; Dubinský, P. Seroprevalence of Toxoplasma gondii in wild boars (Sus scrofa) in the Slovak Republic. *Ann. Agric. Environ. Med.* **2007**, *14*, 71–73.
74. Mashaly, M.; Nabih, N.; Fawzy, I.M.; El Henawy, A.A. Tuberculosis/toxoplasmosis co-infection in Egyptian patients: A reciprocal impact. *Asian Pac. J. Trop. Med.* **2017**, *10*, 315–319. [CrossRef]
75. Hwang, E.H.; Ahn, P.G.; Lee, D.M.; Kim, H.S. Cerebral Toxoplasmosis Combined with Disseminated Tuberculosis. *J. Korean Neurosurg. Soc.* **2012**, *51*, 316–319. [CrossRef]
76. Delibes, M. Feeding ecology of the Spanish lynx in the Coto Doñana. *Acta Theriol. (Warsz).* **1980**, *25*, 309–324. [CrossRef]
77. Carrasco-García, R.; Barroso, P.; Perez-Olivares, J.; Montoro, V.; Vicente, J. Consumption of Big Game Remains by Scavengers: A Potential Risk as Regards Disease Transmission in Central Spain. *Front. Vet. Sci.* **2018**, *5*, 4. [CrossRef] [PubMed]
78. Schley, L.; Roper, T.J. Diet of wild boar Sus scrofa in Western Europe, with particular reference to consumption of agricultural crops. *Mammal Rev.* **2003**, *33*, 43–56. [CrossRef]
79. Lochmiller, R.L.; Deerenberg, C. Trade-offs in evolutionary immunology: Just what is the cost of immunity? *Oikos* **2000**, *88*, 87–98. [CrossRef]
80. Malo, A.F.; Roldan, E.; Garde, J.; Soler, A.; Vicente, J.; Gortazar, C.; Gomendio, M. What does testosterone do for red deer males? *Proc. R. Soc. B Boil. Sci.* **2008**, *276*, 971–980. [CrossRef]
81. Mauget, R.; Campan, R.; Spitz, F.; Dardaillom, M.; Janeau, G.; Pepin, D. Synthise des connaissances actuelles sur la biologie du sanglier, perspectives de recherche. In Proceedings of the Symposium International Sanglier, Les colloques de l'INRA, Tolouse, France, 24–26 April 1984.
82. Rougier, S.; Montoya, J.G.; Peyron, F. Lifelong Persistence of Toxoplasma Cysts: A Questionable Dogma? *Trends Parasitol.* **2017**, *33*, 93–101. [CrossRef] [PubMed]

animals

MDPI

Communication

Intestinal Helminths in Wild Rodents from Native Forest and Exotic Pine Plantations (*Pinus radiata*) in Central Chile

Maira Riquelme [1], Rodrigo Salgado [1], Javier A. Simonetti [2], Carlos Landaeta-Aqueveque [3], Fernando Fredes [4] and André V. Rubio [1,*]

1 Departamento de Ciencias Biológicas Animales, Facultad de Ciencias Veterinarias y Pecuarias, Universidad de Chile, Santa Rosa 11735, La Pintana, Santiago 8820808, Chile; mairariquelme@gmail.com (M.R.); rodrigo.salgadomoya@gmail.com (R.S.)
2 Departamento de Ciencias Ecológicas, Facultad de Ciencias, Universidad de Chile, Casilla 653, Santiago 7750000, Chile; jsimonet@uchile.cl
3 Facultad de Ciencias Veterinarias, Universidad de Concepción, Casilla 537, Chillán 3812120, Chile; clandaeta@udec.cl
4 Departamento de Medicina Preventiva Animal, Facultad de Ciencias Veterinarias y Pecuarias, Universidad de Chile, Santa Rosa 11735, La Pintana, Santiago 8820808, Chile; ffredes@uchile.cl
* Correspondence: arubio@uchile.cl; Tel.: +56-229-780-372

Simple Summary: Land-use changes are one of the most important drivers of zoonotic disease risk in humans, including helminths of wildlife origin. In this paper, we investigated the presence and prevalence of intestinal helminths in wild rodents, comparing this parasitism between a native forest and exotic Monterey pine plantations (adult and young plantations) in central Chile. By analyzing 1091 fecal samples of a variety of rodent species sampled over two years, we recorded several helminth families and genera, some of them potentially zoonotic. We did not find differences in the prevalence of helminths between habitat types, but other factors (rodent species and season of the year) were relevant to explain changes in helminth prevalence. Given that Monterey pine plantations are one of the most important forestry plantations worldwide, and due to the detection of potentially zoonotic helminths, more research should be conducted in this study area and elsewhere in order to better understand the effect of pine plantations on parasites and pathogens in rodents and other wildlife hosts.

Abstract: Native forests have been replaced by forestry plantations worldwide, impacting biodiversity. However, the effect of this anthropogenic land-use change on parasitism is poorly understood. One of the most important land-use change in Chile is the replacement of native forests by Monterey pine (*Pinus radiata*) plantations. In this study, we analyzed the parasitism (presence and prevalence) of intestinal helminths from fecal samples of wild rodents in three habitat types: native forests and adult and young pine plantations in central Chile. Small mammals were sampled seasonally for two years, and a total of 1091 fecal samples from seven small mammal species were analyzed using coprological analysis. We found several helminth families and genera, some of them potentially zoonotic. In addition, new rodent–parasite associations were reported for the first time. The overall helminth prevalence was 16.95%, and an effect of habitat type on prevalence was not observed. Other factors were more relevant for prevalence such rodent species for *Hymenolepis* sp. and season for *Physaloptera* sp. Our findings indicate that pine plantations do not increase helminth prevalence in rodents compared to native forests.

Keywords: Chile; helminths; land conversion; Rodentia; small mammals; zoonosis; wildlife

Citation: Riquelme, M.; Salgado, R.; Simonetti, J.A.; Landaeta-Aqueveque, C.; Fredes, F.; Rubio, A.V. Intestinal Helminths in Wild Rodents from Native Forest and Exotic Pine Plantations (*Pinus radiata*) in Central Chile. *Animals* **2021**, *11*, 384. https://doi.org/10.3390/ani11020384

Academic Editors: Rafael Calero-Bernal and Ignacio Garcia-Bocanegra

Received: 23 January 2021
Accepted: 31 January 2021
Published: 3 February 2021

1. Introduction

Anthropogenic land-use change can impact biodiversity and human health, being a major driver of biodiversity loss and zoonotic disease emergence [1] Land-use change can modify host–parasite interactions through a variety of mechanisms that involve changes in abundance, behavior and immune response of hosts, as well as the composition and structure of host community [2]. Additionally, land-use change can modify abiotic conditions, which, in turn, may influence the transmission and life cycle of parasites such as helminths, which have several life stages with close contact with the environment [3].

Rodents are one of the most important reservoir hosts of zoonotic pathogens [4], and reservoir species are commonly found in high abundance in anthropogenic-modified habitats (e.g., agricultural lands, pasture lands) [5]. Wild rodents can act as definitive, intermediate and paratenic hosts of several endoparasites (helminths and protozoa) with zoonotic potential, such as *Capillaria hepatica* Bancroft 1893, *Cryptosporidium* spp., *Giardia* spp., *Toxoplasma gondii* Nicolle and Manceaux 1908, *Schistosoma* spp., etc. [6,7]. In addition, helminths are the most prevalent group of macroparasites in rodents [8]. Therefore, the study of the effect of anthropogenic land-use change on gastrointestinal helminths in rodents is needed to better understand host–parasite interactions in wildlife with public health implications.

The forest industry constitutes an important economic activity in tropical and temperate regions of developing countries, replacing large areas of native forests and grasslands by plantations of fast-growing exotic trees [9]. Despite the relevance of plantations worldwide, the study of the effect of monoculture plantations on wildlife parasites has been scarce [2], but some studies have found that plantations may increase the abundance of some ectoparasitic mites in rodents [10] as well as parasite richness in several wild mammals [11]. An important plantation in the global forest industry is Monterey pine, *Pinus radiata* (D. Don, 1836), accounting for 32% of productive plantations worldwide [12]. In Chile, Monterey pine plantations are one of the most important land uses in south-central regions [13], covering approximately 1.9 million ha and accounting for 68% of forestry plantations of the country (http://www.corma.cl). This forest plantation modifies biodiversity, including the composition and structure of rodents [14–16], as well as the prevalence and load of mites, *Ornithonyssus* sp. (Mesostigmata), in rodent hosts [10].

Several gastrointestinal helminths have been detected in native and exotic rodents in Chile [17–21]. However, to our knowledge, the effect of land use on intestinal helminths of rodents inhabiting monoculture plantations has not been investigated. Therefore, the aim of this study was to describe the presence and compare the prevalence of gastrointestinal helminths in feces from wild rodents inhabiting forestry plantations and native forests in a highly disturbed landscape from central Chile.

2. Materials and Methods

2.1. Ethical Statement

All procedures for trapping and handling rodents followed the guidelines of the American Society of Mammalogists for the use of wild mammals in research [22] and followed standardized safety guidelines recommended by the Centre for Disease Control and Prevention [23]. Sampling procedures were authorized by the Servicio Agrícola y Ganadero (SAG; Chilean Fish and Wildlife Service) (License No. 6831/2015) and the Ethics Committee of the Faculty of Science, University of Chile.

2.2. Study Area and Sampling Sites

The study was conducted in Tregualemu (35°58′ S, 072°44′ W), located in the coastal range of the Maule Region in central Chile. The landscape contains interspersed remnants of native forest and extensive stands of Monterey pine of different ages [24]. The native forest of the study area includes the Queules National Reserve and contiguous forests that make up an area of 600 ha. The forest is composed of *Nothofagus glauca* (R. Phil.) Krasser 1896 and *N. obliqua* (Mirb.) Oerst, 1871, as dominant species and the evergreen *Cryptocarya alba* (Molina) Looser 1950 and *Peumus boldus* Molina, 1782. Monterey pine plantations in this area are managed by maintaining a developed understory or multiple vegetation strata within monocultures [25]. Understory vegetation in mature pine plantation contains *Aristotelia chilensis* Stunz 1914, *P. boldus* with exotic species such as *Genista monspessulana* (L.) L.A.S.Johnson 1962 and *Rubus ulmifolius* Schott, 1818 [26].

Small mammals were sampled in three dominant habitat types: (1) native forest, (2) adult pine plantations (>15 years old) and (3) young pine plantations (3–4 years old). A total of 12 sampling sites were selected: three sites each in native forest and adult pine plantations and six in young pine plantations. Each site was separated by at least 400 m from each other (mean distance between sites = 2025 m).

2.3. Small Mammals and Fecal Sampling

Small mammals were sampled once each season (summer: January; autumn: April–May; winter: July–August; spring: November) over 2 years (2016–2017). Each site was sampled in all periods, except for one adult pine plantation and one young pine plantation site that were not sampled in autumn 2017 due to logistical constraints. At each site, live Sherman traps were placed separated by 10 m, forming a 7 × 10 grid (70 traps), for four consecutive nights. All traps were baited with rolled oats and vanilla essence and checked daily at dawn. After capture, animals were identified to species [27], measured, weighed, sexed and marked with uniquely numbered ear tags (National Band and Tag Co., Newport, KY). Fecal samples were collected directly from the animal or from the trap and then preserved in 70% ethanol in 2-mL microtubes. After handling, animals were released in the same place they were captured.

2.4. Parasitological Analysis

The presence of helminth eggs and larvae in feces was analyzed using the routine coprological method of modified Telemann [28,29]. Briefly, each stool sample was placed in a 15-mL Falcon tube with 10 mL of 70% ethanol and 2 mL of diethyl ether and centrifuged at 2000 rpm for 10 min at room temperature. After centrifugation, the supernatant was removed and 100 µL of the sample was placed on a glass microscope slide, and then, a 24 × 24-mm cover slip was placed on the surface of the sample. Finally, slides were scanned under 10× and 40× objective lenses of a light microscope. Identification of helminths was based on published taxonomic keys [28] and using references of eggs collected from adult helminths by Landaeta-Aqueveque et al. [30]. Laboratory analyses were performed at the Parasitology Laboratory, Faculty of Veterinary Science, University of Chile, and at the Parasitology Laboratory, Faculty of Veterinary Science, University of Concepcion.

2.5. Data Analysis

Prevalence (the proportion of positive animals divided by the total number tested) was calculated and 95% confidence intervals (CI) were determined using the Clopper–Pearson method [31] for each rodent species. These analyses were performed using the software Quantitative Parasitology v.3.0 [31]. Generalized linear models (GLMs) with binomial distribution and logit function were used to identify variables that may explain prevalence. Models were built using all parasites combined and also using the most frequent helminths separately. The explanatory variables analyzed were habitat type, season and rodent species. For each multivariate model, we calculated Akaike information criterion adjusted for sample size (AICc), differences in AICc and Akaike weights (wr). The best models were based on the lowest AICc values [32]. Selected models were validated by the Hosmer–Lemeshow goodness-of-fit test. We performed these analyses with R software (R Core Team 2017), including packages "MuMIn" [33] and "ResourceSelection" [34].

3. Results

3.1. Small Mammal Sampling

A total of 1962 individuals were captured. Small mammals belonged to seven rodent species and a marsupial, the elegant fat-tailed mouse opossum *Thylamys elegans* (Waterhouse, 1839). The most common species captured were *Abrothrix olivacea* Waterhouse, 1837 (56.5%), *Oligoryzomys longicaudatus* Bennett, 1832 (20.3%) and *Abrothrix longipilis* Waterhouse, 1837 (17.5%), followed by *Phyllotis darwini* Waterhouse, 1837 (2.7%) and the introduced *Rattus rattus* Linnaeus, 1758 (2.3%). Other small mammals captured were

Irenomys tarsalis Philippi, 1900, *Octodon bridgesi* Waterhouse 1844 and *T. elegans* (0.7% all three species pooled).

3.2. Parasitological Analysis

In total, 1091 fecal samples were collected and 185 samples (16.95%) were positive to any helminth egg or larvae and were found in the majority of small mammal species analyzed (Tables 1 and 2). The helminths identified belonged to three phyla: Acanthocephala, Nematoda and Platyhelminthes (Table 1; Figure 1). Another helminth egg was found in several samples from both *Abrothrix* species but could not be identified (Table 1; Figure 1). A larva was also detected in some samples (Table 1) but could not be identified through microscope. The most frequent helminth eggs were *Hymenolepis* sp. (6.1%) and *Physaloptera* sp. (3.5%). Prevalence of helminths in samples of the most abundant rodent species was *A. olivacea* (17.7%), *A. longipilis* (22.2%), *O. longicaudatus* (6.7%) and *P. darwini* (20%) (Table 2). Most helminths were found in both native forest and pine plantations, except Strongylida, which was found only in the native forest, and *Moniliformis* sp., which was only recorded in *Abrothrix* species in both adult and young pine plantations (Table 3).

Table 1. Presence of helminth eggs and larvae from fecal samples of small mammals. N = sample size. Number in parentheses is the number of positive samples.

Hosts	N	Acantocephala	Nematoda	Platyhelminthes	Unidentified
Abrothrix olivacea	583	*Moniliformis* sp. (11)	*Physaloptera* sp. (30) *Syphacia* sp. (6) *Capillaria* sp. (3) Unidentified larva (8)	*Hymenolepis* sp. (40)	Unidentified egg (2)
Abrothrix longipilis	279	*Moniliformis* sp. (3)	*Physaloptera* sp. (3) *Syphacia* sp. (2) *Capillaria* sp. (2) Unidentified larva (4)	*Hymenolepis* sp. (22)	Unidentified egg (20)
Oligoryzomys longicaudatus	180	–	*Physaloptera* sp. (3) *Syphacia* sp. (1) *Capillaria* sp. (1) Strongylida (4)	*Hymenolepis* sp. (3)	–
Phyllotis darwini	30	–	*Physaloptera* sp. (2) *Syphacia* sp. (2)	*Hymenolepis* sp. (2)	–
Rattus rattus	15	–	*Syphacia* sp. (2)	–	–
Thylamys elegans	3	–	–	–	–
Octodon bridgesi	1	–	–	–	–

Table 2. Helminths in fecal samples of small mammals at three habitat types. Sample size (N), number of positive individuals (+), prevalence (P) and 95% confidence intervals (CI) are reported.

	Native Forest			Adult Pine			Young Pine			Overall		
	N (+)	P (%)	95% CI	N (+)	P (%)	95% CI	N (+)	P (%)	95% CI	N (+)	P (%)	95% CI
Abrothrix longipilis	37 (12)	32.4	18–49.8	202 (45)	22.3	16.7–28.6	40 (5)	12.5	4.2–26.8	279 (62)	22.2	17.5–27.6
Abrothrix olivacea	31 (2)	6.5	0.8–21.4	48 (8)	16.7	7.5–30.2	504 (93)	18.5	15.2–22.1	583 (103)	17.7	14.7–21.0
Oligoryzomys longicaudatus	95 (8)	8.4	3.7–15.9	19 (0)	0	-	66 (4)	6.1	1.7–14.8	180 (12)	6.7	3.5–11.4
Phyllotis darwini	0	-	-	0	-	-	30 (6)	20	7.7–38.6	30 (6)	20	7.7–38.6
Rattus rattus	7 (1)	14.3	0.4–57.9	5 (1)	20	0.5–71.6	3 (0)	0	-	15 (2)	13	1.7–40.5
Octodon bridgesi	1 (0)	0	-	0	-	-	0	-	-	1 (0)	0	-
Thylamis elegans	2 (0)	0	-	1 (0)	0	-	0	-	-	3 (0)	0	-
Total	173 (23)	13.3	8.6–19.3	275 (54)	19.7	15.1–24.8	643 (108)	16.8	14–19.9	1091 (185)	16.9	14.8–19.3

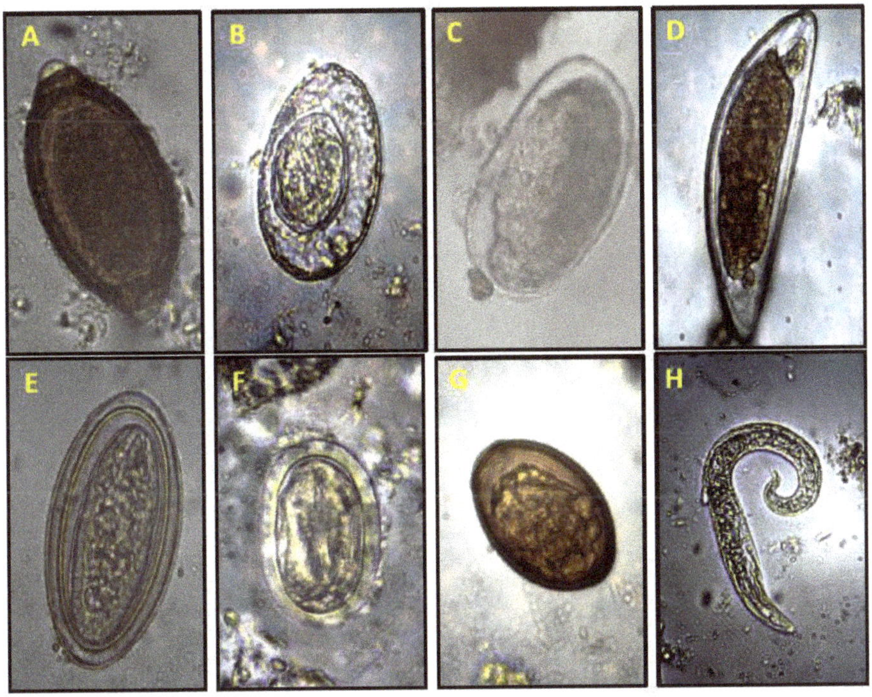

Figure 1. Helminth eggs recorded from fecal samples of rodents. All images were obtained at 40×. (**A**) *Capillaria* sp.; (**B**) *Hymenolepis* sp.; (**C**) Strongylida; (**D**) *Syphacia* sp.; (**E**) *Moniliformis* sp.; (**F**) *Physaloptera* sp.; (**G**) unidentified egg; (**H**) unidentified larva.

Table 3. Number of infected host individuals reported by helminth species, host species (most abundant species) and by habitat type. NF = native forest; AP = adult pine plantation; YP = Young pine plantation; *n* = number sample size. Numbers in parentheses indicate prevalence.

	A. longipilis			***A. olivacea***			***O. longicaudatus***		
	NF *n* **= 37**	**AP** *n* **= 202**	**YP** *n* **= 40**	**NF** *n* **= 31**	**AP** *n* **= 48**	**YP** *n* **= 504**	**NF** *n* **= 95**	**AP** *n* **= 19**	**YP** *n* **= 66**
Hymenolepis sp.	4 (10.8%)	16 (7.9%)	2 (5%)	0	3 (6.2%)	37 (7.3%)	1 (1.1%)	0	2 (3%)
Physaloptera sp.	1 (2.7%)	1 (0.5%)	1 (2.5%)	1 (3.2%)	2 (4.2%)	27 (5.4%)	2 (2.1%)	0	1 (1.5%)
Syphacia sp.	1 (2.7%)	1 (0.5%)	0	0	1 (2.1%)	5 (1%)	1 (1.1%)	0	0
Capillaria sp.	1 (2.7%)	1 (0.5%)	0	0	1 (2.1%)	12 (2.4%)	0	0	1 (1.5%)
Moniliformis sp.	0	2 (1%)	1 (2.5%)	0	1 (2.1%)	10 (2%)	0	0	0
Strongylida	0	0	0	0	0	0	4 (4.2%)	0	0

3.3. Data Analysis

GLM analyses were conducted including the results of the three most abundant rodent species (*A. olivacea*, *A. longipilis* and *O. longicaudatus*). *Phyllotis darwini* was not included because this rodent was only present in young pine plantations [15]. The best

GLM model selected according to AICc values for all helminths included season and host species as explanatory variables (Table S1). This model indicates that prevalence of helminths was higher in spring compared to autumn and summer, and *O. longicaudatus* had lower prevalence compared to both *Abrothrix* species (Table 4). Habitat type was included in the third best GLM model (Table S1), showing that helminth prevalence was not significantly different between native forest and adult pine plantations (estimate = −0.19, standard error (SE) = 0.31, $p = 0.54$) or young pine plantations (estimate = −0.16, SE = 0.30, $p = 0.59$). Regarding GLM analyses using the most common helminth species, the best model for *Hymenolepis* sp. included only host species as an explanatory variable (Table S2). *Oligoryzomys longicaudatus* had lower prevalence of *Hymenolepis* sp. compared to both *Abrothrix* species (Table 5). Habitat type was included in the second best GLM model (Table S2), indicating that *Hymenolepis* sp. prevalence was not significantly different between native forest and adult pine plantations (estimate = 0.25, SE = 0.55, $p = 0.65$) or young pine plantations (estimate = 0.46, SE = 0.55, $p = 0.39$). For *Physaloptera* sp., the best model included the variables season and host species (Table S3), indicating that this parasite had lower prevalence in autumn compared to all other seasons (Table 6). Similar to previous results, habitat type was included in the second best GLM model (Table S3), showing that *Physaloptera* sp. prevalence was not significantly different between native forest and adult pine plantations (estimate = −1.02, SE = 0.78, $p = 0.19$) or young pine plantations (estimate = 0.52, SE = 0.56, $p = 0.35$).

Table 4. Results of the best generalized linear model (GLM) that predicted the probability of positivity for helminths in rodents. Spring and *O. longicaudatus* were used as the reference categories. Hosmer–Lemeshow test: $\chi^2 = 10.96$, $p = 0.20$ ($p > 0.05$ is interpreted as fit).

		Estimate (SE)	*z* Value	*p* Value	Odds Ratio	Lower 95% CI	Upper 95% IC
	Intercept	−2.08 (0.36)	−5.80	<0.001 *			
Season	Autumn	−0.71 (0.24)	−2.98	0.003 *	0.49	0.31	0.78
	Summer	−0.50 (0.23)	−2.18	0.030 *	0.60	0.38	0.95
	Winter	−0.37 (0.25)	−1.50	0.132	0.69	0.43	1.12
Host species	*A. longipilis*	1.21 (0.36)	3.52	<0.001 *	3.36	1.76	6.90
	A. olivacea	0.94 (0.33)	2.87	0.004 *	2.57	1.40	5.13

* *p* values < 0.05.

Table 5. Results of the best GLM that predicted the probability of positivity for *Hymenolepis* sp. in rodents. *O. longicaudatus* was used as the reference category. Hosmer–Lemeshow test: $\chi^2 = 5.16$, $p = 0.74$ ($p > 0.05$ is interpreted as fit).

		Estimate (SE)	*z* Value	*p* Value	Odds Ratio	Lower 95% CI	Upper 95% CI
	Intercept	−2.01 (0.58)	−7.01	<0.001 *			
Host species	*A. longipilis*	1.62 (0.62)	2.60	0.009 *	5.05	1.72	21.56
	A. olivacea	1.47 (0.61)	2.43	0.015 *	4.35	1.56	18.11

* *p* values < 0.05.

Table 6. Results of the best GLM that predicted the probability of positivity for *Physaloptera* sp. in rodents. Autumn and *O. longicaudatus* were used as the reference categories. Hosmer-Lemeshow test: $\chi^2 = 2.43$, $p = 0.97$ ($p > 0.05$ is interpreted as fit).

		Estimate (SE)	*z* Value	*p* Value	Odds Ratio	Lower 95% CI	Upper 95% CI
	Intercept	−4.83 (0.71)	−6.80	<0.001 *			
Season	Spring	1.67 (0.60)	2.78	0.005 *	5.31	1.76	19.80
	Summer	1.21 (0.61)	1.98	0.048 *	3.37	1.08	12.75
	Winter	1.37 (0.60)	2.28	0.030 *	3.93	1.30	14.51
Host species	*A. longipilis*	0.90 (0.85)	−1.06	0.290	0.41	0.07	2.33
	A. olivacea	0.76 (0.65)	1.18	0.240	2.14	0.68	9.42

* *p* values < 0.05.

4. Discussion

In this study, we recorded helminth families or genera that are associated with rodents in Chile and elsewhere [17,35,36]. Furthermore, some of the helminth genera recorded may belong to zoonotic species, such as *Hymenolepis diminuta* Rudolphi 1819, *H. nana* Siebold 1852, *Syphacia obvelata* Rudolphi 1802 and *S. muris* Yamaguti 1941, which have been reported in Chile [30,37]. The following host–parasite associations are registered for the first time: *P. darwini–Physaloptera* sp., *A. longipilis–Moniliformis* sp., *A. longipilis–Physaloptera* sp., *A. longipilis–Capillaria* sp. and *O. longicaudatus–Capillaria* sp. We used traditional routine coprological analyses to detect helminths in feces, which have been successfully employed to address the effect of land use on parasites elsewhere [11,35,38]. However, further studies are needed to identify the helminth species present in the study area which should focus on identification of adult helminths [30] and/or species identification through molecular tools, which has been scarcely used for helminth egg identification from fecal samples of rodents [39].

The findings of some helminths only in native forest (Strongylida in *O. longicaudatus*) or only in pine plantations (*Moniliformis* in *Abrothrix* spp.) might be a consequence of the low prevalence of these parasites (0.37–1.3%, respectively). Our results indicate that helminth prevalence in wild rodents varies among seasons and host species but is not affected by habitat type. Therefore, the replacement of native forest by pine plantations in the study area would not be a driver for increased transmission of helminths with zoonotic potential in rodents. A previous investigation in the study area has shown a similar outcome, in which the prevalence of mites (*Androlaelaps* sp.) parasitizing *A. olivacea* was similar between the native forest and pine plantations [10]. On the other hand, the prevalence of other mite species (*Ornithonyssus* sp.) on the same host was increased in young pine plantations [10]. These contrasting results show that the effect of Monterey pine plantations on parasitism is parasite-dependent, which might be consequence of differences in the ecology, life cycle and other attributes between parasite species and their relation to the environment and their hosts. In fact, the effects of land-use change on helminth parasitism depend on the life history traits of each host–parasite association, as shown in several studies across the world [3,35,40,41].

Hymenolepis sp. was the most common helminth found, with higher prevalence in both *Abrothrix* species compared to *O. longicaudatus*. Most species of the genus *Hymenolepis* have an indirect cycle, including free living and ecto-parasite arthropods as intermediate hosts [42]. Therefore, differences in prevalence among rodents could be related to food habits, where the diets of *A. olivacea* and *A. longipilis* include a higher composition of insects (up to 25–32%) [27] compared to *O. longicaudatus*, for which invertebrates represent a minor component of the diet (5−10%) [27,43]. On the other hand, *Physaloptera* sp., the second most prevalent helminth, presented a seasonal variation, with lower prevalence in autumn compared to other seasons. As some *Physaloptera* species have complex cycles with intermediate hosts such as Coleoptera, Blattodea and Orthoptera, seasonal variation could be related to environmental conditions acting on variations in abundance of intermediate hosts and/or dietary changes of hosts, as suggested by Cawthorn and Anderson [44]. In fact, some invertebrates such as ground beetles (Coleoptera: Carabidae) in the study area have seasonal variations, increasing their abundance and richness in summer [45].

Several studies have investigated the impact of Monterrey pine plantation on biodiversity worldwide [25,46]. However, the effect on parasites and pathogens has only been recently addressed [10,15,47]. In this context, more studies are needed for a better understanding on the effects of pine plantations on parasites and pathogens of wildlife, including an ecological approach using multiple temporal and spatial scales [48]. This information will be useful for addressing the nexus between wildlife and zoonosis risks into land use planning, within the One Health framework [49,50].

5. Conclusions

In this study, we reported several intestinal helminths in fecal samples from wild rodents in the landscape of central Chile, but we did not find differences in helminth prevalence between rodents inhabiting native forests and Monterey pine plantations (adult and young plantations). As Monterey pine plantations are one of the most important forestry plantations worldwide, more research should be conducted in order to better understand the effect of pine plantations on other micro- and macroparasites of public health concern.

Supplementary Materials: The following are available online at https://www.mdpi.com/2076-2615/11/2/384/s1, Table S1: Comparison of models used for helminths; Table S2: Comparison of models used for *Hymenolepis* sp.; Table S3; Comparison of models used for *Physaloptera* sp.

Author Contributions: Conceptualization, J.A.S., F.F., and A.V.R.; methodology, J.A.S., F.F., and A.V.R.; formal analysis, M.R., R.S., C.L.-A., and A.V.R.; investigation, M.R., R.S., and A.V.R.; resources, J.A.S., C.L.-A., F.F., and A.V.R.; writing—original draft preparation, M.R., R.S., and A.V.R.; writing—review and editing, M.R., R.S., J.A.S., C.L.-A., F.F., and A.V.R.; supervision, J.A.S., F.F., and A.V.R.; project administration, A.V.R.; funding acquisition, A.V.R., F.F., and J.A.S. All authors have read and agreed to the published version of the manuscript.

Funding: This research was funded by ANID/FONDECYT No. 3160037 and additional support by FONDECYT No. 1140657 and ANID/PAI, Convocatoria Nacional de Subvención a la Instalación en la Academia, 2018, No. PAI77180009.

Institutional Review Board Statement: The study was conducted according to the guidelines of the Declaration of Helsinki, and approved by the Ethics Committee of University of Chile (Number 0151027 FCS-UCH, October 27, 2015), and by the Servicio Agrícola y Ganadero (SAG; Chilean Fish and Wildlife Service) (License No. 6831/2015).

Informed Consent Statement: Not applicable.

Data Availability Statement: The study did not report any data.

Acknowledgments: We thank M. Barceló, J. Veloso, M. Silva, A. Arzabe, H. Mendoza, C. Muñoz, L. Moreno and other volunteers for their assistance during field sampling. We are grateful to Carlos Reyes (CONAF) and Ronny Zuñiga for logistical support in the field and Claudio Veloso for the loan of traps. We also thank Patricio Toro for his invaluable assistance in the laboratory. We thank Forestal Masisa S.A. and Corporación Nacional Forestal for allowing us to work on their properties.

Conflicts of Interest: The authors declare no conflict of interest.

References

1. Loh, E.H.; Zambrana-Torrelio, C.; Olival, K.J.; Bogich, T.L.; Johnson, C.K.; Mazet, J.A.; Karesh, W.; Daszak, P. Targeting transmission pathways for emerging zoonotic disease surveillance and control. *Vector-Borne Zoonotic Dis.* **2015**, *15*, 432–437. [CrossRef] [PubMed]
2. Gottdenker, N.L.; Streicker, D.G.; Faust, C.L.; Carroll, C. Anthropogenic land use change and infectious diseases: A review of the evidence. *Ecohealth* **2014**, *11*, 619–632. [CrossRef] [PubMed]
3. Werner, C.S.; Nunn, C.L. Effect of urban habitat use on parasitism in mammals: A meta-analysis. *Proc. R. Soc. B* **2020**, *287*, 20200397. [CrossRef] [PubMed]
4. Han, B.A.; Kramer, A.M.; Drake, J.M. Global patterns of zoonotic disease in mammals. *Trends. Parasitol.* **2016**, *32*, 565–577. [CrossRef]
5. Mendoza, H.; Rubio, A.V.; García-Peña, G.E.; Suzán, G.; Simonetti, J.A. Does land-use change increase the abundance of zoonotic reservoirs? Rodents say yes. *Eur. J. Wildl. Res.* **2020**, *66*, 6. [CrossRef]
6. Meerburg, B.G.; Singleton, G.R.; Kijlstra, A. Rodent-borne diseases and their risks for public health. *Crit. Rev. Microbiol.* **2009**, *35*, 221–270. [CrossRef]
7. McCarthy, J.; Moore, T.A. Emerging helminth zoonoses. *Int. J. Parasitol.* **2000**, *30*, 1351–1359. [CrossRef]
8. Weil, Z.M.; Martin, L.B.; Nelson, R.J. Interactions among immune, endocrine, and behavioural response to infection. In *Micromammals and Macroparasites*; Morand, S., Krasnov, B.R., Poulin, R., Eds.; Springer: Tokyo, Japan, 2006; pp. 443–473. ISBN 9784431360247.
9. FAO; UNEP. *The State of the World's Forests 2020. Forests, Biodiversity and People*; FAO: Rome, Italy; UNEP: Nairobi, Kenya, 2020; 214p, ISBN 9789251324196.

10. Veloso-Frías, J.; Silva-De La Fuente, M.C.; Rubio, A.V.; Moreno, L.; González-Acuña, D.; Simonetti, J.A.; Landaeta-Aqueveque, C. Variation in the prevalence and abundance of mites parasitizing *Abrothrix olivacea* (Rodentia) in the native forest and *Pinus radiata* plantations in central Chile. *Hystrix* **2019**, *30*, 107–111. [CrossRef]
11. Chakraborty, D.; Reddy, M.; Tiwari, S.; Umapathy, G. Land use change increases wildlife parasite diversity in Anamalai Hills, Western Ghats, India. *Sci. Rep.* **2019**, *9*, 1–11. [CrossRef]
12. Mead, D.J. *Sustainable Management of Pinus radiata Plantations*; Food and Agriculture Organization of the United Nations (FAO): Rome, Italy, 2013; ISBN 9789251076347.
13. Uribe, S.V.; Estades, C.F.; Radeloff, V.C. Pine plantations and five decades of land use change in central Chile. *PLoS ONE* **2020**, *15*, e0230193. [CrossRef]
14. Moreira-Arce, D.; Vergara, P.M.; Boutin, S.; Simonetti, J.A.; Briceño, C.; Acosta-Jamett, G. Native forest replacement by exotic plantations triggers changes in prey selection of mesocarnivores. *Biol. Conserv.* **2015**, *192*, 258–267. [CrossRef]
15. Rubio, A.V.; Fredes, F.; Simonetti, J.A. Exotic *Pinus radiata* plantations do not increase Andes Hantavirus prevalence in rodents. *EcoHealth* **2019**, *16*, 659–670. [CrossRef] [PubMed]
16. Saavedra, B.; Simonetti, J.A. Small mammals of Maulino forest remnants, a vanishing ecosystem of south-central Chile. *Mammalia* **2005**, *69*, 337–348. [CrossRef]
17. Landaeta-Aqueveque, C.A.; Robles, M.D.; Cattan, P.E. Helmintofauna del roedor *Abrothrix olivaceus* (Sigmodontinae) en áreas sub-urbanas de Santiago de Chile. *Parasitol. Latinoam.* **2007**, *62*, 134–141. [CrossRef]
18. Landaeta-Aqueveque, C.A.; Robles, M.D.; Cattan, P.E. The community of gastrointestinal helminths in the housemouse, *Mus musculus*, in Santiago, Chile. *Parasitol. Latinoam.* **2007**, *62*, 165–169. [CrossRef]
19. Landaeta-Aqueveque, C.; Notarnicola, J.; Correa, J.P.; Yáñez-Meza, A.; Henríquez, A.; Cattan, P.E.; Botto-Mahan, C.; Torres-Pérez, F. First record of *Litomosoides pardinasi* (Nematoda: Onchocercidae) in native and exotic rodents from Chile. *Rev. Mex. Biodivers.* **2014**, *85*, 1032–1037. [CrossRef]
20. Digiani, M.C.; Landaeta-Aqueveque, C.; Serrano, P.C.; Notarnicola, J. Pudicinae (Nematoda: Heligmonellidae) Parasitic in Endemic Chilean Rodents (Caviomorpha: Octodontidae and Abrocomidae): Description of a new species and emended description of *Pudica degusi* (Babero and Cattan) n. comb. *J. Parasitol.* **2017**, *103*, 736–746. [CrossRef]
21. Yáñez-Meza, A.; Landaeta-Aqueveque, C.; Quiroga, N.; Botto-Mahan, C. Helminthic infection in three native rodent species from a semiarid Mediterranean ecosystem. *Rev. Bras. Parasitol. Vet.* **2019**, *28*, 119–125. [CrossRef]
22. Sikes, R.S.; Animal Care Use Committee of the American Society of Mammalogists. 2016 Guidelines of the American Society of Mammalogists for the use of wild mammals in research and education. *J. Mammal.* **2016**, *97*, 663–688. [CrossRef]
23. Mills, J.N.; Childs, J.E.; Ksiazek, T.G.; Peters, C.J.; Velleca, W.M. *Métodos Para Trampeo y Muestreo de Pequeños Mamíferos Para Estudios Virológicos*; Organización Panamericana de la Salud: Washington, DC, USA, 1998; 64p.
24. Bustamante, R.O.; Grez, A.A.; Simonetti, J.A. Efectos de la fragmentación del bosque Maulino sobre la abundancia y diversidad de especies nativas. In *Biodiversidad en Ambientes Fragmentados de Chile: Patrones y Procesos a Diferentes Escalas*; Bustamante, R.O., Grez, A.A., Simonetti, J.A., Eds.; Editorial Universitaria: Santiago, Chile, 2006; pp. 83–97. ISBN 9561119129.
25. Estades, C.F.; Grez, A.A.; Simonetti, J.A. Biodiversity in Monterey pine plantations. In *Biodiversity Conservation in Agroforestry Landscapes: Challenges and Opportunities*; Estades, C.F., Grez, A.A., Simonetti, J.A., Eds.; Editorial Universitaria: Santiago, Chile, 2012; pp. 77–98. ISBN 9789561123847.
26. Poch, T.J.; Simonetti, J.A. Insectivory in *Pinus radiata* plantations with different degree of structural complexity. *For. Ecol. Manag.* **2013**, *304*, 132–136. [CrossRef]
27. Iriarte, A. *Mamíferos de Chile*; Lynx Edicions: Barcelona, España, 2008; 420p, ISBN 9788496553316.
28. Thienpont, D.; Rochette, F.; Vanparijs, O. *Diagnosing Helminthiasis by Coprological Examination*; Janssen Research Foundation: Beerse, Belgium, 1986; pp. 35–36.
29. Riebold, D.; Russow, K.; Schlegel, M.; Wollny, T.; Thiel, J.; Freise, J.; Hüppop, O.; Eccard, J.A.; Plenge-Bönig, A.; Loebermann, M. Occurrence of gastrointestinal parasites in small mammals from Germany. *Vector-Borne Zoonotic Dis.* **2020**, *20*, 125–133. [CrossRef]
30. Landaeta-Aqueveque, C.; del Rosario Robles, M.; Henríquez, A.; Yáñez-Meza, A.; Correa, J.P.; González-Acuña, D.; Cattan, P.E. Phylogenetic and ecological factors affecting the sharing of helminths between native and introduced rodents in Central Chile. *Parasitology* **2018**, *145*, 1570–1576. [CrossRef] [PubMed]
31. Reiczigel, J.; Rózsa, L. Quantitative Parasitology 3.0. 2005. Available online: http://www.zoologia.hu/qp/qp.html (accessed on 9 April 2020).
32. Burnham, K.P.; Anderson, D.R. Multimodel inference: Understanding AIC and BIC in model selection. *Sociol. Methods Res.* **2004**, *33*, 261–304. [CrossRef]
33. Bartoń, K. R Package MuMIn with Tools for Performing Model Selection and Model Averaging. 2019. Available online: https://cran.r-project.org/web/packages/MuMIn/index.html (accessed on 20 April 2020).
34. Lele, S.R.; Keim, J.L.; Solymos, P. ResourceSelection: Resource Selection (Probability) Functions for Use-Availability Data. 2016. Available online: https://cran.r-project.org/web/packages/ResourceSelection/index.html (accessed on 20 April 2020).
35. Wells, K.; Smales, L.R.; Kalko, E.K.; Pfeiffer, M. Impact of rain-forest logging on helminth assemblages in small mammals (Muridae, Tupaiidae) from Borneo. *J. Trop. Ecol.* **2007**, *23*, 35–43. [CrossRef]
36. Skyriene, G.; Ulevicius, A.; Samas, A. Levels of helminth infection of small rodents in two interspersed habitats—The forest and beaver sites. *Balt. For.* **2011**, *17*, 299–307.

37. Ruíz del Río, A. Contribución al Estudio de Las Enfermedades Parasitarias Humanas Transmitidas por Las Ratas en Concepción. Ph.D. Thesis, Escuela de Medicina, Universidad de Chile, Región Metropolitana, Chile, 1938.
38. Rendón-Franco, E.; Muñoz-García, C.I.; Romero-Callejas, E.; Moreno-Torres, K.I.; Suzán, G. Effect of host species diversity on multiparasite systems in rodent communities. *Parasitol. Res.* **2014**, *113*, 447–450. [CrossRef] [PubMed]
39. Gliga, D.S.; Pisanu, B.; Walzer, C.; Desvars-Larrive, A. Helminths of urban rats in developed countries: A systematic review to identify research gaps. *Parasitol. Res.* **2020**, 1–15. [CrossRef] [PubMed]
40. Froeschke, G.; van der Mescht, L.; McGeoch, M.; Matthee, S. Life history strategy influences parasite responses to habitat fragmentation. *Int. J. Parasitol.* **2013**, *43*, 1109–1118. [CrossRef]
41. Cardoso, T.; Simões, R.; Luque, J.; Maldonado, A., Jr.; Gentile, R. The influence of habitat fragmentation on helminth communities in rodent populations from a Brazilian Mountain Atlantic Forest. *J. Helminthol.* **2016**, *90*, 460. [CrossRef]
42. Oldham, J. On the arthropod intermediate hosts of *Hymenolepis diminuta* (Rudolphi 1819). *J. Helminthol.* **1931**, *9*, 21–28. [CrossRef]
43. Silva, S.I. Posiciones tróficas de pequeños mamíferos en Chile: Una revisión. *Rev. Chil. Hist. Nat.* **2005**, *78*, 589–599. [CrossRef]
44. Cawthorn, R.J.; Anderson, R.C. Effects of age, temperature, and previous infection on the development of *Physaloptera maxillaris* (Nematoda: Physalopteroidea) in field crickets (*Acheta pennsylvanicus*). *Can. J. Zool.* **1976**, *54*, 442–448. [CrossRef] [PubMed]
45. Grez, A.A.; Moreno, P.; Elgueta, M. Coleópteros (Insecta: Coleoptera) epígeos asociados al bosque maulino y plantaciones de pino aledañas. *Rev. Chil. Entomol.* **2003**, *29*, 9–18.
46. McFadden, T.N.; Dirzo, R. Opening the silvicultural toolbox: A new framework for conserving biodiversity in Chilean timber plantations. *For. Ecol. Manag.* **2018**, *425*, 75–84. [CrossRef]
47. Rubio, A.V.; Simonetti, J.A. Ectoparasitism by *Eutrombicula alfreddugesi* larvae (Acari: Trombiculidae) on *Liolaemus tenuis* lizard in a Chilean fragmented temperate forest. *J. Parasitol.* **2009**, *95*, 244–245. [CrossRef]
48. Diuk-Wasser, M.A.; VanAcker, M.C.; Fernandez, M.P. impact of land use changes and habitat fragmentation on the eco-epidemiology of tick-borne diseases. *J. Med. Entomol.* **2020**. [CrossRef]
49. Rubio, A.V.; Fredes, F.; Simonetti, J.A. Links between land-sharing, biodiversity, and zoonotic diseases: A knowledge gap. *EcoHealth* **2016**. [CrossRef]
50. McClure, M.; Machalaba, C.; Zambrana-Torrelio, C.; Feferholtz, Y.; Lee, K.D.; Daszak, P.; Karesh, W.B. Incorporating health outcomes into land-use planning. *EcoHealth* **2019**, *16*, 627–637. [CrossRef]

animals

MDPI

Communication

Molecular Characterization of New Haplotype of Genus *Sarcocystis* in Seabirds from Magdalena Island, Southern Chile

Igor C. L. Acosta [1,*], Solange M. Gennari [1,2], Horwald A. B. Llano [1], Sebastián Muñoz-Leal [1,3] and Rodrigo M. Soares [1]

1 Department of Preventive Veterinary Medicine and Animal Health, Faculty of Veterinary Medicine, University of São Paulo, Avenida Prof. Orlando Marques de Paiva, 87, São Paulo CEP 05508-270, Brazil; sgennari@usp.br (S.M.G.); horwald@hotmail.com (H.A.B.L.); sebamunoz@udec.cl (S.M.-L.); rosoares@usp.br (R.M.S.)

2 PhD Program in Medicine, Animal Welfare and Public Health, Faculty of Veterinary Medicine, University Santo Amaro, Rua Prof. Enéas de Siqueira Neto, 340, São Paulo CEP 04829-300, Brazil; sgennari@prof.unisa.br

3 Department of Pathology and Preventive Medicine, Faculty of Veterinary Sciences, University of Concepción, Avenida Vicente Mendez, 595, Chillán CEP 3780000, Chile

* Correspondence: igorclacosta@gmail.com or acostaicl@usp.br

Simple Summary: Sarcocystidae is a family of apicomplexan protozoa highly prevalent in vertebrates. The definitive hosts of sarcocystids eliminate oocysts or sporocysts that infect intermediate hosts. After infection, mature tissue cysts (sarcocysts) develop in intermediate hosts, mostly in muscle and neurological tissues. *Sarcocysts* are infectious for definitive hosts, which acquire them through carnivorous or scavenging habits. Intermediate hosts and definitive hosts are the natural hosts of sarcocystids in which infections are usually mildly or not symptomatic. In 2017, muscular and neurological tissues of 22 birds from Magdalena Islands, southern coast of Chile, were screened for the presence of DNA of sarcocystids. DNA of organisms of the genus *Sarcocystis* was identified in two Chilean skuas (*Stercorarius chilensis*). The genetic makeup of the parasite detected in skuas was unprecedented and probably represent a new species in the genus. It is well known that *Sarcocystis* may cause severe infections in aberrant hosts, which are susceptible animals that do not behave as natural hosts for the parasite and have low resistance to the infection, thus more studies are needed to characterize this parasitosis in skuas and other hosts to understand the epidemiology of the infection and its impact on the health of marine fauna.

Abstract: Evidence of sarcocystid infection was investigated in samples of 16 penguins (*Spheniscus. magellanicus*), four Dominican gulls (*Larus dominicanus*) and two Chilean skuas (*Stercorarius chilensis*) found in Madalenas Islands, Chile, in 2017. Samples of skeletal muscle, cardiac muscle and brain from all birds were screened by a pan-sarcocystid nested-PCR targeting a short fragment of the gene encoding the small ribosomal unit (nPCR-18Sa). The only two positive samples by nPCR-18Sa, both from skuas, were tested by a nested-PCR directed to the internal transcribed spacer 1 (nPCR-ITS1), also a pan-sarcocystidae nested-PCR, and to a nested-PCR directed to the B1 gene (nPCR-B1), for the exclusive detection of *Toxoplasma gondii*. The two nPCR-18Sa-positive samples were nPCR-ITS1-positive and nPCR-B1-negative. The nPCR-ITS1 nucleotide sequences from the two skuas, which were identical to each other, were revealed closely related to homologous sequences of *Sarcocystis halieti*, species found in seabirds of northern hemisphere. Larger fragments of genes encoding 18S and partial sequences of genes coding for cytochrome oxidase subunit 1 were also analyzed, corroborating ITS1 data. The haplotypes found in the skuas are unprecedent and closely related to species that use birds as the definitive host. Further studies need to be carried out to detect, identify and isolate this parasite to understand the epidemiology of the infection and its impact on the health of marine fauna.

Keywords: wild birds; coccidian; molecular; apicomplexa; marine

Citation: Acosta, I.C.L.; Gennari, S.M.; Llano, H.A.B.; Muñoz-Leal, S.; Soares, R.M. Molecular Characterization of New Haplotype of Genus *Sarcocystis* in Seabirds from Magdalena Island, Southern Chile. *Animals* **2021**, *11*, 245. https://doi.org/10.3390/ani11020245

Academic Editors: Rafael Calero-Bernal and Ignacio Garcia-Bocanegra

Received: 16 November 2020
Accepted: 15 January 2021
Published: 20 January 2021

1. Introduction

The phylum Apicomplexa is composed of obligate intracellular parasites that are characterized by having a specialized structure called the apical complex, which is used to invade vertebrate host cells [1] Within this phylum, the Sarcocystidae family comprises more than 196 species of coccidia that form cysts in tissues of intermediate hosts. Although taxonomic controversies still exist, this family has been divided into three subfamilies: Sarcocystinae, represented by the genera *Frenkelia* and *Sarcocystis*; Cystoisosporinae, containing the genus *Cystoisospora*; and Toxoplasmatinae, a subfamily with a few species grouped in the genera *Besnoitia*, *Hammondia*, *Neospora* and *Toxoplasma* [2–5].

Toxoplasma gondii is a coccidian parasite with worldwide distribution. It infects virtually all warm-blooded animals, including humans, but only cats (domestic and wild) act as definitive hosts. Toxoplasmosis has been reported in many avian species; however, little information is available in relation to populations of *Spheniscus magellanicus*, *Stercorarius chilensis* and *Larus dominicanus* [6]. Recently, *T. gondii* antibodies were detected in 57 (43.18%) out of 132 serum samples collected from free-living Magellanic penguins (*Spheniscus magellanicus)* on Magdalena Island, Chile, with titers that ranged from 20 to 320 [7].

The genus *Sarcocystis* has an obligate two-host life cycle. Asexual stages develop in intermediate hosts, usually omnivores, through forming cysts in the musculature and central nervous system. Infection of intermediate hosts occurs through their ingestion of food or water contaminated with sporocysts. Sexual stages only develop in the definitive host, which is typically a carnivore or an omnivore, and infection in this case occurs through ingestion of meat contaminated with cysts [8]. Sarcocystids of the genus *Sarcocystis* may cause severe infections in aberrant hosts, which are susceptible animals that do not behave as natural hosts of the parasite and have low resistance to the infection. Thus, *Sarcocystis* potentially pose risk to human and animal health, depending on the susceptible host behaving as aberrant host or not [8].

A few studies have documented the presence of *Sarcocystis* spp. in wild animals in Chile. Presence of cysts of this parasite has been confirmed in muscle tissues of pudu deer (*Pudu puda*), guanacos (*Lama guanicoe*) and sea lions (*Otaria byronia)* [9–11]. However, *Sarcocystis* has not yet been described in Chilean wild birds.

The Chilean skua *Stercorarius chilensis* is a large predatory seabird that inhabits shore ecosystems along the southern cone of South America from central Peru to northern Argentina, with occasional occurrence on the coasts of Ecuador, Brazil, Uruguay and Antarctica [12]. Skuas belong to the order Charadriiformes and are considered to be opportunist feeders, preying on a wide diversity of animals such as small seabirds, fish, scraps and carrion [13,14]. Populations of skuas may be small, but they do not approach the thresholds for vulnerable classification following a population-size criterion (<10,000 mature individuals) [15].

Considering other coastal birds' species, the Kelp Gull (*Larus dominicanus*) is an opportunistic feeder like numerous Laridae and consumes a wide variety of fishes, invertebrates and fisheries waste [16,17]. A high diversity in the use of habitat types has been recorded throughout its distributional range in the Southern Hemisphere, including Argentina, Brazil, Chile, Peru and Uruguay, and the breeding population has been estimated at least 160,000 pairs [17]. In contrast, the Magellanic penguin (*S. magellanicus*) has approximately 1.1 to 1.6 million breeding pairs that nest along the eastern and western coasts of South America, in Argentina Chile and the Malvinas/Falkland Islands [18]. *Spheniscus magellanicus* has a primarily piscivorous diet with the presence of some cephalopods and crustaceans [19].

To date, more than 25 species of *Sarcocystis* have been found to use birds as intermediate hosts [8,20]. *Sarcocystis falcatula*, *Sarcocystis calchasi* and the recently described unnamed species *Sarcocystis* sp. Chicken-2016-DF-BR, which can possibly be interpreted as *Sarcocystis wenzeli* [21] are species that may be pathogenic for intermediate hosts [22–24].

Focusing on *T. gondi* and the genus *Sarcocystis*, the aim of this study was to screen for the evidence of new species or species genotypes of Sarcocystidae in seabird carcasses from southern Chile, a region with scarce data on the occurrence of this group of parasites.

Molecular evidence of a unique haplotype of genus *Sarcocystis* was found in two Chilean skuas (*S. chilensis*).

2. Materials and Methods

2.1. Ethical Considerations

Sample collections on Magdalena Island were conducted under license no. 039/2016 issued by the National Forestry Corporation (Corporación Nacional Forestal; CONAF), and permit no. 2799 issued by the National Fisheries Service (Servicio Nacional de Pesca; SERNAPESCA), Chile. This study was approved by the Ethics Committee on Animal Use (CEUA-no. 9701041113) of the School of Veterinary Medicine, University of São Paulo (FMVZ-USP).

2.2. Collection of Samples

In January 2017, fragments from the pectoral muscle, heart and brain, comprising approximately 5–10 g each, were collected from fresh seabird carcasses on Magdalena Island. This island is located in the Strait of Magellan, near the city of Punta Arenas, in southern Chile (52°55′10.0″ S; 70°34′37.7″ W), and constitutes a natural reserve named "Monumento Natural Los Pinguinos". Necropsies were performed in situ and samples were stored in sterile microtubes at −20 °C until the time of analysis. Samples were collected from 22 birds: 16 penguins (*S. magellanicus*), four Dominican gulls (*L. dominicanus*) and two Chilean skuas (*S. chilensis*), totaling 66 samples (22 from pectoral muscles, 22 from hearts and 22 from brains).

2.3. Molecular Identification

Tissue samples of 25–50 mg were subjected to DNA extraction using the DNeasy Blood and Tissue kit (Qiagen, Hilden, Germany) following the manufacturer's recommendations, with the exception of final elution of the product into 50 μL of elution buffer from Qiagen Kit. As internal control for the evaluation of the successfulness of the DNA extraction, DNA samples were tested by conventional PCR targeting mitochondrial hypervariable region in the penguins derived samples and by PCR directed to mitochondrial 16S rRNA gene in the tissues from the other seabirds [25–27].

Initial screening targeting the Sarcocystidae family was performed using a pan-sarcocystid nested PCR based on primers [28] directed to a short fragment of 18S rDNA gene (nPCR-18Sa). The nPCR-18Sa positive samples were further investigated for the presence of DNA of *T. gondii* to amplify partial fragments of gene B1 (nPCR-B1) using the primers described by [29]. The nPCR-18Sa positive samples were also tested by a second pan-sarcocystid nested PCR directed to internal transcribed spacer 1 (nPCR-ITS1) [30,31] The nPCR-ITS1 were used in order to obtain genetic sequences capable of differentiating the species of the Sarcocystid screened with nPCR-18Sa. The nPCR-ITS1 positive samples were further tested with a third pan-sarcocystid nested PCR, now targeting a larger fragment of 18S rDNA gene (nPCR-18Sb) using primers described by [32], as well with a *Sarcocystis* specific nested PCR [30] directed to cytochrome oxidase subunit I (nPCR-CO1). The primers are depicted in Table 1.

The first round of nPCR-18Sa were performed with 3.0 μL of extracted DNA, 1.8 μL of 10× PCR Buffer (KCL 50 mM; Tris HCl 10 mM; pH 9.0) (Life Technologies Corporation, Carlsbad, CA 92008 USA), 0.7 μL of $MgCL_2$ (50 mM), 1.4 μL of mixed dNTPs (10 mM), 0.1 μL of each primer (25 μM), 0.14 μL of Platinum™ Taq DNA Polymerase (5 U/μL) (Life Technologies Corporation, Carlsbad, CA 92008 USA) and ultrapure autoclaved water to a volume of 18 μL per reaction. The PCR thermal cycling consisted of an initial incubation at 94 °C for 30 sec, followed by 30 cycles (94 °C for 25 sec, 55 °C for 1 min, 72 °C for 1.5 min) and a final extension at 72 °C for 10 min. For the second rounds: 1 μL of template derived from the first reactions, 2.5 μL of 10× PCR Buffer (KCL 50 mM; Tris HCl 10 mM; pH 9.0) (Life Technologies Corporation, Carlsbad, CA 92008 USA), 2.5 μL of $MgCL_2$ (50 mM), 4.0 μL of mixed dNTPs (10 mM), 1.25 μL of each primer (10μM), 0.15 μL of Platinum™ Taq DNA Polymerase (5 U/μL) (Life Technologies Corporation, Carlsbad, CA 92008 USA)and ultrapure autoclaved water to a volume of 25 μL per reaction. The PCR thermal cycling consisted of an initial incubation at 94 °C for 4 min, followed by 30 cycles (94 °C for 30 s, 55 °C for 1 min, 72 °C for 2.0 min) and a final extension at 72 °C for 10 min. For the second rounds the same quantities of the reagent mixture with primers at 50 μM, using 2 μL of the product of the PCR diluted in ultra-pure water (1:2). The nPCR thermal cycling consisted of an initial incubation at 94 °C for 4 min, followed by 35 cycles (94 °C for 30 s, 55 °C for 1 min, 72 °C for 2.0 min) and a final extension at 72 °C for 10 min.

Table 1. Primers for the detection of Sarcocystidae in tissues of seabirds from Magdalena Island, Chile.

PCR	Primers	Sequences	PCR Step [a]	Reference
nPCR-18S	Tg18s48F	CCATGCATGTCTAAGTATAAGC	1	[28]
	Tg18s359R	GTTACCCGTCACTGCCAC	1	[28]
	Tg18s58F	CTAAGTATAAGCTTTTATACGGC	2	[28]
	Tg18s348R	TGCCACGGTAGTCCAATAC	2	[28]
nPCR-B1	T1	AGCGTCTCTCTTCAAGCAGCGTA	1	[29]
	T2	TCCGCAGCGACTTCTATCTCTGT	1	[29]
	T3	TGGGAATGAAAGAGACGCTAATGTG	2	[29]
	T4	TTAAAGCGTTCGTGGTCAACTATCG	2	[29]
nPCR-18Sb	18S9L	GGATAACCTGGTAATTCTATG	1 + 2	[32]
	18S1H	GGCAAATGCTTTCGCAGTAG	1 + 2	[32]
nPCR-CO1	COX1-227F25	GTTTTGGTAACTACTTTGTACCGAT	1	[31]
	COX1-885R25	GAAATATGCACGAGTATCTACCTCT	1	[31]
	COX1-275F22	TGTACCCACGAATTAATGCAGT	2	[31]
	COX1-844R21	GTGTGCCCATACTAGAGAACC	2	[31]
nPCR-ITS1	JS4	CGAAATGGGAAGTTTGAAC	1	[33]
	CT2c	CTGCAATTCACATTCGC	1	[30]
	JS4b	AGTCGTAACAAGGTTTCCGTAGG	2	[30]
	CT2b	TTGCGCGAGCCAAGACATC	2	[30]

[a] Primers used in the first round of amplification (1); primers used in the second round of amplification (2); primers used in both first and second round of amplification (1 + 2).

The first round of nPCR-B1 were performed with 1.0 μL of extracted DNA, 2.5 μL of 10× PCR Buffer (KCL 50 mM; Tris HCl 10 mM; pH 9.0), (Life Technologies Corporation, Carlsbad, CA 92008 USA), 0.75 μL of $MgCL_2$ (50 mM), 4.0 μL of mixed dNTPs (10 mM), 1.25 μL of each primer (10 μM), 0.15 μL of Platinum™ Taq DNA Polymerase (5 U/μL) (Life Technologies Corporation, Carlsbad, CA 92008 USA)and ultrapure autoclaved water to a volume of 25 μL per reaction. The PCR thermal cycling consisted of an initial incubation at 94 °C for 3 min, followed by 25 cycles (94 °C for 25 s, 55 °C for 1 min, 72 °C for 1.5 min) and a final extension at 72 °C for 10 min. For the second rounds: 1 μL of template derived from the first reactions, 2.5 μL of 10× PCR Buffer (KCL 50 mM; Tris HCl 10 mM; pH 9.0) (Life Technologies Corporation, Carlsbad, CA 92008 USA), 2.5 μL of $MgCL_2$ (50 mM), 4.0 μL of mixed dNTPs (10 mM), 1.25 μL of each primer (10 μM), 0.15 μL of Platinum™ Taq DNA Polymerase (Life Technologies Corporation, Carlsbad, CA 92008 USA) and ultrapure autoclaved water to a volume of 25 μL per reaction. The nPCR thermal cycling consisted of an initial incubation at 94 °C for 3 min, followed by 35 cycles (94 °C for 25 s, 55 °C for 1 min, 72 °C for 1.5 min) and a final extension at 72 °C for 10 min.

The first round of nPCR-18Sb, nPCR-ITS1 and nPCR-CO1 were performed with 4 μL of extracted DNA, 2.5 μL of 10× PCR Buffer (KCL 50 mM; Tris HCl 10 mM; pH 9.0) (Life Technologies Corporation, Carlsbad, CA 92008 USA), 1.0 μL of $MgCL_2$ (50 mM), 0.5 μL of mixed dNTPs (10 mM), 1.0 μL of each primer (10 μM), 0.2 μL of Platinum™ Taq DNA Polymerase (Life Technologies Corporation, Carlsbad, CA 92008 USA) (5 U/μL) (Termofischer Scientific) and ultrapure autoclaved water to a volume of 25 μL per reaction. The PCR thermal cycling consisted of an initial incubation at 94 °C for 3 min, followed by 35 cycles (94 °C for 30 s, 56 °C for 30 s, 72 °C for 50 s) and a final extension at 72 °C for 5 min. For the second rounds: 2 μL of template derived from the first reactions, 2.5 μL of 10× PCR Buffer (KCL 50 mM; Tris HCl 10 mM; pH 9.0) (Life Technologies Corporation, Carlsbad,

CA 92008 USA), 1.0 μL of $MgCL_2$ (50 mM), 0.5 μL of mixed dNTPs (10 mM), 2.5 μL of each primer (10 μM), 0.2 μL of Platinum™ Taq DNA) (Life Technologies Corporation, Carlsbad, CA 92008 USA) (5 U/μL) and ultrapure autoclaved water to a volume of 25 μL per reaction. The thermal cycling was the same used in the first round.

DNA of *Sarcocystis neurona*, *Neospora caninum* and *Hammondia hammondi* was used as positive controls and ultrapure DNAse-free water as the negative control for all reactions.

PCR products were resolved on 2.0% agarose gels and viewed through UV transillumination. Amplicons of the expected sizes were treated with ExoSAP-IT (Affimetryx/Thermo Fisher Scientific, Santa Clara, CA, USA), prepared for sequencing using the Big Dye Terminator cycle sequencing kit (Applied Biosystems, Foster City, CA, USA) and sequenced in an ABI automated sequencer (ABI 3500 Genetic Analyzer, Applied Biosystems). Sequencing was performed using the same primers as in the nPCR consensus. Sequence edition and contig assemblies were done by using the software Codoncode aligner, Codoncode Corporation. Final sequences were compared with homologous available in GenBank, using the BLASTn algorithm (Table S1) (https://blast.ncbi.nlm.nih.gov/Blast.cgi).

For the phylogenies, sequences were aligned using the Clustal W program, as implemented in the BioEdit Sequence Alignment Editor [34]. The phylogenetic tree based on ITS1 was inferred using MEGA X [35], through the maximum likelihood method and T92 model of evolutionary distances. Branch supports were tested through 1000 bootstrap replications.

3. Results

3.1. Molecular Identification

Sixty-six tissue samples from 22 seabirds were screened by nPCR-18Sa and only two samples of pectoral muscle from two Chilean skuas were positive. None of these two samples were positive for the *T. gondii*-specific nested-PCR (nPCR-B1). The two nPCR-18Sa-positive samples were also positive by nPCR-ITS1, nPCR-18Sb and nPCR-CO1 (Figure 1, left panel). After sequencing nPCR-ITS1, nPCR-18Sb and nPCR-CO1 amplicons and removal of primer-derived sequences, 861, 783 and 547 base pairs were obtained, respectively. Fragments of the sequences obtained are shown in Figure 1, right panel. The homologous sequences of the two samples were identical to each other, thus only one set were submitted to the GenBank, under the accession numbers MW160469, MW161469, MW157378. Through BLAST searches, ITS1, CO1 and 18S genetic sequences were compared with sequences producing the most significant alignments, with query coverage $\geq$ 99% and percentage similarities $\geq$ 99.00% in the cases of CO1 and 18S. All ITS1 sequences with query cover $\geq$ 96% were used for analyses of the genetic sequence of the skuas.

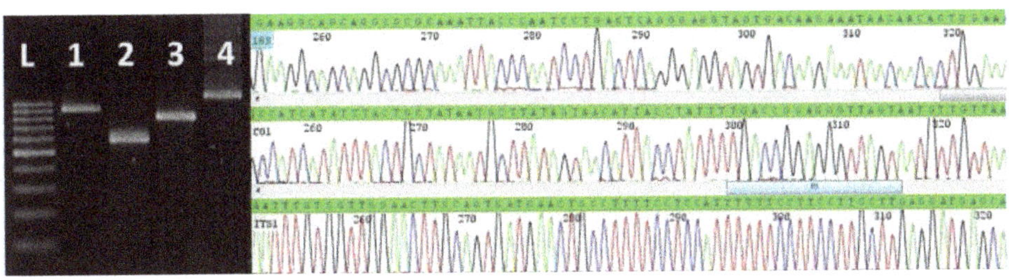

Figure 1. Left panel: Agarose gel electrophoresis of nPCR-ITS1 (**1**), nPCR-CO1 (**2**), and nPCR-18Sb (**3**) amplicons from *Sarcocystis* sp. ex *Stercorarius chilensis*, nPCR-ITS1 amplicons (**4**) from *Sarcocystis neurona* and Ladder Scada 100 bp, Sinapse, Inc. (**L**). Right panel: segments of the electropherogram obtained after sequencing nPCR-18Sb (**top**), nPCR-CO1 (**middle**), and nPCR-ITS1 (**bottom**) amplicons from *Sarcocystis* sp. ex *Stercorarius chilensis*.

The ITS1 fragment from the skuas showed 96.14–96.28% identity to sequences of *Sarcocystis halieti* from herring gulls (*Larus argentatus*) (MN450340–MN450356), great cormorants (*Phalacrocorax carbo*) (MH130209, JQ733513) and white-tailed sea-eagles (*Haliaeetus albicilla*) (MF946589–MF946596). *Sarcocystis* sp. from Cooper's hawk (*Accipiter cooperii*)

(KY348755), *Sarcocystis columbae* from common woodpigeons (*Columba palumbus*) (GU253885, HM125052) and herring gulls (*Larus argentatus*) (MN450338–MN450339) and *Sarcocystis corvusi* from Eurasian jackdaws (*Corvus monedula*) (JN256119) showed less than 94% sequence identity with homologous sequences from the Chilean skuas at ITS1 locus.

In contrast to ITS1, much less genetic variability was observed within the CO1 and 18S coding genes. The haplotype obtained from the skuas was named *Sarcocystis* sp. ex *Stercorarius chilensis*.

The CO1 sequences from *Sarcocystis* sp. ex *Stercorarius chilensis* were 100% identical with homologous sequences of *S. corvusi* (MH138314), *S. columbae* (MH138312) and *S. halieti* (MH138308-09; MF946583). *Sarcocystis fulicae* (MH138316), *Sarcocystis wobeseri* (MH138315), *Sarcocystis cornixi* (MH138313) and *Sarcocystis* sp. ex *Accipiter cooperii* (KY348756) differed by only one nucleotide substitution from *Sarcocystis* sp. ex *Stercorarius chilensis* at this locus. Regarding the 18S rRNA gene, the maximum percentage identity was found between *Sarcocystis* sp. ex *Stercorarius chilensis* and *S. halieti* (99.74%).

3.2. Phylogeny

The ITS1-based phylogeny demonstrated that the species that shares the most recent ancestral commonality with *Sarcocystis* sp. ex *Stercorarius chilensis* was *S. halieti* (Figure 2).

These sequences were separated with high support from a major clade comprising the species *Sarcocystis* sp. ex *Columba livia* (FJ232948), *Sarcocystis calchasi* (KC733715–KC733718) and *Sarcocystis wobeseri* (MN450365–MN450373, HM159421, JN256121), which exploit Anseriformes, Charadriiformes, Columbiformes, Psittaciformes and other birds as intermediate hosts.

Altogether, the *Sarcocystis* species that were most similar to *Sarcocystis* sp. ex *Stercorarius chilensis* used birds as intermediate hosts.

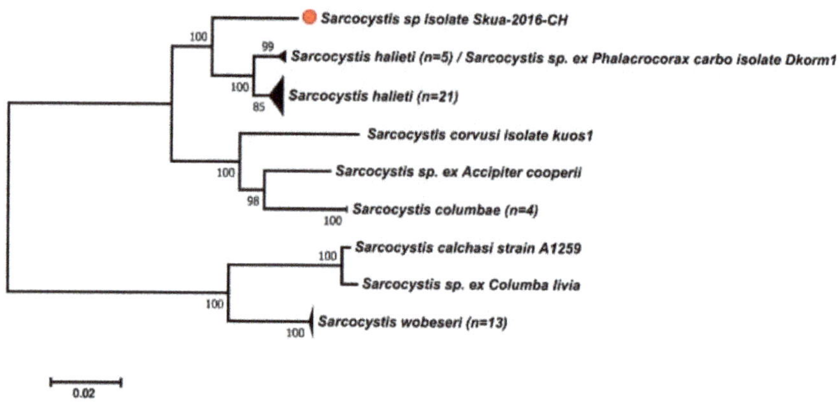

Figure 2. Phylogenetic tree of *Sarcocystis* species based on ITS1 sequences. The tree was constructed using the maximum likelihood method and Tamura 3 parameters nucleotide substitution model. The final alignment contained 49 sequences and 814 aligned nucleotide positions. All positions containing gaps and missing data were eliminated (complete deletion option). Numbers in branches represent bootstrap values after 1000 replicates. The red dot identifies the sequence of *Sarcocystis* sp. ex *Stercorarius chilensis*.

4. Discussion

Toxoplasma gondii has high genotypic diversity and several new genotypes have been described in wildlife [36–38], which has aided to understand the shape of molecular evolution and the epidemiology of the infection [38]. Similarly, several species descriptions have been made for the genus *Sarcocystis*, most of them with the aid of molecular methods [39–43]. This study presents the results of a molecular screening of Sarcocystidae focusing on animal species and geographical areas where these parasites have rarely or not yet been identified. DNA of organisms of the genus *Sarcocystis* was identified in two Chilean skuas, whereas DNA of *T. gondii* were not found in any sample. The *Sarcocystis* haplotype detected in skuas was named *Sarcocystis* sp. ex *Stercorarius chilensis*.

Although all the samples tested negative for the presence of *T. gondii*, antibodies against this parasite were detected previously in 57 (43.18%) of the 132 serum samples from free-living Magellanic penguins from the same region, with titers that ranged from 20 to 320 [7]. Herein, infected animals were not encountered possibly because the sampling was insufficient to find at least one positive animal, as the prevalence of the infection in seabirds in the sampled area are not known and the sampling might have not been representative of the population surveyed. In addition, the mass of tissue that was tested might have been insufficient because of the very sparse and focal distribution of *T. gondii* cysts in the tissues of the HI, thus, digestion of samples previously to the DNA extraction and subsequent DNA detection would be more appropriate than direct DNA extraction, as used in this study [44].

Oocysts of *T. gondii* can sporulate and survive in seawater for months [45,46]. Marine mammals in different groups (cetaceans, pinnipeds and sirenians) and seabirds might become infected through consumption of water containing the oocysts. Thus, *T. gondii* oocysts from felidae feces might enter the marine environment and contaminate both the water and several invertebrate species, which could act as vectors of infection for mammals and seabirds [47]. Mice can be experimentally infected when fed with *T. gondii*-contaminated oysters (*Crassostrea virginica*) [45] proving that *T. gondii* was able to survive for several months in these mollusks [48]. Anchovies and Pacific sardines can be experimentally contaminated with *T. gondii* oocysts, which indicates that migratory filter feeders may serve as biotic vectors for this parasite [49]. Another study proved that freshwater crustaceans were able to bioaccumulate *T. gondii* oocysts. It should be noted that crustaceans are part of penguins' and many seabirds' food chain [50,51]. Thus, although the birds screened here were found not infected by *T. gondii*, marine fauna are at risk of acquiring the infection, by ingesting oocysts carried by transport hosts (oysters, fish and other) or through predation of intermediate hosts in the marine or in the coastal environment.

Based on molecular data, *Sarcocystis* sp. ex *Stercorarius chilensis* is an undescribed *Sarcocystis* species, closely related to *S. halieti*. The molecular identification based on ITS1, CO1 and 18S rRNA gene sequences showed a closed relationship between *Sarcocystis* sp. from Chilean skuas and other *Sarcocystis* spp. that use birds as intermediate hosts and predatory birds as definitive hosts. As expected, the most variable locus was ITS1, and phylogenies based on 18S rRNA and CO1 genes showed insufficient discrimination power to differentiate between species within the genus [39,41].

The most similar sequences to ITS1 of *Sarcocystis* sp. ex *Stercorarius chilensis* are those from *Sarcocystis* spp. that use hawks as definitive hosts. *Sarcocystis* sp. ex *Stercorarius chilensis* grouped together with *S. halieti*, a species that uses white-tailed sea-eagles (*Haliaeetus albicilla*) and Eurasian sparrowhawks (*Accipiter nisus*) as definitive hosts [39]. Other taxa found through ITS1-based BLAST searches encompass *Sarcocystis* spp. that also use hawks as definitive hosts (*Accipiter cooperii*, *Accipiter nisus*), except for *S. corvusi*, for which this information remains unknown [52,53]. Accipiter hawks (*Accipiter gentilis*, *Accipiter nisus*) are definitive hosts for *S. calchasi* [54–56].

Several studies have expanded the knowledge on the host specificity of *Sarcocystis*, as unequivocal identification of the parasite can be achieved after identifying sarcocysts and oocysts to species level using molecular methods. *Sarcocystis halieti* and *Sarcocystis lari* were found to have formed oocysts in the intestine of white-tailed sea eagle (*Haliaeetus albicilla*), showing for the first time the potential role of sea eagle as definitive host of those species of *Sarcocystis* [53]. Likewise, european seabirds were found to harbor several species of *Sarcocystis* after DNA of *Sarcocystis lari*, *S. wobeseri*, *S. columbae* and *S. halieti* were detected in sarcocysts infecting muscle of herring gulls (*Larus argentatus*), great black-backed gulls (*Larus marinus*) and great cormorants (*Phalacrocorax carbo*) in Lithuania [40–42].

The four morphologically indistinguishable *Sarcocystis* species, *Sarcocystis lari, S. wobeseri, S. columbae* and *S. halieti,* could only be differentiated in *L. argentatus* by means of ITS1 sequence analysis [42]. Likewise, only ITS1 clearly discriminated *Sarcocystis* sp. ex *Stercorarius chilensis* from *S. halieti,* which reinforces that molecular characterization using this marker is of paramount importance to distinguish closed related species within the genus.

It is well known that a single animal can host more than one *Sarcocystis* species [40]. Here, sarcocysts were not individually excised and subjected to molecular examination, notwithstanding, the possibility of mixed infected samples of skuas was discarded because single peaks and clean sequence throughout the chromatograms were obtained for each sequence. Thus, a haplotype could be confidently assigned to the samples.

Although screening *Sarcocystis* by using molecular methods without morphological characterization of parasitic structures is obviously not enough to name a new species, this procedure may provide subsides to future studies on the epidemiology of the infection and its impact on the health of marine fauna. To our knowledge, *Sarcocystis* in south American seabirds were identified only once [43], which suggests that a wide field of research on diversity of sarcocystidae can be explored on this continent.

5. Conclusions

Although few animals have been screened in this study and morphological characterization of the parasites was not carried out, evidence of an unprecedented haplotype of *Sarcocystis* was found in skuas from Chile, which demonstrate that molecular screening of *Sarcocystis* can be a valuable tool to prospect for new species, contributing to knowledge on the epidemiology of sarcocystosis and life cycle of *Sarcocystis*. Sporocysts shed with feces, sarcocysts in tissues or rapid dividing structures in acute sarcocystosis (schyzonts and merozoites) can be more easily and accurately identified as data on *Sarcocystis* genetic sequences increases. Nevertheless, a complete study encompassing aspects of life cycle and morphological data is necessary to fully describe *Sarcocystis* sp. ex *Stercorarius chilensis* and additional studies are needed to better understand the epidemiology of the infection and its impact on the health of marine fauna.

Supplementary Materials: The following are available online at https://www.mdpi.com/2076-2615/11/2/245/s1, Table S1. Results from BLAST search using sequences of ITS1, COX1 and 18S of *Sarcocystis* sp. ex *Stercorarius chilensis*.

Author Contributions: Conceptualization, I.C.L.A., S.M.G. and R.M.S. writing—original draft preparation, I.C.L.A. and R.M.S.; review and editing, I.C.L.A., S.M.G., H.A.B.L., S.M.-L. and R.M.S.; visualization, I.C.L.A., S.M.G., H.A.B.L., S.M.-L. and R.M.S. All authors have read and agreed to the published version of the manuscript.

Funding: This research was funded by FAPESP (Fundação de Amparo à Pesquisa do Estado de São Paulo), Brazil and by Coordenação de Aperfeiçoamento de Pessoal de Nível Superior (CAPES), Brazil. R.M.S. and S.M.G. received a fellowship from CNPq (Conselho Nacional de Desenvolvimento Científico e Tecnológico).

Institutional Review Board Statement: This study was approved by the Ethics Committee on Animal Use (CEUA-no. 9701041113 - 2014) of the School of Veterinary Medicine, University of São Paulo (FMVZ-USP). Sample collections on Magdalena Island were conducted under license no. 039/2016 issued by the National Forestry Corporation (Corporación Nacional Forestal; CONAF), and permit no. 2799 issued by the National Fisheries Service (Servicio Nacional de Pesca; SERNAPESCA), Chile.

Informed Consent Statement: Not applicable.

Data Availability Statement: Not applicable.

Acknowledgments: To Roberto Fernández and Ricardo Cid (CONAF) for their valuable assistance during the sample collections on Magdalena Island.

Conflicts of Interest: The authors declare no conflict of interest.

References

1. Sam-Yellowe, T.Y. Rhoptry organelles of the apicomplexa: Their role in host cell invasion and intracellular survival. *Parasitol. Today* **1996**, *12*, 308–316. [CrossRef]
2. Morrison, D.A.; Bornstein, S.; Thebo, P.; Wernery, U.; Kinne, J.; Mattsson, J.G. The current status of the small subunit rRNA phylogeny of the coccidia (Sporozoa). *Int. J. Parasitol.* **2004**, *34*, 501–514. [CrossRef] [PubMed]
3. Mugridge, N.B.; Morrison, D.A.; Johnson, A.M.; Luton, K.; Dubey, J.P.; Votýpka, J.; Tenter, A.M. Phylogenetic relationships of the genus *Frenkelia*: A review of its history and new knowledge gained from comparison of large subunit ribosomal ribonucleic acid gene sequences. *Int. J. Parasitol.* **1999**, *29*, 957–972. [CrossRef]
4. Samarasinghe, B.; Johnson, J.; Ryan, U. Phylogenetic analysis of Cystoisospora species at the rRNA ITS1 locus and development of a PCR-RFLP assay. *Exp. Parasitol.* **2008**, *118*, 592–595. [CrossRef] [PubMed]
5. Tenter, A.M.; Barta, J.R.; Beveridge, I.; Duszynski, D.W.; Mehlhorn, H.; Morrison, D.A.; Andrew Thompson, R.C.; Conrad, P.A. The conceptual basis for a new classification of the coccidian. *Int. J. Parasitol.* **2002**, *32*, 595–616. [CrossRef]
6. Dubey, J.P. *Toxoplasmosis of Animals and Humans*, 2nd ed.; CRC Press: Boca Raton, FL, USA, 2010.
7. Acosta, I.C.L.; Souza-Filho, A.F.; Muñoz-Leal, S.; Soares, H.S.; Heinemann, M.B.; Moreno, L.; González-Acuña, D.; Gennari, S.M. Evaluation of antibodies against *Toxoplasma gondii* and *Leptospira* spp. in Magellanic penguins (*Spheniscus magellanicus*) on Magdalena Island, Chile. *Vet. Parasitol. Reg. Stud. Rep.* **2019**, *16*, 100282. [CrossRef]
8. Dubey, J.P.J.; Calero-Bernal, R.; Rosenthal, B.M.B.; Speer, C.A.C.; Fayer, R. *Sarcocystosis of Animals and Humans*; CRC Press: Boca Raton, FL, USA, 2015.
9. Rioseco, H.; Cubillos, V.; González, H.; Díaz, L. Sarcosporidiosis en pudues (*Pudu pudu*, Molina, 1782) Primera comunicación en Chile. *Arch. Med. Vet.* **1976**, *8*, 122–123.
10. Gorman, T.R.; Alcaíno, H.A.; Muñuz, H.; Cunazza, C. *Sarcocystis* sp. in guanaco (*Lama guanicoe*) and effect of temperature on its viability. *Vet. Parasitol.* **1984**, *15*, 95–101. [CrossRef]
11. Sepúlveda, M.A.; Seguel, M.; Alvarado-Rybak, M.; Verdugo, C.; Muñoz-Zanzi, C.; Tamayo, R. Postmortem findings in four South American sea lions (*Otaria byronia*) from an urban colony in Valdivia, Chile. *J. Wildl. Dis.* **2015**, *51*, 279–282. [CrossRef]
12. Tavares, D.C.; Moura, J.F.; de Amorim, C.E.; Boldrini, M.A.; Siciliano, S. Aves, Stercorariidae, Chilean Skua *Stercorarius chilensis* Bonaparte, 1857: First documented record for the state of Espírito Santo, southeastern Brazil. *Check List* **2012**, *8*, 560. [CrossRef]
13. Furness, R.W. *The Skuas*; Poyser, Poyser Monographs; Bloomsbury Collections: London, UK, 1987.
14. Robinson, I. Seabirds. In *Handbook of Avian Medicine*; Elsevier Ltd.: Amsterdam, The Netherlands, 2009; pp. 377–403.
15. BirdLife International. Species Factsheet: Catharacta Chilensis. Available online: http://www.birdlife.org (accessed on 16 July 2020).
16. Ludynia, K.; Garthe, S.; Luna-Jorquera, G. Seasonal and regional variation in the diet of the Kelp Gull in northern Chile. *Waterbirds* **2005**, *28*, 359–365. [CrossRef]
17. Yorio, P.; Branco, J.O.; Lenzi, J.; Luna-Jorquera, G.; Zavalaga, C. Distribution and Trends in Kelp Gull (*Larus dominicanus*) Coastal Breeding Populations in South America. *Waterbirds* **2016**, *39*, 114–135. [CrossRef]
18. Boersma, P.D.; Frere, E.; Kane, O.; Pozzi, L.M.; Pütz, K.; Rey, A.R. Magellanic penguins. In *Penguins: Natural History and Conservation*; Borboroglu, P.G., Boersma, P.D., Eds.; University of Washington Press: Seattle, WA, USA, 2013; pp. 233–263.
19. Frere, E.; Gandini, P.; Lichtschein, V. Variacion latitudinal en la dieta del pinguino de Magallanes (*Spheniscus magellanicus*) en la Costa Patagonica, Argentina. *Ornitol. Neotrop.* **1996**, *7*, 35–41.
20. Odening, K. The present state of species-systematics in *Sarcocystis* Lankester, 1882 (Protista, Sporozoa, Coccidia). *Syst. Parasitol.* **1998**, *41*, 209–233. [CrossRef]
21. Pan, J.; Ma, C.; Huang, Z.; Ye, Y.; Zeng, H.; Deng, S.; Hu, J.; Tao, J. Morphological and molecular characterization of *Sarcocystis wenzeli* in chickens (*Gallus gallus*) in China. *Res. Sq.* **2020**, *13*, 1–7. [CrossRef]
22. Konradt, G.; Bianchi, M.V.; Leite-Filho, R.V.; da Silva, B.Z.; Soares, R.M.; Pavarini, S.P.; Driemeier, D. Necrotizing meningoencephalitis caused by *Sarcocystis falcatula* in bare-faced ibis (*Phimosus infuscatus*). *Parasitol. Res.* **2017**, *116*, 809–812. [CrossRef]
23. Olias, P.; Maier, K.; Wuenschmann, A.; Reed, L.; Armién, A.G.; Shaw, D.P.; Gruber, A.D.; Lierz, M. *Sarcocystis calchasi* has an expanded host range and induces neurological disease in cockatiels (*Nymphicus hollandicus*) and North American rock pigeons (*Columbia livia* f. dom.). *Vet. Parasitol.* **2014**, *200*, 59–65. [CrossRef]
24. Wilson, T.M.; Sousa, S.K.H.; Paludo, G.R.; de Melo, C.B.; Llano, H.A.B.; Soares, R.M.; Castro, M.B. An undescribed species of *Sarcocystis* associated with necrotizing meningoencephalitis in naturally infected backyard chickens in the Midwest of Brazil. *Parasitol. Int.* **2020**, *76*, 102098. [CrossRef]
25. Roeder, A.D.; Ritchie, P.A.; Lambert, D.M. New DNA markers for penguins. *Conserv. Genet.* **2002**, *3*, 341–344. [CrossRef]
26. Pons, J.M.; Hassanin, A.; Crochet, P.A. Phylogenetic relationships within the Laridae (Charadriiformes, Aves) inferred from mitochondrial markers. *Mol. Phylogenet. Evol.* **2005**, *37*, 686–699. [CrossRef]
27. Han, Y.D.; Baek, Y.S.; Kim, J.H.; Choi, H.G.; Kim, S. Complete mitochondrial genome of the South Polar Skua *Stercorarius maccormicki* (Charadriiformes, Stercorariidae) in Antarctica. *Mitochondrial DNA* **2016**, *27*, 1783–1784.
28. Su, C.; Shwab, E.K.; Zhou, P.; Zhu, X.Q.; Dubey, J.P. Moving towards an integrated approach to molecular detection and identification of *Toxoplasma gondii*. *Parasitology* **2010**, *137*, 1–11. [CrossRef] [PubMed]
29. Yai, L.E.O.; Cañon-Franco, W.A.; Geraldi, V.C.; Summa, M.E.L.; Camargo, M.C.G.O.; Dubey, J.P.; Gennari, S.M. Seroprevalence of *Neospora caninum* and *Toxoplasma gondii* antibodies in the South American opossum (*Didelphis marsupialis*) from the city of São Paulo, Brazil. *J. Parasitol.* **2003**, *89*, 870–871. [CrossRef] [PubMed]

30. Soares, R.M.; Lopes, E.G.; Keid, L.B.; Sercundes, M.K.; Martins, J.; Richtzenhain, L.J. Identification of *Hammondia heydorni* oocysts by a heminested-PCR (hnPCR-AP10) based on the *H. heydorni* RAPD fragment AP10. *Vet. Parasitol.* **2011**, *175*, 168–172. [CrossRef] [PubMed]
31. Gondim, L.F.P.; Soares, R.M.; Tavares, A.S.; Silva, W.B.; de Jesus, R.F.; Llano, H.A.B.; Gondim, L.Q. *Sarcocystis falcatula*-like derived from opossum in Northeastern Brazil: In vitro propagation in avian cells, molecular characterization and bioassay in birds. *Int. J. Parasitol. Parasites Wildl.* **2019**, *10*, 132–137. [CrossRef]
32. Li, Z.Q.; Yang, Y.X.; Zuo, S.W.; Attwood, X.W.; Chen, Y.P.; Zhang, A. PCR-based RFLP analysis of *Sarcocystis cruzi* (Protozoa: Sarcocystidae) in Yunnan Province, PR China, reveals the water buffalo (*Bubalus bubalis*) as a natural intermediate host. *J. Parasitol.* **2002**, *88*, 1259–1261. [CrossRef]
33. Šlapeta, J.R.; Koudela, B.; Votýpka, J.; Modrý, D.; Hořejš, R.; Lukeš, J. Coprodiagnosis of Hammondia heydorni in dogs by PCR based amplification of ITS 1 rRNA: Differentiation from morphologically indistinguishable oocysts of *Neospora caninum*. *Vet. J.* **2002**, *163*, 147–154. [CrossRef]
34. Thompson, J.D.; Gibson, T.J.; Plewniak, F.; Jeanmougin, F.; Higgins, D.G. The CLUSTAL_X windows interface: Flexible strategies for multiple sequence alignment aided by quality analysis tools. *Nucleic Acids Res.* **1997**, *25*, 4876–4882. [CrossRef] [PubMed]
35. Kumar, S.; Stecher, G.; Li, M.; Knyaz, C.; Tamura, K. MEGA X: Molecular evolutionary genetics analysis across computing platforms. *Mol. Biol. Evol.* **2018**, *35*, 1547–1549. [CrossRef] [PubMed]
36. Sibley, L.D.; Khan, A.; Ajioka, J.W.; Rosenthal, B.M. Genetic diversity of *Toxoplasma gondii* in animals and humans. *Philos. Trans. R. Soc. Lond. B Biol. Sci.* **2009**, *364*, 2749–2761. [CrossRef] [PubMed]
37. Vitaliano, S.N.; Soares, H.S.; Minervino, A.H.H.; Santos, A.L.Q.; Werther, K.; Marvulo, M.F.V.; Siqueira, D.B.; Pena, H.F.J.; Soares, R.M.; Su, C.; et al. Genetic characterization of *Toxoplasma gondii* from Brazilian wildlife revealed abundant new genotypes. *Int. J. Parasitol. Parasites Wildl.* **2014**, *3*, 276–283. [CrossRef] [PubMed]
38. Aguirre, A.A.; Longcore, T.; Barbieri, M.; Dabritz, H.; Hill, D.; Klein, P.N.; Lepczyk, C.; Lilly, E.L.; Milcarsky, R.M.J.; Su, C.; et al. The one health approach to toxoplasmosis: Epidemiology, control, and prevention strategies. *EcoHealth* **2019**, *16*, 378–390. [CrossRef] [PubMed]
39. Gjerde, B.; Vikøren, T.; Hamnes, I.S. Molecular identification of *Sarcocystis halieti* n. sp., *Sarcocystis lari* and *Sarcocystis truncata* in the intestine of a white-tailed sea eagle (*Haliaeetus albicilla*) in Norway. *Int. J. Parasitol. Parasites Wildl.* **2018**, *7*, 1–11. [CrossRef] [PubMed]
40. Prakas, P.; Butkauskas, D.; Juozaitytė-Ngugu, E. Molecular identification of four *Sarcocystis* species in the herring gull, *Larus argentatus*, from Lithuania. *Parasit Vectors* **2020**, *13*, 1–6. [CrossRef]
41. Prakas, P.; Butkauskas, D.; Švažas, S.; Stanevičius, V. Morphological and genetic characterisation of *Sarcocystis halieti* from the great cormorant (*Phalacrocorax carbo*). *Parasitol. Res.* **2018**, *117*, 3663–3667. [CrossRef]
42. Prakas, P.; Kutkiene, L.; Butkauskas, D.; Sruoga, A.; Zalakevicius, M. Description of *Sarcocystis lari* sp. n. (Apicomplexa: Sarcocystidae) from the great black-backed gull, *Larus marinus* (Charadriiformes: Laridae), on the basis of cyst morphology and molecular data. *Folia Parasitol.* **2014**, *61*, 11–17. [CrossRef]
43. Acosta, I.C.L.; Soares, R.M.; Mayorga, L.F.S.P.; Alves, B.F.; Soares, H.S.; Gennari, S.M. Occurrence of tissue cyst forming coccidia in Magellanic penguins (*Spheniscus magellanicus)* rescued on the coast of Brazil. *PLoS ONE* **2018**, *13*, e0212467. [CrossRef]
44. Dubey, J.P. Long-term persistence of *Toxoplasma gondii* in tissues of pigs inoculated with *T. gondii* oocysts and effect of freezing on viability of tissue cysts in pork. *Am. J. Vet. Res.* **1988**, *49*, 910–913.
45. Lindsay, D.S.; Collins, M.V.; Mitchell, S.M.; Cole, R.A.; Flick, G.J.; Wetch, C.N.; Lindquist, A.; Dubey, J.P. Sporulation and survival of *Toxoplasma gondii* oocysts in seawater. *J. Eukaryot. Microbiol.* **2003**, *50*, 687–688. [CrossRef]
46. Fayer, R.; Dubey, J.P.; Lindsay, D.S. Zoonotic protozoa: From land to sea. *Trends Parasitol.* **2004**, *20*, 531–536. [CrossRef]
47. Cole, R.A.; Lindsay, D.S.; Howe, D.K.; Roderick, C.L.; Dubey, J.P.; Thomas, N.J.; Baeten, L.A. Biological and molecular characterization of *Toxoplasma gondii* strains obtained from southern sea otters (*Enhydra lutris nereis*). *J. Parasitol.* **2000**, *86*, 526–530. [CrossRef]
48. Lindsay, D.S.; Phelps, K.K.; Smith, S.A.; Flick, G.; Sumner, S.S.; Dubey, J.P. Removal of *Toxoplasma gondii* oocysts from sea water by eastern oysters (*Crassostrea virginica*). *J. Eukaryot. Microbiol.* **2001**, *48*, 197S–198S. [CrossRef] [PubMed]
49. Lindsay, D.S.; Collins, M.V.; Mitchell, S.M.; Wetch, C.N.; Rosypal, A.C.; Flick, G.J.; Zajac, A.M.; Lindquist, A.; Dubey, J.P. Survival of *Toxoplasma gondii* oocysts in Eastern oysters (*Crassostrea virginica*). *J. Parasitol.* **2004**, *90*, 1054–1057. [CrossRef] [PubMed]
50. Massie, G.N.; Ware, M.W.; Villegas, E.N.; Black, M.W. Uptake and transmission of *Toxoplasma gondii* oocysts by migratory, filter-feeding fish. *Vet. Parasitol.* **2010**, *169*, 296–303. [CrossRef] [PubMed]
51. Bigot-Clivot, A.; Ladeiro, M.P.; Lepoutre, A.; Bastien, F.; Bonnard, I.; Dubey, J.P.; Villena, I.; Aubert, D.; Geffard, O.; François, A.; et al. Bioaccumulation of *Toxoplasma* and *Cryptosporidium* by the freshwater crustacean *Gammarus fossarum*: Involvement in biomonitoring surveys and trophic transfer. *Ecotoxicol. Environ. Saf.* **2016**, *133*, 188–194. [CrossRef]
52. Mayr, S.L.; Maier, K.; Müller, J.; Enderlein, D.; Gruber, A.D.; Lierz, M. *Accipiter hawks* (Accipitridae) confirmed as definitive hosts of *Sarcocystis turdusi*, *Sarcocystis cornixi* and *Sarcocystis* sp. ex *Phalacrocorax carbo*. *Parasitol. Res.* **2016**, *115*, 3041–3047. [CrossRef]
53. Lindsay, D.S.; Verma, S.K.; Scott, D.; Dubey, J.P.; von Dohlen, A.R. Isolation, molecular characterization, and in vitro schizogonic development of *Sarcocystis* sp. ex *Accipiter cooperii* from a naturally infected Cooper's hawk (*Accipiter cooperii*). *Parasitol. Int.* **2017**, *66*, 106–111. [CrossRef]
54. Olias, P.; Olias, L.; Krücken, J.; Lierz, M.; Gruber, A.D. High prevalence of *Sarcocystis calchasi* sporocysts in European *Accipiter hawks*. *Vet. Parasitol.* **2011**, *175*, 230–236. [CrossRef]

55. Olias, P.; Gruber, A.D.; Hafez, H.M.; Heydorn, A.O.; Mehlhorn, H.; Lierz, M. *Sarcocystis calchasi* sp. nov. of the domestic pigeon (*Columba livia* f. domestica) and the Northern goshawk (*Accipiter gentilis*): Light and electron microscopical characteristics. *Parasitol. Res.* **2010**, *106*, 577–585. [CrossRef]
56. Olias, P.; Olias, L.; Lierz, M.; Mehlhorn, H.; Gruber, A.D. *Sarcocystis calchasi* is distinct to *Sarcocystis columbae* sp. nov. from the wood pigeon (*Columba palumbus*) and Sarcocystis sp. from the sparrow hawk (*Accipiter nisus*). *Vet. Parasitol.* **2010**, *171*, 7–14. [CrossRef]

animals

MDPI

Brief Report

Zoonotic Microsporidia in Wild Lagomorphs in Southern Spain

Anabel Martínez-Padilla [1,†], Javier Caballero-Gómez [2,3,*,†], Ángela Magnet [4], Félix Gómez-Guillamón [5], Fernando Izquierdo [4], Leonor Camacho-Sillero [5], Saúl Jiménez-Ruiz [2,6], Carmen del Águila [4] and Ignacio García-Bocanegra [2]

1 Departamento de Biología Molecular y Bioquímica, Universidad de Málaga (UMA), 29010 Málaga, Spain; anabelmp@uma.es
2 Grupo de Investigación en Sanidad Animal y Zoonosis (GISAZ), Departamento de Sanidad Animal, Universidad de Córdoba (UCO), 14014 Córdoba, Spain; sauljimenezruiz@gmail.com (S.J.-R.); nacho.garcia@uco.es (I.G.-B.)
3 Unidad de Enfermedades Infecciosas, Grupo de Virología Clínica y Zoonosis, Instituto Maimónides de Investigación Biomédica de Córdoba (IMIBIC), Universidad de Córdoba (UCO), 14004 Córdoba, Spain
4 Facultad de Farmacia, Universidad San Pablo-CEU, CEU Universities, Urbanización Montepríncipe, Boadilla del Monte, 28660 Madrid, Spain; angela.magnetdavila@ceu.es (Á.M.); ferizqui@ceu.es (F.I.); cagupue@ceu.es (C.d.Á.)
5 Programa de Vigilancia Epidemiológica de la Fauna Silvestre en Andalucía (PVE), Consejería de Agricultura, Ganadería, Pesca y Desarrollo Sostenible, Junta de Andalucía, 29006 Málaga, Spain; felixj.gomezguillamon@juntadeandalucia.es (F.G.-G.); leonorn.camacho@juntadeandalucia.es (L.C.-S.)
6 Grupo Sanidad y Biotecnología (SaBio), Instituto de Investigación en Recursos Cinegéticos (IREC-CSIC-JCCM), Universidad de Castilla-la Mancha (UCLM), 13005 Ciudad Real, Spain
* Correspondence: javiercaballero15@gmail.com; Tel.: +34-957218725
† Equally contributed as first author.

Received: 9 November 2020; Accepted: 23 November 2020; Published: 26 November 2020

Simple Summary: A cross-sectional study was carried out to assess the presence of zoonotic microsporidia in organ meats of European wild rabbits and Iberian hares consumed by humans in Spain. Between July 2015 and December 2018, kidney samples from 383 wild rabbits and kidney and brain tissues from 79 Iberian hares in southern Spain were tested by species-specific polymerase chain reactions (PCRs) for the detection of microsporidia DNA. We confirmed the presence of *Enterocytozoon bieneusi* in three wild rabbits and *Encephalitozoon intestinalis* in one wild rabbit and three Iberian hares. However, none of the 462 sampled wild lagomorphs showed *Encephalitozoon hellem* nor *Encephalitozoon cuniculi* infection. This is the first report of *E. intestinalis* infection in wild rabbits and Iberian hares. The presence of *E. bieneusi* and *E. intestinalis* in organ meats from wild lagomorphs can be of public health concern. Additional studies are required to determine the real prevalence of these parasites in European wild rabbit and Iberian hare.

Abstract: Microsporidia are obligate intracellular protist-like fungal pathogens that infect a broad range of animal species, including humans. This study aimed to assess the presence of zoonotic microsporidia (*Enterocytozoon bieneusi, Encephalitozoon intestinalis, Encephalitozoon hellem,* and *Encephalitozoon cuniculi*) in organ meats of European wild rabbit (*Oryctolagus cuniculus*) and Iberian hare (*Lepus granatensis*) consumed by humans in Spain. Between July 2015 and December 2018, kidney samples from 383 wild rabbits and kidney and brain tissues from 79 Iberian hares in southern Spain were tested by species-specific PCR for the detection of microsporidia DNA. *Enterocytozoon bieneusi* infection was confirmed in three wild rabbits (0.8%; 95% CI: 0.0–1.7%) but not in hares (0.0%; 95% CI: 0.0–4.6%), whereas *E. intestinalis* DNA was found in one wild rabbit (0.3%; 95% CI: 0.0–0.8%) and three Iberian hares (3.8%; 95% CI: 0.0–8.0%). Neither *E. hellem* nor *E. cuniculi* infection were detected in the 462 (0.0%; 95% CI: 0.0–0.8%) lagomorphs analyzed. The absence of *E. hellem* and *E. cuniculi* infection suggests a low risk of zoonotic foodborne transmission from these wild lagomorph species in southern Spain. To the authors' knowledge, this is the first report of *E. intestinalis* infection in wild rabbits and Iberian hares. The presence of *E. bieneusi* and *E. intestinalis* in organ meats from wild lagomorphs can be of public health concern. Additional studies are required to determine the real prevalence of these parasites in European wild rabbit and Iberian hare.

Keywords: *Encephalitozoon intestinalis*; *E. hellem*; *E. cuniculi*; *Enterocytozoon bieneusi*; European wild rabbit; Iberian hare; zoonotic; foodborne

1. Introduction

The European wild rabbit (*Oryctolagus cuniculus*) and the Iberian hare (*Lepus granatensis*) are two endemic species in the Iberian Peninsula. Both lagomorph species play a key role in the ecology of Mediterranean ecosystems [1,2]. They are the staple prey for a large number of predators, including endangered species such as the Iberian lynx (*Lynx pardinus*) and the Spanish imperial eagle (*Aquila adalberti*) [3]. These lagomorphs are also among the main small game species in Spain. About 5.3 million wild rabbits and 890 thousand Iberian hares are harvested annually in this country and are generally consumed without sanitary inspection since they are intended for small-scale retail sale or personal consumption [4]. Public health concerns indicate the need for epidemiological studies on zoonotic diseases affecting wildlife species that are a source of food for humans [5]. In this respect, the role of wild lagomorphs as reservoirs of zoonotic parasites has been widely documented [6–8].

The phylum Microsporidia comprises more than 1500 obligate intracellular spore-forming parasites classified as fungi that can infect a wide range of vertebrate and invertebrate hosts through the fecal–oral route [9]. Among the 17 zoonotic microsporidia species, *Enterocytozoon bieneusi, Encephalitozoon intestinalis, E. hellem,* and *E. cuniculi* are the most common species reported in humans [10,11]. Microsporidiosis in humans is usually characterized by chronic diarrhea and wasting syndrome, although keratoconjunctivitis, hepatitis, myositis, kidney and urogenital infection, ascites, and/or cholangitis have also been reported, particularly in immunocompromised individuals [12].

Foodborne transmission has previously been evidenced as a zoonotic route of microsporidia infection [13]. In this regard, in European countries, zoonotic microsporidia species have been detected in the Eastern cottontail rabbit (*Sylvilagus floridanus*), the European brown hare (*Lepus europaeus*), and the European rabbit [14–17]. Even though microsporidia infections in these lagomorph species are usually asymptomatic, microsporidiosis has been reported in domestic rabbit and European brown hare, causing neurological and/or renal disorders and even death [17–19]. However, knowledge about the role of wild lagomorphs in the epidemiology of microsporidia remains scarce [20]. Hence, the present study aimed to assess the presence of *E. bieneusi, E. intestinalis, E. hellem,* and *E. cuniculi* in organ meats from European wild rabbit and Iberian hare, the most frequent wild lagomorph species consumed by humans in Spain.

2. Materials and Methods

A cross-sectional study was carried out on wild lagomorphs in Andalusia (southern Spain) (36° N–38°60′ N, 1°75′ W–7°25′ W, 87,268 km^2) between July 2015 and December 2018. The sample size for wild rabbits was calculated assuming a prevalence of 50% (the highest sample size for studies with unknown prevalence) with a 95% confidence level (95% CI) and a desired precision of ±5% [21], which gave 384 wild rabbits to be sampled. Whenever possible, a minimum of 49 animals were sampled in each of the eight provinces of Andalusia in order to detect microsporidia infection with a 95% probability and an assumed minimum within-province prevalence of 6%. Kidney samples were collected from 383 wild rabbits in 40 hunting areas (Figure 1). In the same study region and period, kidney and brain samples were also obtained from 79 Iberian hares in 43 hunting or protected areas (Figure 1). Sampled animals were necropsied in the field. Samples were placed in sterile bottles

(25 mL), preserved at 4 °C while transported to the laboratory, and stored at −20 °C until analysis. Data on species, age, gender, sampling site, province, and sampling year were gathered from each wild lagomorph sampled, whenever possible. Data analyses were carried out using the SPSS statistical software package version 25.0 (IBM Corporation, Somers, NY, USA).

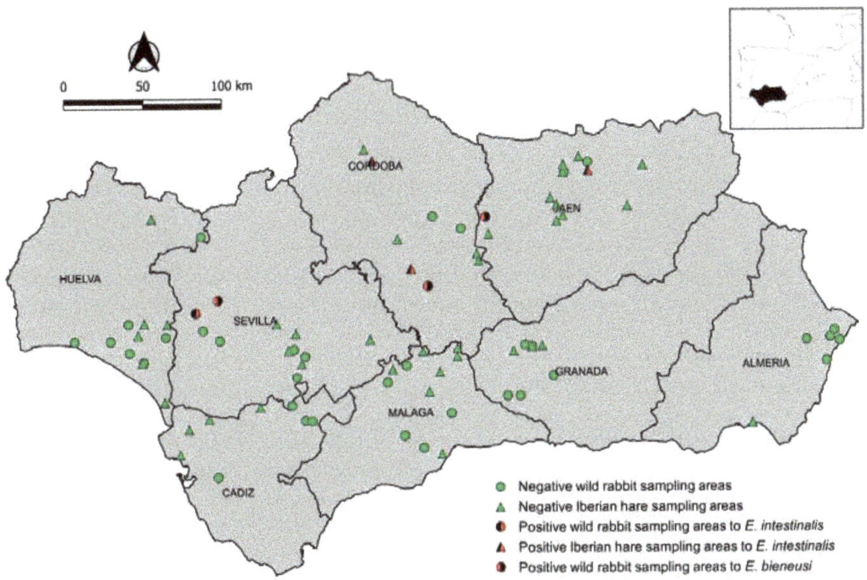

Figure 1. Map of Andalusia (southern Spain) showing the locations of European wild rabbits (dots) and Iberian hares (triangles) sampled. Green and red symbols indicate negative and positive animals, respectively, detected in these hunting areas.

This study did not involve purposeful killing of animals. The collection of samples was performed by personnel of the Epidemiological Surveillance Program in Wildlife of the Government of Andalusia. Samples were collected from legally hunted animals during the hunting seasons or by passive surveillance in complete agreement with Andalusian, Spanish, and European regulations. No ethical approval by an Institutional Animal Care and Use Committee was deemed necessary.

Genomic DNA was extracted from the kidney and brain samples using the G-spin™ total DNA extraction kit (Intron Biotechnology, Seongnam-Si, Korea), following the manufacturer's instructions. PCR amplifications were performed using the GenAmp kit (Perkin-Elmer Cetus, Norwalk, CT, USA) to detect small-subunit rRNA (SSU rRNA) coding regions of the microsporidia using species-specific primers previously described (Table 1) in a Gene Amp PCR system 9700 thermocycler (Perkin Elmer). The final concentration was 0.2 mM of each dNTP, 0.2 μM of each primer, buffer with $MgCl_2$ (1.5 mM), and 1.25 U of Taq polymerase. Conditions for PCR reactions were as follows: denaturing at 94 °C for 30 s followed by 35 cycles of annealing at 45 °C for 30 s for *E. intestinalis* primers or 55 °C for 30 s for the remaining microsporidia species and extension at 72 °C for 90 s. PCR products were analyzed by electrophoresis in 2% agarose gel stained with ethidium bromide and examined under UV light.

Table 1. Primers used for the detection of *Enterocytozoon bieneusi*, *Encephalitozoon intestinalis*, *Encephalitozoon hellem*, and *Encephalitozoon cuniculi*.

Microspodia Species	Primers	Forward Sequence (5′-3′)	Reverse Sequence (5′-3′)	Obtained Amplicon Size	Reference
E. bieneusi	EBIEF1/EBIER1	GAAACTTGTCCACTCCTTACG	CCATGCACCACTCCTGCCATT	607	[22]
E. intestinalis	SINTF/SINTR	TTTCGAGTGTAAAGGAGTCGA	TGCCATGCACTCACAGGCATC	822	[23]
E. hellem	EHELF/EHELR	TGAGAAGTAAGATGTTTAGCA	GTAAAAACACTCTCACACTCA	547	[24]
E. cuniculi	ECUNF/ECUNR	ATGAGAAGTGATGTGTGTGCG	TGCCATGCACTCACAGGCATC	549	[25]

3. Results

Microsporidia infection was detected in 7 of the 462 (1.5%; 95% CI: 0.4–2.6) wild lagomorphs analyzed. The distribution of prevalence according to species, province, year, age, and sex is shown in Table 2. Four of the 383 (1.0%; 95% CI: 0.0–2.1) wild rabbits and 3 of the 79 (3.8%; 95% CI: 0.0–8.0) Iberian hares tested positive for microsporidia infection. Positive animals were found in 7 of 83 (8.4%) sampling areas and 3 of 8 (37.5%) provinces (Figure 1). *E. bieneusi* infection was detected in wild rabbits (3/383; 0.8%; 95% CI: 0.0–1.7) but not in Iberian hares (0/79; 0.0%; 95% CI: 0.0–4.6). *E. intestinalis* was confirmed in one wild rabbit (0.3%; 95% CI: 0.0–0.8) and in two brain and one kidney samples from three Iberian hares (3.8%; 95% CI: 0.0–8.0). Co-infections with *E. bieneusi* and *E. intestinalis* were not observed. None of the 462 (0.0%; 95% CI: 0.0–0.8) wild lagomorphs showed positive results for *E. hellem* or *E. cuniculi* infection in the examined organs.

Table 2. Prevalence of zoonotic microsporidia (*Enterocytozoon bieneusi*, *Encephalitozoon intestinalis*, *Encephalitozoon hellem*, and *Encephalitozoon cuniculi)* in organ meats from wild lagomorph species in Andalusia, southern Spain.

Variable	Categories	Prevalence (No. Positives/Overall) †				
		Microsporidia	***E. bieneusi***	***E. intestinalis***	***E. hellem***	***E. cuniculi***
Species	Iberian hare	3.8 (3/79)	0 (0/79)	3.8 (3/79)	0 (0/79)	0 (0/79)
	Wild rabbit	1.0 (4/383)	0.8 (3/383)	0.3 (1/383)	0 (0/383)	0 (0/383)
Province	Cadiz	0 (0/56)	0 (0/56)	0 (0/56)	0 (0/56)	0 (0/56)
	Cordoba	7.0 (3/43)	2.3 (1/43)	4.6 (2/43)	0 (0/43)	0 (0/43)
	Jaen	2.4 (2/82)	1.2 (1/82)	1.2 (1/82)	0 (0/82)	0 (0/82)
	Almeria	0 (0/52)	0 (0/52)	0 (0/52)	0 (0/52)	0 (0/52)
	Malaga	0 (0/67)	0 (0/67)	0 (0/67)	0 (0/67)	0 (0/67)
	Seville	3.7 (2/54)	1.8 (1/54)	1.8 (1/54)	0 (0/54)	0 (0/54)
	Granada	0 (0/49)	0 (0/49)	0 (0/49)	0 (0/49)	0 (0/49)
	Huelva	0 (0/59)	0 (0/59)	0 (0/59)	0 (0/59)	0 (0/59)
Year	2015	1.2 (4/349)	0.9 (3/349)	0.3 (1/349)	0 (0/349)	0 (0/349)
	2016	0 (0/1)	0 (0/1)	0 (0/1)	0 (0/1)	0 (0/1)
	2017	0 (0/33)	0 (0/33)	0 (0/33)	0 (0/33)	0 (0/33)
	2018	3.8 (3/79)	0 (0/79)	3.8 (3/79)	0 (0/79)	0 (0/79)
Age	Yearling	5.0 (1/20)	0 (0/20)	5.0 (1/20)	0 (0/20)	0 (0/20)
	Subadult	1.6 (1/62)	1.6 (1/62)	0 (0/62)	0 (0/62)	0 (0/62)
	Adult	1.1 (4/366)	0.6 (2/366)	0.6 (2/366)	0 (0/366)	0 (0/366)
Sex	Female	0.9 (2/235)	0.4 (1/235)	0.4 (1/235)	0 (0/235)	0 (0/235)
	Male	1.9 (4/212)	0.9 (2/212)	0.9 (2/212)	0 (0/212)	0 (0/212)

† Prevalences are depicted as percentages. Missing values omitted.

4. Discussion

This is the first large-scale study on zoonotic microsporidia species, including *E. bieneusi*, *E. intestinalis*, *E. hellem*, and *E. cuniculi*, in wild rabbits in the Iberian Peninsula and also the first report of microsporidia infection in Iberian hares worldwide. The presence of *E. bieneusi* infection in wild rabbits is consistent with that previously documented [26] and confirms the susceptibility of European wild rabbits to this pathogen. *E. bieneusi* infection has also been detected in farmed rabbits [27–29]. The low prevalence of this parasite detected in organ meats from wild rabbits and the absence of positivity in Iberian hares suggest that these mammal species could act as a spillover host rather than a true reservoir of this emerging pathogen. However, additional studies including fecal samples are required to determine the real prevalence of *E. bieneusi* in wild lagomorphs in the study area. In Spanish Mediterranean ecosystems, *E. bieneusi* infection has been detected in different sympatric species, including wild boar [30], red foxes (*Vulpes vulpes*), beech martens (*Martes foina*), and European badgers (*Meles meles*) [11,27], some of which can be predator species of wild lagomorphs. The high homology between *E. bieneusi* isolates obtained from humans, wild boar, and wild carnivores provides evidence of the zoonotic potential of wildlife [11,30]. In this regard, *E. bieneusi* infection has been reported in HIV/AIDS patients, organ transplant recipients, and immunocompetent individuals in Spain [31–33]. Further molecular studies are needed to assess the risk of zoonotic transmission of this pathogen from wild rabbits.

To the best of our knowledge, this is the first report of *E. intestinalis* infection in wild rabbits and Iberian hares, which increases the number of animal species susceptible to this microsporidian species. There is only one previous report of *E. intestinalis* infection in wild lagomorphs, particularly European brown hares [17]. *E. intestinalis* infection has been reported in humans in several European countries [34–36]. Even though this parasite has not been detected in humans in Spain so far, and the frequency of positive organ meat samples detected in the present study was low (0.3% and 3.8% in wild rabbits and Iberian hares, respectively), our results reveal *E. intestinalis* circulation in this country and, therefore, the risk of transmission of these food-borne zoonotic pathogen cannot be ruled out. Based on the number of animals infected with *E. bieneusi* or *E. intestinalis* found in the present study and hunting bag records for wild lagomorphs [4], around 53,000 wild rabbits and 33,800 Iberian hares infected with these microsporidia species may be consumed every year in Spain.

E. hellem or *E. cuniculi* DNA was not found in any of the 383 wild rabbits or 79 Iberian hares analyzed. These results are consistent with previous studies that also failed to detect these parasites in feces in wild and domestic rabbits [26–29] and in kidney and brain samples in domestic rabbits [35]. In contrast, DNA of *E. hellem* was confirmed in the kidneys of a free-living European brown hare with chronic interstitial nephritis [17]. In addition, *E. cuniculi* infections have been confirmed in kidney and brain tissues of free-living Eastern cottontail rabbits [15] as well as in farmed and pet European rabbits [37,38]. Serological evidence of *E. cuniculi* exposure has also been detected in wild rabbits in other European countries, with seroprevalence values ranging between 3.9% in France and 44.7% in Slovakia [16,39]. The absence of *E. cuniculi* infection in Iberian hares in our study is in accordance with previous observations in other hare species [37,40]. Likewise, seroprevalence values found in European brown hares ranged between 0.0% in Italy and 2.9% in the Czech Republic [14,41]. Further serosurvey studies are warranted to assess *E. hellem* and *E. cuniculi* circulation in the wild lagomorph populations in Spain.

Our study has some limitations that should be noted. Even though our objective was focused on assessing the risk of zoonotic foodborne transmission of the selected microsporidia species, fecal or duodenal tissue could be more appropriate samples for detecting infection of some of these pathogens. On the other hand, *E. intestinalis* DNA was detected in brain samples from Iberian hares, whose kidney tissues tested negative. Unfortunately, brain samples could not be collected from wild rabbits in the present study. While kidney tissue has been shown to be suitable for detection of microsporidia infection in wildlife species [42], including lagomorphs [17], previous observations have shown different sensitivities using brain and kidney samples [15,41,42]. Our results indicate that the prevalence of *E. intestinalis* obtained in wild rabbits in the present study, as well as those previously reported in other species, may be underestimated. To increase the sensitivity of microsporidia detection in organ meat samples, kidney and brain tissues should be tested in parallel.

5. Conclusions

The absence of positivity to *E. hellem* or *E. cuniculi* denotes a limited role of wild rabbits and Iberian hares in the zoonotic transmission of these microsporidia species in southern Spain. The detection of *E. bieneusi* in wild rabbits and *E. intestinalis* in both wild lagomorph species could be of public health concern. Further studies are warranted to elucidate *E. bieneusi* and *E. intestinalis* infection levels in meat and products derived from wild lagomorphs and the risk of transmission of these food-borne zoonotic pathogens.

Author Contributions: Conceptualization, A.M.-P. and I.G.-B.; Methodology, A.M.-P., J.C.-G., Á.M., F.I., L.C.-S., S.J.-R. and C.d.Á.; Formal Analysis, A.M.-P., J.C.-G., Á.M. and I.G.-B.; Investigation, A.M.-P. and J.C.-G.; Resources, Á.M., I.G.-B. and F.G.-G.; Writing—Original Draft Preparation, A.M.-P., J.C.-G. and I.G.-B.; Writing—Review and Editing, A.M.-P., J.C.-G. and I.G.-B.; Supervision, I.G.-B.; Project Administration, I.G.-B.; Funding Acquisition, Á.M. and I.G.-B. All authors have read and agreed to the published version of the manuscript.

Funding: This work has benefited from the financial aid of research grants funded by the Spanish Ministry of Science and Innovation (PID2019-111080RB-C21) and by the University of Córdoba (UCO-FEDER-1264967).

Acknowledgments: We gratefully acknowledge the help of E. Rayas and V. Talavera of the Epidemiological Surveillance Program in Wildlife (Regional Government of Andalusia) in the collection of samples and epidemiological information. J. Caballero-Gómez is supported by an FPU grant from the Spanish Ministry of Science, Innovation, and Universities (FPU17/01319) and S. Jiménez-Ruiz holds a Ph.D. contract from the UCLM co-supported by the European Social Fund (2018/12504).

Conflicts of Interest: The authors declare there are no conflicts of interest.

References

1. Delibes-Mateos, M.; Redpath, S.M.; Angulo, E.; Ferreras, P.; Villafuerte, R. Rabbits as a keystone species in southern Europe. *Biol. Conserv.* **2007**, *137*, 149–156. [CrossRef]
2. Gortazar, C.; Millán, J.; Acevedo, P.; Escudero, M.A.; Marco, J.; De Luco, D.F. A large-scale survey of brown hare *Lepus europaeus* and Iberian hare *L. granatensis* populations at the limit of their ranges. *Wild. Biol.* **2007**, *13*, 244–250. [CrossRef]
3. Ferrer, M.; Negro, J.J. The near extinction of two large European predators: Super specialists pay a price. *Conserv. Biol.* **2004**, *18*, 344–349. [CrossRef]
4. Ministerio de Agricultura, Pesca y Alimentación (MAPA). Available online: https://www.mapa.gob.es/es/desarrollo-rural/estadisticas/caza_pesca.aspx (accessed on 2 November 2020).
5. Wang, L.F.; Crameri, G. Emerging zoonotic viral diseases. *Rev. Sci. Tech.* **2014**, *33*. [CrossRef]
6. Jiménez, M.; González, E.; Martín-Martín, I.; Hernández, S.; Molina, R. Could wild rabbits (*Oryctolagus cuniculus*) be reservoirs for *Leishmania infantum* in the focus of Madrid, Spain? *Vet. Parasitol.* **2014**, *202*, 296–300. [CrossRef]
7. Fernández-Aguilar, X.; Alzaga, V.; Villanúa, D.; Cabezón, O.; García-Bocanegra, I.; Dubey, J.P.; Almería, S. Epidemiology and prevalence of *Toxoplasma gondii* infection in the Iberian hare (*Lepus granatensis*). *Vet. Parasitol.* **2013**, *196*, 194–198. [CrossRef]
8. Molina, R.; Jiménez, M.I.; Cruz, I.; Iriso, A.; Martín-Martín, I.; Sevillano, O.; Melero, S.; Bernal, J. The hare (*Lepus granatensis*) as potential sylvatic reservoir of *Leishmania infantum* in Spain. *Vet. Parasitol.* **2012**, *190*, 268–271. [CrossRef]
9. Snowden, K. Microsporidia in higher vertebrates. In *Microsporidia: Pathogens of Opportunity*; Weiss, L., Becnel, J., Eds.; John Wiley & Sons: Hoboken, NJ, USA, 2014; pp. 469–491. [CrossRef]
10. Qiu, L.; Xia, W.; Li, W.; Ping, J.; Ding, S.; Liu, H. The prevalence of microsporidia in China: A systematic review and meta-analysis. *Sci. Rep.* **2019**, *9*, 1–9. [CrossRef]
11. Santín, M.; Calero-Bernal, R.; Carmena, D.; Mateo, M.; Balseiro, A.; Barral, M.; Barbero, F.L.; Habela, M.Á. Molecular characterization of *Enterocytozoon bieneusi* in wild carnivores in Spain. *J. Eukaryot. Microbiol.* **2018**, *65*, 468–474. [CrossRef]
12. Weiss, L. Microsporidiosis. In *Hunter's Tropical Medicine and Emerging Infectious Disease*, 10th ed.; Magill, A.J., Ryan, E.T., Eds.; Elsevier: Amsterdam, The Netherlands, 2020; pp. 825–831. [CrossRef]

13. Didier, E.S.; Stovall, M.E.; Green, L.C.; Brindley, P.J.; Sestak, K.; Didier, P.J. Epidemiology of microsporidiosis: Sources and modes of transmission. *Vet. Parasitol.* **2004**, *126*, 145–166. [CrossRef]
14. Bártová, E.; Marková, J.; Sedlak, K. Prevalence of antibodies to *Encephalitozoon cuniculi* in European hares (*Lepus europaeus*). *Ann. Agric. Environ. Med.* **2015**, *22*, 674–676. [CrossRef] [PubMed]
15. Zanet, S.; Palese, V.; Trisciuoglio, A.; Alonso, C.C.; Ferroglio, E. *Encephalitozoon cuniculi, Toxoplasma gondii* and *Neospora caninum* infection in invasive Eastern Cottontail Rabbits *Sylvilagus floridanus* in Northwestern Italy. *Vet. Parasitol.* **2013**, *197*, 682–684. [CrossRef] [PubMed]
16. Künzel, F.; Joachim, A. Encephalitozoonosis in rabbits. *Parasitol. Res.* **2010**, *106*, 299–309. [CrossRef] [PubMed]
17. De Bosschere, H.; Wang, Z.; Orlandi, P.A. First diagnosis of *Encephalitozoon intestinalis* and *E. hellem* in a European brown hare (*Lepus europaeus*) with kidney lesions. *Zoonoses Public Health* **2007**, *54*, 131–134. [CrossRef]
18. Morsy, E.A.; Salem, H.M.; Khattab, M.S.; Hamza, D.A.; Abuowarda, M.M. *Encephalitozoon cuniculi* infection in farmed rabbits in Egypt. *Acta Vet. Scand.* **2020**, *62*, 1–11. [CrossRef]
19. Wasson, K.; Peper, R.L. Mammalian microsporidiosis. *Vet. Pathol.* **2000**, *37*, 113–128. [CrossRef]
20. Hinney, B.; Sak, B.; Joachim, A.; Kváč, M. More than a rabbit's tale–*Encephalitozoon* spp. in wild mammals and birds. *Int. J. Parasitol. Parasites Wildl.* **2016**, *5*, 76–87. [CrossRef]
21. Thrusfield, M. *Veterinary Epidemiology*, 4th ed.; Wiley Backweel: Hoboken, NJ, USA, 2018.
22. Da Silva, A.J.; Schwartz, D.A.; Visvesvara, G.S.; De Moura, H.; Slemenda, S.B.; Pieniazek, N.J. Sensitive PCR diagnosis of Infections by *Enterocytozoon bieneusi* (microsporidia) using primers based on the region coding for small-subunit rRNA. *J. Clin. Microbiol.* **1996**, *34*, 986–987. [CrossRef]
23. Da Silva, A.J.; Slemenda, S.B.; Visvesvara, G.S.; Schwartz, D.A.; Wilcox, C.M.; Wallace, S.; Pieniazek, N.J. Detection of *Septata intestinalis* (Microsporidia) using polymerase chain reaction primers targeting the small subunit ribosomal RNA coding region. *J. Mol. Diagn.* **1997**, *2*, 47–52. [CrossRef]
24. Visvesvara, G.S.; Leitch, G.J.; da Silva, A.J.; Croppo, G.P.; Moura, H.; Wallace, S.; Slemenda, S.B.; Schwartz, D.A.; Moss, D.; Bryan, R.T. Polyclonal and monoclonal antibody and PCR-amplified small-subunit rRNA identification of a microsporidian, *Encephalitozoon hellem*, isolated from an AIDS patient with disseminated infection. *J. Clin. Microbiol.* **1994**, *32*, 2760–2768. [CrossRef]
25. De Groote, M.A.; Visvesvara, G.; Wilson, M.L.; Pieniazek, N.J.; Slemenda, S.B.; da Silva, A.J.; Gordon, J.L.; Bryan, R.T.; Reves, R. Polymerase chain reaction and culture confirmation of disseminated *Encephalitozoon cuniculi* in a patient with AIDS: Successful therapy with albendazole. *J. Infect. Dis.* **1995**, *171*, 1375–1378. [CrossRef]
26. Del Aguila, C.; Izquierdo, F.; Navajas, R.; Pieniazek, N.J.; Miro, G.; Alonso, A.I.; Da Silva, A.J.; Fenoy, S. *Enterocytozoon bieneusi* in animals: Rabbits and dogs as new hosts. *J. Eukaryot. Microbiol.* **1999**, *46*, 8–9.
27. Askari, Z.; Mirjalali, H.; Mohebali, M.; Zarei, Z.; Shojaei, S.; Rezaeian, T.; Rezaeian, M. Molecular detection and identification of zoonotic microsporidia spore in fecal samples of some animals with close-contact to human. *Iran. J. Parasitol.* **2015**, *10*, 381. [CrossRef] [PubMed]
28. Galván-Díaz, A.L.; Magnet, A.; Fenoy, S.; Henriques-Gil, N.; Haro, M.; Gordo, F.P.; Miró, G.; Del águila, C.; Izquierdo, F. Microsporidia detection and genotyping study of human pathogenic *E. bieneusi* in animals from Spain. *PLoS ONE* **2014**, *9*, e106017. [CrossRef] [PubMed]
29. Lores, B.; Del Aguila, C.; Arias, C. *Enterocytozoon bieneusi* (microsporidia) in faecal samples from domestic animals from Galicia, Spain. *Mem. Inst. Oswaldo Cruz* **2002**, *97*, 941–945. [CrossRef]
30. Dashti, A.; Rivero-Juarez, A.; Santín, M.; López-López, P.; Caballero-Gómez, J.; Frías-Casas, M.; Köster, P.C.; Bailo, B.; Briz, V.; Carmena, D. *Enterocytozoon bieneusi* (Microsporidia): Identification of novel genotypes and evidence of transmission between sympatric wild boars (*Sus scrofa ferus*) and Iberian pigs (*Sus scrofa domesticus*) in Southern Spain. *Transbound. Emerg. Dis.* **2020**. [CrossRef]
31. Abreu-Acosta, N.; Lorenzo-Morales, J.; Leal-Guio, Y.; Coronado-Álvarez, N.; Foronda, P.; Alcoba-Florez, J.; Izquierdo, F.; Batista-Díaz, N.; Del águila, C.; Valladares, B. Enterocytozoon bieneusi (microsporidia) in clinical samples from immunocompetent individuals in Tenerife, Canary Islands, Spain. *Trans. R. Soc. Trop. Med. Hyg.* **2005**, *99*, 848–855. [CrossRef]
32. Galván, A.L.; Sánchez, A.M.; Valentín, M.P.; Henriques-Gil, N.; Izquierdo, F.; Fenoy, S.; Del Aguila, C. First cases of microsporidiosis in transplant recipients in Spain and review of the literature. *J. Clin. Microbiol.* **2011**, *49*, 1301–1306. [CrossRef]

33. Lores, B.; López-Miragaya, I.; Arias, C.; Fenoy, S.; Torres, J.; Del Aguila, C. Intestinal Microsporidiosis Due to *Enterocytozoon bieneusi* in Elderly Human Immunodeficiency Virus–Negative Patients from Vigo, Spain. *Clin. Infect. Dis.* **2002**, *34*, 918–921. [CrossRef]
34. Bednarska, M.; Bajer, A.; Siñski, E.; Wolska-Kusnierz, B.; Samoliñski, B.; Graczyk, T.M. Occurrence of intestinal microsporidia in immunodeficient patients in Poland. *Ann. Agric. Environ. Med.* **2014**, *21*, 244–248. [CrossRef]
35. Talabani, H.; Sarfati, C.; Pillebout, E.; van Gool, T.; Derouin, F.; Menotti, J. Disseminated infection with a new genovar of *Encephalitozoon cuniculi* in a renal transplant recipient. *J. Clin. Microbiol.* **2010**, *48*, 2651–2653. [CrossRef] [PubMed]
36. Mathis, A.; Weber, R.; Deplazes, P. Zoonotic potential of the microsporidia. *Clin. Microbiol. Rev.* **2005**, *18*, 423–445. [CrossRef] [PubMed]
37. Espinosa, J.; Ferreras, M.C.; Benavides, J.; Cuesta, N.; Pérez, C.; García Iglesias, M.J.; García Marín, J.F.; Pérez, V. Causes of Mortality and Disease in Rabbits and Hares: A Retrospective Study. *Animals* **2020**, *10*, 158. [CrossRef] [PubMed]
38. Csokai, J.; Joachim, A.; Gruber, A.; Tichy, A.; Pakozdy, A.; Künzel, F. Diagnostic markers for encephalitozoonosis in pet rabbits. *Vet. Parasitol.* **2009**, *163*, 18–26. [CrossRef]
39. Bálent, P.; Halanova, M.; Sedlakova, T.; Valencakova, A.; Cislaková, L. *Encephalitozoon cuniculi* infection in rabbits and laboratory mice in eastern Slovakia. *Bull. Vet. Inst. Pulawy* **2004**, *48*, 113–116.
40. Pipia, A.P.; Giobbe, M.; Mula, P.; Varcasia, A.; Sanna, G.; Walochnik, J.; Lavazza, A.; Scala, A. Epidemiological and Biomolecular Updates on *Encephalitozoon cuniculi* in Lagomorpha of Sardinia (Italy). In *Veterinary Science: Current Aspects in Biology, Animal Pathology, Clinic, and Food Hygien*; Pugliese, A., Gaiti, A., Eds.; Springer: Berlin/Heidelberg, Germany, 2012; pp. 47–50. [CrossRef]
41. Ebani, V.V.; Poli, A.; Rocchigiani, G.; Bertelloni, F.; Nardoni, S.; Papini, R.A.; Mancianti, F. Serological survey on some pathogens in wild brown hares (*Lepus europaeus*) in Central Italy. *Asian Pac. J. Trop. Med.* **2016**, *9*, 465–469. [CrossRef]
42. Tsukada, R.; Tsuchiyama, A.; Sasaki, M.; Park, C.H.; Fujii, Y.; Takesue, M.; Hatai, H.; Kudo, N.; Ikadai, H. *Encephalitozoon* infections in *Rodentia* and *Soricomorpha* in Japan. *Vet. Parasitol.* **2013**, *198*, 193–196. [CrossRef]

animals MDPI

Brief Report

Broad Range Screening of Vector-Borne Pathogens in Arctic Foxes (*Vulpes lagopus*) in Iceland

Sándor Hornok [1,*], Kristin Mühldorfer [2], Nóra Takács [1], Regina Hofmann-Lehmann [3], Marina L. Meli [3], Miklós Gyuranecz [4], Ester R. Unnsteinsdóttir [5], Alex D. Greenwood [2,6] and Gábor Á. Czirják [2,*]

1 Department of Parasitology and Zoology, University of Veterinary Medicine, 1078 Budapest, Hungary; takacs.nora@univet.hu
2 Department of Wildlife Diseases, Leibniz Institute for Zoo and Wildlife Research, 10315 Berlin, Germany; muehldorfer@izw-berlin.de (K.M.); greenwood@izw-berlin.de (A.D.G.)
3 Clinical Laboratory, Department of Clinical Diagnostics and Services, and Center for Clinical Studies, Vetsuisse Faculty, University of Zurich, 8057 Zurich, Switzerland; rhofmann@vetclinics.uzh.ch (R.H.-L.); mmeli@vetclinics.uzh.ch (M.L.M.)
4 Institute for Veterinary Medical Research, Centre for Agricultural Research, 1143 Budapest, Hungary; m.gyuranecz@gmail.com
5 Ecology and Consultancy Department, Icelandic Institute of Natural History, 210 Garðabær, Iceland; Ester.R.Unnsteinsdottir@ni.is
6 Department of Veterinary Medicine, Freie Universität Berlin, 14163 Berlin, Germany
* Correspondence: hornok.sandor@univet.hu (S.H.); czirjak@izw-berlin.de (G.Á.C.)

Received: 18 September 2020; Accepted: 29 October 2020; Published: 4 November 2020

Simple Summary: The arctic fox is the only native terrestrial mammal in Iceland. The population comprises both "coastal" and "inland" fox ecotypes, with regard to food resources. Because of the relatively low biodiversity within arctic ecosystems and the involvement of the species in both marine and terrestrial ecosystems, the Icelandic arctic fox population could serve as sentinels for the overall ecosystem health of Iceland. After screening the samples from 60 foxes for tick-/vector-borne pathogens, this study reports the near-absence (very low prevalence) of these pathogens in Icelandic arctic foxes in 2011–2012. Taking into account the broad range of target microorganisms analyzed here, as well as the warming climate and increasing presence of the vector *Ixodes ricinus* in Iceland, our results will be very useful as baseline data for comparison in future monitoring of the emergence of ticks and tick-borne diseases in this country.

Abstract: The arctic fox (*Vulpes lagopus*) is the only native terrestrial mammal in Iceland. While red foxes (*V. vulpes*) are known to be epidemiologically important carriers of several vector-borne pathogens in Europe, arctic foxes have never been evaluated in a similar context on this continent. This has become especially relevant in the last decade, considering the establishing populations of the tick species *Ixodes ricinus* in Iceland. In this study, liver DNA extracts of 60 arctic foxes, hunted between 2011–2012, were molecularly screened for vector-borne protozoan parasites (*Trypanosomatidae*, *Babesia*, *Theileria*, *Hepatozoon*) and bacteria (*Anaplasma*, *Ehrlichia*, *Rickettsia*, *Borrelia*, hemotropic *Mycoplasma*). One sample was real-time qPCR positive for *Anaplasma phagocytophilum*, though this positivity could not be confirmed with sequencing. Samples were negative for all other tested vector-borne pathogens. Results of this study indicate that, except for *A. phagocytophilum*, Icelandic arctic foxes were apparently "not yet infected" with vector-borne pathogens in 2011–2012, or their infections were "below the detection limit" of applied methods. Taking into account the broad range of target microorganisms analyzed here, as well as the warming climate and increasing presence of the vector *I. ricinus* in Iceland, our results will be very useful as baseline data for comparison in future monitoring of the emergence of ticks and tick-borne diseases in this country.

Keywords: arctic fox; Iceland; vector-borne bacteria; vector-borne protozoan parasites; climate change

1. Introduction

The arctic fox (*Vulpes lagopus*) is a small size (ca. 3–5 kg) generalist predator species with circumpolar distribution, associated primarily with arctic and tundra habitats. In Iceland, the arctic fox is the only native terrestrial mammal, divided into coastal and inland ecotypes [1]. While coastal ecotype foxes mainly feed on sea birds and eggs, invertebrates, and marine mammal carcasses, the inland foxes feed on ptarmigans, migrating waterfowl, eggs, reindeer and livestock carcasses, and wood mice [2].

The closest relative of the arctic fox in the Palearctic region is the red fox (*V. vulpes*), and in northern countries, these two species show partly overlapping geographical range and niche [3]. The red fox is widespread in Europe and is known to harbor a broad range of pathogens transmitted by hard ticks (Acari: *Ixodidae*) and other blood-sucking arthropod vectors [4–6]. This implies that red foxes participate in the natural transmission cycle of vector-borne pathogens but may also be used as sentinels for epidemiological surveillance of associated disease risks. On the other hand, a similar epidemiological role has not been postulated for the arctic fox, most likely because they were not found to be tick-infested in early studies [7]. However, this scenario may change with the warming climate, as increasing tick abundance has recently been experienced in Iceland [8].

Eight hard tick species have been reported in Iceland [9], among which *Ixodes ricinus* has the highest veterinary-medical importance in Europe [10]. Since the first record of *I. ricinus* in Iceland, dating back to 1967, its growing significance has been reported [8,9]. Apart from the import of this tick species by migratory birds, its host-questing activity on the vegetation is a likely indicator of established populations [8]. Importantly, arctic foxes are among the possible wildlife reproduction hosts of *I. ricinus* (i.e., feeding adults), while rodents, such as wood mice (*Apodemus sylvaticus*), can support larvae and nymphs.

Between 2012–2013, vector-borne pathogens were examined in 28 arctic foxes in Canada, and *Bartonella henselae*, *Mycoplasma haemocanis*, and *Ehrlichia canis* were detected [11]. However, despite the emergence of *I. ricinus* in Iceland, arctic foxes have not been screened for vector-borne pathogens in this country. Thus, the aim of the present study was to screen liver samples of 60 arctic foxes from Iceland for DNA of vector-borne protozoa (Euglenozoa: *Trypanosomatidae*; Apicomplexa: *Babesia*, *Theileria*, *Hepatozoon*) and bacteria (Proteobacteria: *Anaplasma*, *Ehrlichia*, *Rickettsia*, *Bartonella*; *Spirochaetes*: *Borrelia*; Tenericutes: hemotropic *Mycoplasma*), with emphasis on tick-borne pathogens.

2. Materials and Methods

All experimental procedures described in the materials and methods section were approved by the Internal Committee for Ethics and Animal Welfare of the Leibniz Institute for Zoo and Wildlife Research (permit #2010-01-03). All experiments were carried out in compliance with the approved guidelines of the Leibniz Institute for Zoo and Wildlife Research and the laws of Germany, Hungary, and Iceland.

Liver samples of 60 arctic foxes hunted between 2011 and 2012 in various inland and coastal regions of Iceland were used in this study. Hunting was covered by the government and managed by The Environmental Agency, under the wildlife act (64/1994) and animal welfare act (55/2013). Scientific use of the harvest was organized by the Icelandic Institute of Natural History, courtesy of The Icelandic Ministry for the Environment and Natural Resources. The animals included 40 adults (23 males, 17 females) and 20 juveniles (11 males, 9 females). Age determination of the arctic foxes was performed by X-ray measurements based on dental pulp cavity volume and cementum analysis. Juveniles were defined as younger than 1 year and adults as older than 1 year of age [12]. Carcasses were stored at −20 °C until necropsy was performed.

During necropsy, approximately 10 g of liver tissue was sampled and then stored at −80 °C until further analysis. The liver was chosen as the target organ, because it contains blood from both the portal and systemic circulation and is rich in cells of the reticuloendothelial system, i.e., components that are the most likely to contain the DNA of hemotropic, vector-borne microorganisms. Ectoparasites were not seen on the foxes post-mortem at the time of sample collection.

DNA was extracted from liver samples, using the NucleoSpin Tissue kit (Macherey-Nagel GmbH, Düren, Germany), individually from each fox. A total of 100 μL of homogenized tissue was digested with Proteinase-K overnight at 56 °C and prepared as described in the user manual. Quality and quantity of extracted DNA were assessed by ND-1000 UV-Vis Spectrophotometer (Nanodrop Technologies, Wilmington, DE, USA) measurements.

All 60 DNA extracts were screened for a broad range of tick-/vector-borne pathogens with PCRs summarized in the Supplementary Material (Table S1). For *Anaplasma phagocytophilum*, the assay consisted of 40 cycles and the results were regarded as positive if the threshold cycle (Ct) value was below 39. The detection limit of this qPCR was a 0.125 ratio (one-eighths) of an *A. phagocytophilum* infected cell [13]. For Sanger sequencing of *A. phagocytophilum* from the real-time qPCR-positive sample, two conventional PCRs were attempted, the first by using the msp2 primers and annealing temperature, as in method "A" (Supplementary Materials, Table S1), and the second to amplify part of the heat shock chaperonin GroEL gene [14]. The latter PCR was repeated, but no sequenceable product was obtained.

3. Results and Discussion

Out of 60 samples, one adult female arctic fox from the costal ecotype was positive for *A. phagocytophilum*-specific real-time qPCR (Ct = 36.71; sequencing was repeatedly not successful). In line with this finding, red foxes have been reported to carry *A. phagocytophilum* DNA in at least 10 European countries [4,5]. *Anaplasma phygocytophilum* (Rickettsiales: Anaplasmataceae) is the causative agent of granulocytic anaplasmosis in multiple hosts (including humans) and tick-borne fever in ruminants [15]. In Europe, *A. phagocytophilum* is transmitted by *I. ricinus*, the tick-species increasingly recognized to affect animals in Iceland [8]. Nevertheless, to the best of our knowledge, this is the first molecular evidence of *A. phagocytophilum* in arctic foxes and the first autochthonous occurrence of this tick-borne species in Iceland. However, since this real-time qPCR positivity for *A. phagocytophilum* could not be confirmed with sequencing, probably because of the low amount of DNA, as reflected by the high Ct value in the more sensitive real-time qPCR, further studies are needed to verify the presence of this species in arctic foxes in Iceland. Although seropositivity to *A. phagocytophilum* has already been reported in humans in Iceland, in the absence of published data on the autochthonous occurrence of its vector in the country, this finding was explained by exposure during travel abroad [16].

All 60 samples were negative or contained DNA below the PCR detection limit for other tested vector-borne pathogens. Taking into account that red foxes are known to harbor a broad range of the vector-borne protozoa and bacteria targeted here (e.g., in northern Europe [5]; in southern Europe [4,6]), PCR-negativity of most samples implies that the warming climate of Iceland has not entailed changes of similar magnitude in the distribution of tick-borne pathogens in 2011–2012, as reported in the same period from Scandinavia [17]. This is further confirmed by the absence of any ticks on the 60 foxes analyzed here, as well as on 315 arctic foxes examined in 2015–2016 [8].

Unlike what is shown here, arctic foxes were reported to harbor *Bartonella* spp., *M. haemocanis*, and *E. canis* in northern Canada [11,18]. However, in the latter case, the majority of positive samples were flea-borne bartonellae and not tick-borne pathogens. At least three rodent species are known to occur in Iceland [19], but their mite and flea species are usually not shared with foxes [20]. Despite this, during prey consumption, the role of temporary infestation of arctic foxes with these non-host specific ectoparasites, in the transmission of vector-borne pathogens, should not be discounted [18].

Nevertheless, results of the present study suggest that different spectra of vector-borne bacteria might emerge in arctic foxes in Iceland compared to Canada.

4. Conclusions

The above results provide baseline data on the near-absence (very low prevalence) of tick-/vector-borne pathogens in Icelandic arctic foxes in 2011–2012. Already observed and foreseeable climate changes in Iceland (milder winter, warmer spring and autumn [8]) will most likely promote multi-focal establishment of *I. ricinus* and possibly other tick species in this country, with increased risks of autochthonous transmission of tick-borne pathogens [8]. This warrants continuous monitoring of emerging vector-borne pathogens in Iceland, similar to all arctic ecosystems. Taking into account the broad range of target microorganisms analyzed here (for which red foxes are known to be susceptible), as well as the warming climate and increasing presence of *I. ricinus* in Iceland, our results will be very useful as baseline data for comparison in future monitoring of the emergence of ticks and tick-borne diseases in this country.

Supplementary Materials: The following are available online at http://www.mdpi.com/2076-2615/10/11/2031/s1. Table S1: Details of the screening methods used in this study, listed according to target groups of vector-borne pathogens.

Author Contributions: S.H., K.M., and G.Á.C. designed the study; M.G., E.R.U., A.D.G., and G.Á.C. coordinated different parts of the study; E.R.U. collected the samples; S.H., K.M., N.T., R.H.-L., and M.L.M. performed laboratory analysis; S.H. wrote the manuscript with the contributions of all the co-authors. All authors have read and agreed to the published version of the manuscript.

Funding: This research was supported by funds from the Leibniz Institute for Zoo and Wildlife Research, Germany.

Acknowledgments: We would like to thank Katja Pohle and Nadine Jahn for their help with curating the samples and conducting the laboratory analyses. The laboratory work was partly performed using the logistics of the Center for Clinical Studies at the Vetsuisse Faculty of the University of Zurich. This manuscript benefited from comments of six anonymous referees. The publication of this article was funded by the Open Access Fund of the Leibniz Association.

Conflicts of Interest: The authors declare no conflict of interest. The funders had no role in the design of the study; in the collection, analyses, or interpretation of data; in the writing of the manuscript, or in the decision to publish the results.

References

1. Carbonell Ellgutter, J.A.; Ehrich, D.; Killengreen, S.T.; Ims, R.A.; Unnsteinsdóttir, E.R. Dietary variation in Icelandic arctic fox (*Vulpes lagopus*) over a period of 30 years assessed through stable isotopes. *Oecologia* **2020**, *192*, 403–414. [PubMed]
2. Bocharova, N.; Treu, G.; Czirják, G.Á.; Krone, O.; Stefanski, V.; Wibbelt, G.; Unnsteinsdóttir, E.R.; Hersteinsson, P.; Schares, G.; Doronina, L.; et al. Correlates between feeding ecology and mercury levels in historical and modern arctic foxes (*Vulpes lagopus*). *PLoS ONE* **2013**, *8*, e60879. [CrossRef] [PubMed]
3. Elmhagen, B.; Tannerfeldt, M.; Angerbjorn, A. Food-niche overlap between arctic and red foxes. *Can. J. Zool.* **2002**, *80*, 1274–1285. [CrossRef]
4. Hodžić, A.; Alić, A.; Fuehrer, H.P.; Harl, J.; Wille-Piazzai, W.; Duscher, G.G. A molecular survey of vector-borne pathogens in red foxes (*Vulpes vulpes*) from Bosnia and Herzegovina. *Parasites Vectors* **2015**, *8*, 88. [CrossRef] [PubMed]
5. Sakalauskas, P.; Lipatova, I.; Radzijevskaja, J.; Paulauskas, A. Pathogen screening in the red fox (*Vulpes vulpes*) from Lithuania. *Biologija* **2019**, *65*, 283–295. [CrossRef]
6. Sukara, R.; Juwaid, S.; Ćirović, D.; Penezić, A.; Mihaljica, D.; Veinović, G.; Radojičić, S.; Hodžić, A.; Duscher, G.G.; Tomanović, S. *Candidatus Neoehrlichia* sp. (FU98) and *Borrelia burgdorferi sensu lato* in red foxes (*Vulpes vulpes*) from Serbia. *Acta Vet-Beogr.* **2019**, *69*, 312–324. [CrossRef]
7. Hersteinsson, P. (Ed.) *Íslensk Spendýr [Icelandic Mammals]*; Vaka-Helgafell: Reykjavík, Iceland, 2004; p. 344. (In Icelandic)
8. Alfredsson, M.; Olafsson, E.; Eydal, M.; Unnsteinsdóttir, E.R.; Hansford, K.; Wint, W.; Alexander, N.; Medlock, J.M. Surveillance of *Ixodes ricinus* ticks (Acari: Ixodidae) in Iceland. *Parasites Vectors* **2017**, *10*, 466. [CrossRef] [PubMed]

9. Richter, S.H.; Eydal, M.; Skírnisson, K.; Ólafsson, E. Tick species (Ixodida) identified in Iceland. *Icel. Agric. Sci.* **2013**, *26*, 3–10.
10. Gray, J.S. The development and seasonal activity of the tick *Ixodes ricinus*: A vector of Lyme borreliosis. *Rev. Med. Vet. Entomol.* **1991**, *79*, 323–333.
11. Mascarelli, P.E.; Elmore, S.A.; Jenkins, E.J.; Alisauskas, R.T.; Walsh, M.; Breitschwerdt, E.B.; Maggi, R.G. Vector-borne pathogens in arctic foxes, *Vulpes lagopus*, from Canada. *Res. Vet. Sci.* **2015**, *99*, 58–59. [CrossRef]
12. Treu, G.; Krone, O.; Unnsteinsdóttir, E.R.; Greenwood, A.D.; Czirják, G.Á. Correlations between hair and tissue mercury concentrations in Icelandic arctic foxes (*Vulpes lagopus*). *Sci. Total Environ.* **2018**, *619–620*, 1589–1598. [CrossRef]
13. Courtney, J.W.; Kostelnik, L.M.; Zeidner, N.S.; Massung, R.F. Multiplex real-time PCR for detection of *Anaplasma phagocytophilum* and *Borrelia burgdorferi*. *J. Clin. Microbiol.* **2004**, *42*, 3164–3168. [CrossRef] [PubMed]
14. Alberti, A.; Zobba, R.; Chessa, B.; Addis, M.F.; Sparagano, O.; Pinna Parpaglia, M.L.; Cubeddu, T.; Pintori, G.; Pittau, M. Equine and canine *Anaplasma phagocytophilum* strains isolated on the island of Sardinia (Italy) are phylogenetically related to pathogenic strains from the United States. *Appl. Environ. Microbiol.* **2005**, *71*, 6418–6422. [CrossRef]
15. Woldehiwet, Z. The natural history of *Anaplasma phagocytophilum*. *Vet. Parasitol.* **2010**, *167*, 108–122. [CrossRef]
16. Skarpheðinsson, S.; Søgaard, P.; Pedersen, C. Seroprevalence of human granulocytic ehrlichiosis in high-risk groups in Denmark. *Scand. J. Inf. Dis.* **2001**, *33*, 206–210.
17. Jaenson, T.G.; Jaenson, D.G.; Eisen, L.; Petersson, E.; Lindgren, E. Changes in the geographical distribution and abundance of the tick *Ixodes ricinus* during the past 30 years in Sweden. *Parasites Vectors* **2012**, *5*, 8. [CrossRef] [PubMed]
18. Buhler, K.J.; Maggi, R.G.; Gailius, J.; Galloway, T.D.; Chilton, N.B.; Alisauskas, R.T.; Samelius, G.; Bouchard, É.; Jenkins, E.J. Hopping species and borders: Detection of *Bartonella* spp. in avian nest fleas and arctic foxes from Nunavut, Canada. *Parasites Vectors* **2020**, *13*, 469. [CrossRef]
19. Bengtson, S.A.; Brinck-Lindroth, G.; Lundqvist, L.; Nilsson, A.; Rundgren, S. Ectoparasites on small mammals in Iceland: Origin and population characteristics of a species-poor insular community. *Holarct. Ecol.* **1986**, *9*, 143–148. [CrossRef]
20. Brinck-Lindroth, G.; Smit, F.G.A.M. *The Fleas (Siphonaptera) of Fennoscandia and Denmark*; Brill Academic Publishers: Leiden, The Netherlands, 2007.

Publisher's Note: MDPI stays neutral with regard to jurisdictional claims in published maps and institutional affiliations.

MDPI
St. Alban-Anlage 66
4052 Basel
Switzerland
Tel. +41 61 683 77 34
Fax +41 61 302 89 18
www.mdpi.com

Animals Editorial Office
E-mail: animals@mdpi.com
www.mdpi.com/journal/animals

www.ingramcontent.com/pod-product-compliance
Lightning Source LLC
Chambersburg PA
CBHW061619210926
51850CB00009B/2376
* 9 7 8 3 0 3 6 5 6 8 5 2 2 *